Protozoa
and other protists

Protozoa
and other protists

Michael A. Sleigh

The right of the
University of Cambridge
to print and sell
all manner of books
was granted by
Henry VIII in 1534.
The University has printed
and published continuously
since 1584.

CAMBRIDGE UNIVERSITY PRESS

Cambridge
New York   Port Chester
Melbourne   Sydney

Published by the Press Syndicate of the University of Cambridge
The Pitt Building, Trumpington Street, Cambridge CB2 1RP
40 West 20th Street, New York, NY 10011-4211, USA
10 Stamford Road, Oakleigh, Victoria 3166, Australia

First published by Edward Arnold 1973 as *The biology of protozoa*
This edition first published by Edward Arnold 1989 and first
published by Cambridge University Press 1991

Printed in Great Britain by
Butler & Tanner Ltd, Frome and London

A catalogue record of this book is available from the British Library

ISBN 0 521 41751 1 hardback
ISBN 0 521 42805 X paperback

Contents

Preface to *The Biology of Protozoa*

Research using Protozoa has flourished in recent years not only because of studies directed at increasing our knowledge of the Protozoa themselves, but also because many people have recognized that these organisms provide excellent subjects for studies of general biological phenomena at the cellular level — the possibility of obtaining large numbers of identical cells, uncontaminated, and even at the same stage in their growth and division cycle is particularly valuable in biochemical studies; studies on Protozoa have therefore made a substantial contribution to modern Cell Biology. This flowering of research activity has principally been directed towards elucidating the structure and understanding the functioning of Protozoa as cells – their organization, biochemistry and physiology, but new research in such fields as ecology and genetics has also revealed more about the importance of Protozoa in the economy of nature and about the variety of methods evolved by living things to solve problems of living. We are therefore at an exciting stage in the development of the study of Protozoa.

The growth of Protozoology as a substantial discipline has been linked with the appearance of several excellent textbooks; in recent years the two volumes edited by Grassé, and books by Grell, Mackinnon and Hawes, Sandon, Dogiel and Kudo have given coverage to various aspects of the study of Protozoa, but only bring the reader to the threshold of modern studies in Cell Biology. Several recent texts are valuable but are more limited in scope, covering ultrastructure in the book by Pitelka, only parasitic species in the book by Baker and only a few species in the book by Vickerman and Cox. Accounts of recent research are therefore not generally available in textbooks, and one has to depend upon specialist reviews and original papers to provide a reasonable coverage of the subject. Comparison of the recent review articles in the four volumes of *Research in Protozoology* with their predecessor *Protozoa in Biological Research* shows the enormous development of many branches of the subject, but all of these articles and those in the three volumes of *Biochemistry and Physiology of Protozoa* have provided much information for this book. The increase in research activity has led to an enormous literature about Protozoa, so that the references quoted have been restricted to major works, reviews and other recent papers that refer to earlier studies and provided an introduction to literature that could not be listed in the bibliography.

It is the main aim of this book to emphasize new insights into cell functioning and ultrastructure as they apply to Protozoa, and to illustrate the wealth of diversity in organization and physiology that has been revealed among the various groups of organisms classed as Protozoa. The background of systematic

Protozoology is related with modern developments in the hope that this book will provide students with a guide to many aspects of new research on Protozoa.

It is a pleasure to acknowledge the help given by many people in the preparation of this book. I owe a particular debt of gratitude to Mrs. Margaret Attwood whose skilful drawings illustrate the diversity of flagellates, amoebae and ciliates in Chapters 5, 7 and 8; almost every drawing incorporates information taken from several sources, so that in few cases is it possible to acknowledge a single origin for the figure. I should also like to thank the following who generously gave me copies of electron micrographs for illustrations that appear in the book: Drs. L. H. Bannister, C. F. Bardele, J. and M. Cachon, R. M. Crawford, J. D. Dodge, A. E. and A. G. H. Dorey, R. Gliddon, H. Jørgen Hansen, R. H. Hedley, A. Hollande, B. S. C. Leadbeater, A. C. Macdonald, J. W. Murray, B. Nisbet, E. B. Small. J. B. Tucker and K. Vickerman. The permission from authors and publishers for the reproduction of figures previously published elsewhere is also acknowledged with gratitude. A number of secretaries have assisted at various stages, and I am particularly grateful to Miss Gaye McCrindle, whose accurate work, patience and attention to detail greatly aided the preparation of the final manuscript; Miss Jeanette Date and Miss Margaret David have also given valuable help. I wish to express my thanks to all those colleagues who have helped by answering questions, discussing evidence and reading sections of this book; their comments have added to my understanding and have contributed substantially to the accuracy of many statements. The wise guidance of Professor E. J. W. Barrington, the General Editor, is gratefully acknowledged.

Bristol 1973   M.A.S.

Preface to the second edition

This volume is based on my earlier book on *The Biology of Protozoa*, and incorporates many changes that reflect both the sustained growth in research on protists over the last 15 years and some new attitudes to these organisms. The fact that protists generally function as independent cells has made them convenient and important objects for study by cell biologists, and as in the earlier book the approach followed here is essentially that of the cell biologist. Research on the ultrastructure of a much greater range of protists, together with the application of techniques of molecular biology, has led to a flourishing of phylogenetic speculation both about the origin of eukaryote protists and about possible relationships among the groups. It has become clear that the protozoa and algae, formerly regarded as taxonomic groups, are two assemblages of organisms each comprising members drawn from various parts of a wide evolutionary radiation of protists. The prime algal characteristic of the possession of chloroplasts has proved to be a secondary character gained at various times by protists of various types. For these reasons, all protists must be considered together to understand their relationships to one another, even if, as here, the macroscopic photosynthetic forms are only briefly mentioned in passing. Significant progress has also been made in ecological studies on protists, where the important roles played by unicellular eukaryotes in soils, sediments and most aquatic habitats are now being demonstrated by quantitative studies of improving reliability. Advances in studies on parasitic protists have also been rapid; indeed to do justice to the details about important protistan parasites required by a professional parasitologist would require a book to itself, so here are given only outlines of the structure, relationships, physiological features and life cycles of examples of the main parasitic groups at the same level as for other protists. It is in phylogenetic, systematic and ecological areas that this book will be found to differ most from its predecessor, although the chapters on the cellular characteristics of protists have also been extensively revised. I believe this broad approach was responsible for the very gratifying success of the earlier book, and I hope that this new book will retain its appeal in spite of the incorporation of much new material. The extensive referencing found popular in the earlier book has also been retained, with the bulk of references being new.

A large proportion of the drawings made by Margaret Attwood are used again in this book, as are some of the original micrographs from biologists whose generosity in providing illustrations was previously acknowledged. It is a pleasure to thank C. F. Bardele, D. Barlow, G. Brugerolle, J. and M. Cachon, R. M. Crawford, K Hausmann, J.-P. Mignot, D. J. Patterson and F. E. Round for kindly providing copies of figures for this book. The permission of authors and publishers for the reproduction of figures previously published elsewhere is gratefully acknowledged. I am also grateful to many colleagues who have kindly commented on various sections of this manuscript or on parts of the earlier book, and have thereby improved the accuracy of the information; responsibility for the shortcomings which remain rests with me.

Southampton 1988   *M. A. Sleigh*

1

Protozoa as members of the Protista

During the last ten years there has been much debate about the origins of eukaryotes from prokaryotes, and about relationships amongst lower eukaryotes. Attention was drawn in the first edition of this book to the artificial boundary drawn between algae and protozoa. The position is now clearer, and there is much support for the view that eukaryotes are best divided into four kingdoms: Animalia or multicellular animals (Parazoa, Mesozoa and Metazoa), Plantae or green land plants (Bryophyta and Tracheophyta), Fungi (the non-flagellate groups) and Protista, comprising eukaryote groups formerly classed as algae, protozoa and flagellate fungi. The name Protoctista is used by some authors, e.g. Margulis (1981), for this last kingdom, but there seem compelling reasons to prefer the simpler name Protista (see Corliss, 1984).

This book is principally concerned with those sections of the kingdom Protista traditionally recognised as protozoa, but will extend to consider relevant features of related unicellular protists and some features of the multicellular algae. The view is taken here, as outlined in the next section, that the taxon Protozoa no longer exists, although the concept of protozoa as animal-like protists remains valuable (Corliss, 1981). Similarly, there are plants described as algae both within the prokaryotes (Cyanobacteria = blue-green algae) and in various sections of the kingdom Protista (red algae, brown algae, green algae, etc.), but several protist groups contain both protozoa and algae (notably euglenids and dinoflagellates), so the taxon Algae can not exist as a single proper taxonomic entity. It seems quite likely that both the kingdom Protista and the prokaryote kingdom Monera represent levels of organisation rather than single major branches of the evolutionary tree, and are not therefore comparable with say the kingdom Plantae, which appears to be essentially monophyletic. The Protista includes at least several evolutionary branches with distinct phylogenies, the characteristics of whose members are the subject of this book. As Sogin *et al*. (1986) point out 'the diversity within the Kingdom Protista dwarfs that of the Plantae, Fungi and Animalia combined'. Since the taxonomic concept of a kingdom may have two different meanings within the eukaryotes, we may expect at some future date a change in rank of the assemblage of protists (perhaps a confederation ?) to recognise the fact that it encompasses several branches of the evolutionary tree.

The origins and evolution of protists

The themes discussed have been reviewed recently by Margulis, 1981; Sleigh, 1986a; Woese, 1987.

There have been recent improvements in both the quality and quantity of information that can shed light on the origins and early evolution of eukaryotes, based both on traditional methods of comparative anatomy and palaeontology, and on more modern techniques of organic geochemistry, comparative biochemistry and molecular biology. The last of these may indeed provide quantitative indications of the closeness of relationships between organisms by comparison of the sequence of components within certain highly conserved but widely distributed molecules, like the sequence of amino acids in cytochromes or tubulins or of nucleotides in certain ribosomal RNAs (rRNAs). Comparisons of nucleotide sequences in rRNA molecules of three different types (5S rRNA, 16–18S rRNA and 23–28S rRNA) derived from a wide range of organisms, including a diversity of protists, tend to suggest similar patterns of relationships between protistan groups (Hori and Osawa, 1986; Sogin *et al.*, 1986; Adoutte *et al.*, 1988), although the less reliable 5S rRNA results deviate in some respects. The trends indicated in these studies, as well as other evidence, are incorporated in phylogenetic trees outlined in Figs. 1.1 and 5.34. This approach is likely to yield even more convincing data in the near future as the rRNA nucleotide sequences of more organisms are studied. Present data are consistent with the view that the ancestral eukaryote evolved as a phagotrophic anaerobe living among and feeding upon the diverse forms of prokaryote that dominated life on earth before about 1.5×10^9 years ago.

The atmosphere of the early earth following its formation about 4.5×10^9 years ago lacked oxygen and allowed synthesis of a variety of organic compounds using energy from ultra-violet light and electric storms. Progressive polymerisation and aggregation of these compounds led eventually to the development of simple organisms. These were characterised by the presence of a limiting lipoprotein membrane regulating internal concentrations and exchange of materials with the environment, and by the property of self-replication through inherited nucleic acid molecules that could direct protein synthesis. Such primeval organisms or 'progenotes' probably first appeared on earth before 3.5×10^9 years ago, and lived as heterotrophs deriving their energy for maintenance, growth and reproduction from the fermentation of prebiologically synthesised organic matter. The diversification of these organisms to exploit different forms of nutrition led to an adaptive radiation of prokaryotes. At least three primary lines are thought to have evolved, leading to Archaebacteria, Eubacteria and an Urkaryote ancestor of the eukaryote lineage (Woese, 1987); this divergence probably took place early since it is revealed by fundamental differences in the cell membrane lipids and properties of the protein synthesis mechanism between the three groups. Lake *et al.* (1984) suggest that a fourth primary line, the Eocytes, also separated early from the Archaebacteria.

The driving force for evolution is the need to capture energy more efficiently than one's competitors. The abiologically synthesised organic matter was in limited supply, but energy could also be derived from light and from the promotion of certain inorganic reactions. Certain bacteria developed compounds which could absorb light energy and use it in photosynthesis. Early forms used bacteriochlorophyll to trap energy for driving the reaction of carbon dioxide with hydrogen derived from organic compounds or from hydrogen sulphide, but eventually the absorption of light by the combination of chlorophyll *a* and phycobilic proteins (p. 101) allowed the development of coupled photosystems that mobilised sufficient energy to split water to provide hydrogen for carbon dioxide fixation and release oxygen. The release of oxygen by these cyano-bacteria (and also by other chlorophyll-containing prokaryotes, like chloroxy-bacteria, with different pigment combinations) eventually transformed the earth's atmosphere from a reducing one to an oxidising one – an event that can be traced geologically by the appearance of oxidised rocks at about 2×10^9 years ago. It also allowed the development of oxidative metabolic pathways using atmospheric oxygen as the final electron acceptor. In particular, certain bacteria adapted cytochrome chains, already used to produce ATP in photosynthesis, for the generation of ATP by oxidative phosphorylation driven by energy from coenzymes reduced in Krebs cycle reactions (p. 58) in an aerobic respiratory system that is energetically more efficient than anaerobic respiration.

The adaptive radiation of bacteria produced a great variety of anaerobic heterotrophs (fermenters), aerobic heterotrophs, chemo-autotrophs and photo-autotrophs, a wide range of which persist to this day in appropriate habitats. In parallel with this bacterial evolution the Urkaryote is assumed to have developed as an anaerobic phagotroph, feeding upon bacteria and organic debris. It is speculated that this mode of life required a naked surface, internal membrane systems for phagotrophic digestion and cytoskeletal systems for prey capture and for internal transport of food vacuoles and lysosomes. During the development of these eukaryotic features the nuclear material became isolated within a nuclear envelope, possibly to remove it from an unstable anchorage at a naked plasma membrane and from the digestive activities of the cytoplasm, although the separation of nucleic acid replication and transcription in the nucleus from RNA code translation at protein synthesis in the cytoplasm also permitted the development of RNA processing in eukaryotes (p. 53). The presence of multiple chromosomes presumably led to the need for eukaryotic mitosis involving microtubules, and eukaryote sex and meiosis probably also developed early, since they have similar features in all eukaryotes.

The resulting ancestral eukaryote is assumed to have been essentially amoe-boid in form, and dependent for food on rare local concentrations of bacteria on surfaces. By developing flagella from projecting bundles of microtubules this organism could more efficiently capture the bacteria and migrate to new food sites (Sleigh, 1986b). However, as anaerobic heterotrophs they were energe-tically inefficient, and therefore poor competitors. The acquisition of the ability to perform oxidative phosphorylation in mitochondria, presumably by develo-ping a stable symbiosis with certain aerobic bacteria (Yang *et al.*, 1985), possibly of several types (p. 18), allowed these heterotrophs to extract more energy from each bacterium captured, and resulted in an adaptive radiation of eukar-yotes. Representatives of several lines developed stable symbiotic relations with different photosynthetic prokaryotes or eukaryotes and became photoauto-trophic eukaryotes (see p. 103), some of which lost the phagotrophic habit whilst others retained it. The increase in availability of eukaryotic prey allowed further diversification of the phagotrophic heterotrophs, whilst other heterotrophs became saprotrophs, losing phagotrophic ability and obtaining organic mole-cules by absorption of products of external breakdown of organic matter.

Some of the diversity of protist lineages believed to have evolved by these routes is outlined in Fig. 1.1, and more details of parts of the scheme are given in

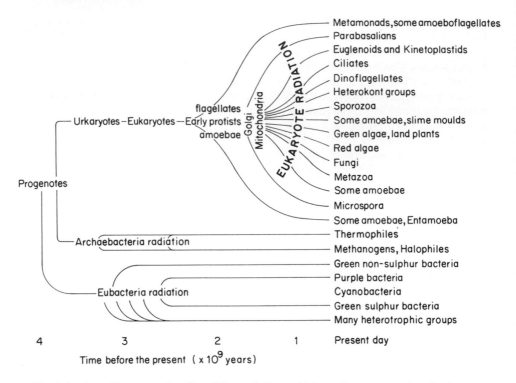

Fig. 1.1   An outline reconstruction of the evolutionary history of groups mentioned in the text.

Fig. 5.34 (p. 152). Evidence provided in many sections of this book is taken to support the general form of this evolutionary scheme, though new information added each year is allowing its progressive refinement. The different lineages and their sub-divisions are characterised by unique combinations of fundamental features of nuclear organisation, nuclear division, pattern of life cycle, the arrangement, appendages, movement and roots of flagella, the presence, structure and pigmentation of plastids, the nature and position of food stores, the presence and nature of cell walls, pellicles, tests or other secreted surface structures or internal skeletons, the presence of mitochondria and the form of their cristae and special features of other internal membranous or fibrous organelles. The fundamental nature of the features that characterise the different lineages indicate that they separated early in evolution and represent ancient independent branches of the evolutionary tree. Many such branches have therefore come to be recognised by some authors as divisions of phylum (or botanical division) rank within the Protista, but agreement about the taxonomic rank to be accorded to many of these groups has yet to be reached. In addition to the major branches of the lines shown in Fig. 1.1, there are many smaller groups, some known at present only from a single species or genus (which may have been thoroughly or only superficially studied), which appear to have fundamental differences from all other groups and so, until we know more about them and any relatives, must be classified separately.

Protists that lack mitochondria, and could have separated from the eukaryote stock in the anaerobic period before the acquisition of mitochondrial symbionts, include metamonad flagellates and certain amoebae (Pelobiontea), which lack Golgi membranes, Microspora, in which possible Golgi membranes but not dictyosomes have been identified, and parabasalian flagellates, that possess numerous dictyosomes (p. 16). It is interesting that non-mitochondrial symbionts and hydrogenosomes, both of which may serve some mitochondrial functions, are often present in these groups. Most other groups contain mitochondria, but in some parasitic groups that lack mitochondria their absence may be secondary because these forms are closely related to groups that possess mitochondria. The wide distribution of mitochondria in both heterotrophic and autotrophic groups suggests that mitochondria were acquired before plastids.

It is not possible in a book of this length to provide a full separate account of the many phyla of protists that are currently recognised, so they have been divided for systematic description into four sets in a rather traditional way. These are:

1. those phyla in which the main life stage is flagellate, or is clearly derived from flagellate forms

2. phyla that are essentially amoeboid, whether or not they sometimes have flagellate stages

3. the ciliates

4. those phyla that reproduce by spores, are primarily parasitic in animals and possess flagella only on gametes, if at all.

It is strongly emphasised that these are not regarded as taxonomic groupings of phylogenetic significance (except Ciliophora), but merely groups of phyla whose shared features allow their diversity to be described in a reasonably economical fashion. The red algae do not have flagella and have a special form of sexual reproduction; they are peripheral to the subject of this book, but will be mentioned alongside flagellate phyla since that is where the other photosynthetic phyla occur.

Protists came to dominate life on earth about 1.5×10^9 years ago and have clearly continued to evolve. Some of them presumably became larger and more complex as time passed, and the other three eukaryotic kingdoms evolved from among the Protista, so that by 7×10^8 years ago the multicellular animals, for example, had become sufficiently complex for the skeletons of members of several phyla to be present in the fossil record. The green land plants clearly evolved from certain flagellated green algae. The true fungi (Zygomycetes, Ascomycetes and Basidiomycetes) never possess flagella and show features that may be regarded as primitive, especially among yeasts, but appear from studies on nucleotide sequences of rRNA to have evolved relatively recently and to be fairly close to higher plants and red algae; these fungi diverged from algae and took up a saprotrophic mode of life. The origins of higher animals are also obscure (Hanson, 1977); favoured suggestions are that a planuloid form (like a larval cnidarian or primitive flatworm) could have been derived from a colonial flagellate or by cellularization of a primitive ciliated form (before nuclear dimorphism?); sponges are assumed to have had an origin from collar flagellates, and may not have been ancestral to any metazoan group. Results of rRNA sequencing suggest that metazoans are closer to higher plants than to ciliates, and thus a flagellate origin for metazoa seems likely.

Features of the four sets of protistan phyla

The flagellated phyla, discussed in Chapter 5, contain photoautotrophic members with a variety of plastid characters as well as a diversity of free-living and parasitic heterotrophs. Multicellular, often non-motile, members are found in many groups, especially the photosynthetic phyla. Flagellate cells usually contain most components of the full range of eukaryotic organelles, as can be seen from the simple example in Fig. 1.2. Many of their features take specific forms in different phyla; for example, although in this case the cell surface is simple, with only a little secreted material (not shown), many flagellates have

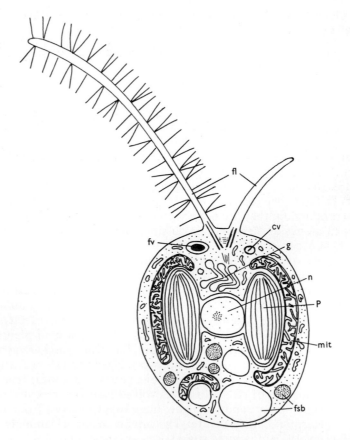

Fig. 1.2 Diagram of structures found in **Ochromonas**, a flagellate member of the Chrysophyta, showing the contractile vacuole *cv*, flagella *fl*, food storage bodies *fsb*, food vacuole *fv*, Golgi body *g*, mitochondion *mit*, nucleus *n* and plastid *p*; length of cell body about 10 μm.

complete cell walls, extensive coats of secreted scales or well-developed internal pellicles. All phagotrophic cells retain an area of unthickened membrane for food intake, which may be restricted to a specialised cytostome. The most common number of flagella appears to be two, but flagellar multiplication has occurred in some phyla. Many flagellates have contractile vacuoles.

Members of the amoeboid groups, discussed in Chapter 6, typically possess some form of pseudopodium, usually broadly-lobed, needle-like or reticulate. Little structural organisation is seen (at least at light microscope level) in many amoebae (Fig. 1.3) in comparison with that in flagellates and ciliates. This is consistent with the fact that much of the cytoplasm of many amoeboid forms is capable of flowing fairly freely into cellular projections, often carrying with it a variety of inclusions including contractile and food vacuoles. Some forms have

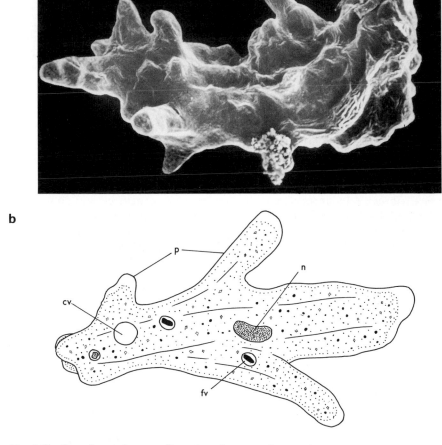

Fig. 1.3 *Amoeba proteus*. **a**, Scanning electron micrograph (reproduced from Small and Marszalek, 1969). **b**, Diagram showing contractile vacuole *cv*, food vacuole *fv*, nucleus *n* and pseudopodia *p*; cell length about 500 μm.

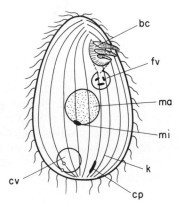

Fig. 1.4 Diagram of the ciliate *Tetrahymena* showing buccal cavity *bc* with feeding cilia, cytoproct *cp*, contractile vacuole *cv* with two pores, forming food vacuole *fv*, kinety rows of cilia *k*, macronucleus *ma* and micronucleus *mi*; cell length about 50 μm.

permanent or temporary internal skeletons, and others secreted shells, scales or spines, which provide a more regular structure. These groups contain phago-trophic forms, but there are a few amoeboid taxa that have closer affinities with flagellates and betray this by photoautotrophic nutrition or by the permanent possession of flagella as well as pseudopodia.

The Ciliophora (ciliates), that form the subject of Chapter 7, clearly form a natural group isolated from other protists by a number of specialised features of their pellicle and nuclei. Common features of ciliates are illustrated in Fig. 1.4. The body surface bears many cilia in rows called kineties, and in many species specialised feeding cilia in the buccal cavity around the cytostome may be grouped into compound structures. The infraciliature (ciliary bases and attached fibres) is incorporated with membranous structures and sometimes other material in a specialised superficial zone of cytoplasm to form the pellicle. Within this the cytoplasm is more fluid and mobile, and contains nuclei of two different types as well as food vacuoles and other cell organelles and inclusions. Contractile vacuoles are present in most ciliates and are closely associated with the pellicle.

The groups in the final set, considered in Chapter 8, share the features of reproduction by spores and parasitism within other eukaryotes, usually animals. The largest group are the Sporozoa (also known as the Apicomplexa) which possess a characteristic apical complex of organelles (typically present in infective cells of the life cycle, see p. 222), and produce spores, following sexual reproduction, which germinate to release haploid sporozoites (Fig. 1.5a). Many sporozoa have flagellated male gametes, but flagella are absent from the other groups. The spores of the Microspora and the Myxospora, formerly classed together as the Cnidospora, both contain eversible filaments, like those of cnidarian nematocysts (Fig. 1.5b,c). In both cases the spores release amoeboid sporoplasms, but in microspora the single sporoplasm escapes through the everted polar filament whilst in myxosporans one or more sporoplasms escape after breakdown of the spore wall. Microspora are intracellular parasites

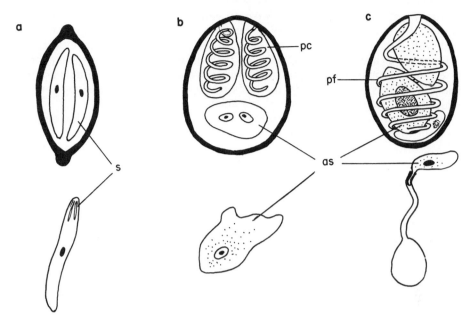

Fig. 1.5 Comparative diagrams of spores of members of **a** Sporozoa, **b** Myxospora and **c** Microspora. In each case there is a thick spore wall enclosing an infective cell that hatches as a sporozoite *s* in sporozoans, and as an amoeboid sporoplasm *as* in the other groups. The spore of a myxosporan usually contains several polar capsules *pc* while the microsporan spore contains a single eversible polar filament *pf*. Spores **a** and **b** are commonly 10–25 μm long and those of **c** are usually less than 5 μm long.

lacking mitochondria and producing unicellular spores, but myxospora form extensive plasmodia containing mitochondria and produce multicellular spores. The spores of Haplosporidia, which also differentiate within a plasmodial mass containing mitochondria, are unicellular with an apical opening and a single uninucleate sporoplasm (Fig. 8.13b); they never possess polar capsules. The paramyxids have closed spores and probably represent a separate group.

The cellular organisation of protists

The simplest organisms of most phyla of protists are clearly unicellular. Some members of most groups have become complex, following a variety of routes. Colonies are often formed, much as they are in prokaryotes, when cells remain associated after division; these cells may all be similar, and in most cases possess a full range of eukaryotic cell constituents, but gain advantage by creating larger feeding currents or extending into better illuminated areas for photosynthesis. In more complex members of some groups the cells within a colony may develop different specialisations and become more interdependent. Complexity can also develop within cells which grow so large that additional nuclear material is apparently required, in the form of replicate sets of genes in polytene chromosomes, polyploid nuclei or multiple nuclei; a large ciliate may have 10^6 or more times the volume of a small flagellate. There has been much semantic discussion about the cellular nature of the protozoan body (Hyman, 1940; Mackinnon and Hawes, 1961; Pitelka, 1963); largely because of the existence of multinucleate and multicellular forms within the same phyla, some biologists have described protozoans as acellular or non-cellular. It seems more logical to regard protists as essentially unicellular organisms which have adapted their cellular components in various ways in a great diversity of evolutionary experiments, some of which have produced the most complex cells known.

Every protistan cell is a highly integrated unit of life. Its functioning depends on a turnover of chemical materials, a store of biochemical information and a through-flow of information. All three are normally necessary for maintenance as well as growth and other characteristic activities of the organism. Organic molecules which are incorporated into the structural elements and many other vital components of the cell may be built up within autotrophic organisms from such simple molecules as carbon dioxide, water and salts containing nitrogen, phosphorus, sulphur and other minor elements; the synthesis of organic compounds from such simple precursors occurs in association with the fixation of light energy in the plastids of photosynthetic cells, but similar synthesis may utilize energy derived from the breakdown of simple organic compounds (Chapter 3). Heterotrophic protists take in organic molecules both as a source of raw materials for cell construction and as a source of energy; usually their powers of synthesis of small organic molecules are limited. While the small molecules required by autotrophic organisms may enter the cell by diffusion, the uptake of complex molecules requires an active process and usually involves taking the food substance into vacuoles in the cell, and the subsequent enzymic breakdown of macromolecules into smaller units which can be passed through the vacuole membrane. The cell membrane forms the zone of contact between a protistan cell and the environment; everything that enters or leaves the organism must pass through this membrane, so that it is a vital controlling region exerting an important influence on intracellular concentrations. Agents which act on the protist must either act at the cell membrane or must pass through the membrane to an intracellular site of action.

Energy derived from light or the breakdown of organic molecules is stored in the cell in the chemical bonds of organic compounds. Some of these compounds are labile and carry a relatively large amount of readily available energy; the best known example is adenosine triphosphate (ATP), whose stored energy is released by removal of one or two terminal phosphate groups, but certain reduced coenzymes are also energy carriers. Many other compounds which serve as stores of energy are inert and the energy they contain is mobilised by the enzymic oxidation processes associated with respiration and is transferred to energy carriers like ATP. Aerobic respiration involves the utilization of oxygen and is associated with mitochondria. Anaerobes can only perform partial oxidations in the cytoplasm and mobilise less energy from the same substrates. Energy given up by ATP and similar carrier molecules is used for synthesis and for movement by the formation of chemical bonds; this is obvious in the case of chemical synthesis, and in movement it is thought that the changes in shape or position of molecules are the result of the formation of temporary chemical bonds. Most cellular reactions involve the loss of some energy to entropy and little or none of the energy released on the breakage of many types of chemical bond is recoverable.

The information store necessary for molecular synthesis within protists as within other cells is provided by 'blueprints' carried in the genes of the nucleus in the form of coded sequences of nucleotide subunits in their DNA (deoxyribonucleic acid) chains. The information encoded in DNA finds expression in the synthesis of proteins; the information is first copied to RNA (ribonucleic acid) in a transcription process and then used to determine the sequence of amino acids of a polypeptide chain in a process of translation which takes place in the cytoplasm at ribonucleoprotein bodies called ribosomes. Some of the proteins synthesised have structural roles, but the majority are enzymes functioning in the promotion of a variety of cell reactions, including the synthesis of molecules of many types. The molecules synthesised are mainly used within the cells, but most protists produce some forms of extracellular products, including the walls or external envelopes of cysts or other cells, extrusive organelles and secretions of other sorts.

The character of a protistan cell depends on the types of proteins it contains. Individuals within a species may show minor variations in their total protein constitution; if the differences between the protein constitutions of two populations of a species become extensive, it is likely that the two populations will no longer interbreed, and therefore that they become isolated as two distinct species. The differences between the protein constitutions of two species will be greater or less according to the remoteness or closeness of their phylogenetic relationship. Most protist cells change in character, because of changes in the pattern of protein synthesis, in different phases of their life cycle; e.g. there is usually a period in which shells, skeletons or other new organelles are formed, in species of many groups there is a cyst stage, and in sporozoans the organism goes through a sequence of different morphological stages. A cell carries the same gene pattern in the nucleus throughout its life cycle, but utilises different assortments of these genes at different times so that a different character will temporarily be imparted to the cell by the different proteins synthesised. A complex interaction of internal and environmental factors determines which genes are used for protein synthesis at a particular time.

The provision of a complete information store for each daughter cell at the division of a protist is made possible by the exact replication of the DNA molecules before division of the nucleus. The precise replication and high degree of stability of the DNA molecules are vital for the inheritance of specific features and the conservation of the characteristics of a species. However, DNA molecules are not absolutely stable and errors in replication do occur. Variations (mutations) can occur within the gene pattern of the cell and can be inherited to provide variants in cell character which form the raw material of the process of natural selection. Variation in gene patterns is necessary for evolution to take place. Some protists contain only a single set of chromosomes (haploid cells), and in these forms every gene may find expression and be 'tested' for its usefulness to the organism; other forms contain two or more sets of chromosomes (diploid or polyploid cells), and a mutant gene may or may not find expression depending on the nature of the corresponding (allelic) gene(s) on the other homologous chromosome(s). In haploid organisms only successful mutations survive, but unsuccessful genes may persist in diploid organisms. Many genes are only useful when they occur in suitable combinations with other genes, e.g. many enzymes are involved in chain reactions where each step is essential for completion of the sequence. The occurrence of several suitable mutations simultaneously, so as to make possible the establishment of a new reaction sequence, is a very unlikely event, and in haploid organisms demands very large populations or results in slow evolution. The chance of the occurrence of suitable combinations of mutations may be increased in diploid or polyploid organisms, but the most important means of bringing about increased variation in the gene content of organisms are those processes associated with sexual fusion (Chapter 4). The segregation of genes in meiosis and the recombination of genes at fertilisation provide great plasticity of the gene pattern within an interbreeding population of organisms. The evolution of sexually reproducing protists can be much more rapid than that in species which only reproduce asexually.

In subsequent chapters this introduction to cellular organisation of protists will be expanded by discussions of the structure and functions of the component organelles of protistan cells, the nutrition and the reproduction of protists. The characteristics and range of form of the various phyla of Protista will be described, with particular emphasis on protozoan forms, and in the final chapter the ecology and biological and economic importance of members of the kingdom Protista will be discussed.

2
Features of protistan organization

The major features of body structure in unicellular protists are at the organelle level (Pitelka, 1963). Most of the cytoplasmic structures which are found in other eukaryote cells are also recognizable in protists. These structures are principally either membrane elements or fibrous elements; the nuclear components and extracellular products of cells do not fall strictly within these categories.

Membranous structures

Membrane elements in cells (*see* Allen, 1978) are composed of lipoprotein structures about 7.5 nm thick, which appear in electron micrographs of sectioned membranes as a pair of dense lines each about 2 nm thick separated by a lighter space 3–4 nm thick. Current evidence suggests that such membranes are fluid mosaics comprised of protein molecules embedded in a lipid bilayer (Singer and Nicholson, 1972); the hydrophobic fatty acid 'tails' of the lipids and hydrophobic regions of integral membrane proteins occupy the centre of the bilayer, with polar end-groups at either surface. Both lipid and protein molecules diffuse freely within their membrane layer unless restrained by attachments. Lipids and proteins differ at the two sides of the membrane and in membranes from different sources. The fluidity of a membrane depends on the types of lipids it contains, and the proportions of lipids may vary with temperature (to keep fluidity optimal) (Nozawa and Thompson, 1979); the layer facing the cytoplasm tends to have negatively charged lipids and the outer layer often has glycolipids with sugar groups extending outwards. Transmembrane proteins may act as receptors, transport molecules, enzymes or structural links across the membrane; they may be linked to cytoskeletal proteins within the cell, which may thereby stabilise the membrane, and/or may carry polypeptide or more often carbohydrate end groups outside the cell as anchorage, protective or recognition components in the cell coat (glycocalyx). It is not clear whether protein molecules ever merely insert part-way into a membrane. Membrane proteins are recognised as intramembranous particles (IMPs) in freeze fracture preparations where the membrane is cleaved between the two lipid layers (e.g. Fig. 2.10, p. 32).

The lipid component of such a 'unit membrane' makes it rather impermeable; to take advantage of this property, the cytoplasm of every protist is totally enclosed in a unit membrane called the plasma membrane. Through this membrane must pass all substances which enter or leave the cell, so that the properties of the membrane, and particularly its proteins, determine the character of the cell by controlling the concentration of substances within it. The membrane is very permeable to some substances which may enter or leave readily, and is very impermeable to other substances, which may not be allowed to enter or may be kept within the cell; some substances to which the membrane has a low permeability may be actively transported across the membrane with the use of metabolic energy. Such differential permeability of the cell membrane makes it possible for the cell to accumulate certain substances so that a suitable chemical environment may be provided for intracellular processes, and the semipermeability of the membrane often also necessitates some regulation of the osmotic concentration of the organism, particularly in fresh water. The plasma membrane forms a most important part of the homeostatic machinery of protists. The low permeability of the membrane to certain ions and the ability to transport ions actively have made it possible for many cells, including protists, to maintain an electrical potential across the plasma membrane and some other membranes in cells, and to use these membrane potentials in a number of physiological activities.

It is also important that since the plasma membrane is the region of contact with the outside world, external agents which exert an effect on the organism do so at the membrane or after penetrating through the membrane. The membrane probably has little control over electromagnetic or other radiations, but a large proportion of other influences or substances probably 'distort' the membrane physically or chemically in order to achieve their characteristic effect.

Binding sites on the plasma membranes of cells can move in the plane of the membrane to transport bound molecules or larger structures. By this means secreted structures, such as scales (p. 129) or mastigonemes (p. 35), produced within Golgi or other vesicles, are carried from their site of release to their final destination on the cell surface or flagellum. Similarly, bound lectins, such as concanavalin A, are capped (carried together into a cluster) on a diversity of amoeboid and other cells. The migration of such binding sites, or groups of them that form 'membrane domains', also seem responsible for the transport of extrusomes (p. 17) along the inside and bacteria along the outside of axopod membranes of centrohelids (p. 177) (Bardele, 1976). Gliding movements of various flagellates (p. 107) and the transport of polystyrene beads along *Chlamydomonas* flagellar surfaces appear to use the same mechanism (Bloodgood, 1981), and the gliding movements of gregarines (p. 226) (King, 1981; Walker *et al.*, 1979) and diatoms (Edgar and Pickett-Heaps, 1983) also probably depend upon it, in association with actin filaments.

The various internal membranes of cells are a little thinner than the plasma membrane and do not show such an obvious polarity. The general form of some of the internal membranous organelles of protists is shown in Fig. 2.1. The endoplasmic reticulum (ER), which is developed to a greater or lesser extent in all eukaryotes, forms a system of membranes separating off closed channels or vesicles whose interior is not in continuity with the cytoplasm. This system provides at least two compartments in the cell, the contents of which may be maintained under different conditions, and it thus allows the separation of substances from each other. It also serves as a surface upon which enzyme systems may be arranged; in particular, the cytoplasmic surface of the ER is a major site of activity of ribosomes during protein synthesis (p. 52). Polypeptides synthesised at these ribosomes may be passed directly through the membrane into the

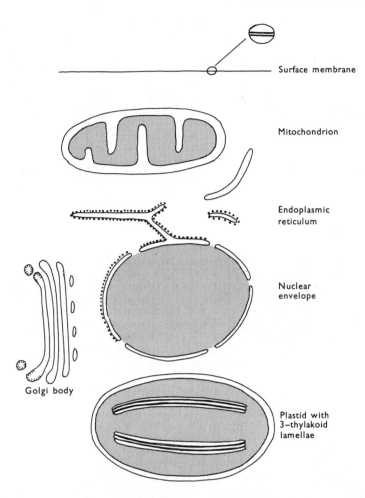

Surface membrane

Mitochondrion

Endoplasmic
reticulum

Nuclear
envelope

Golgi body

Plastid with
3–thylakoid
lamellae

Fig. 2.1 Diagrams to show the arrangement of membranes in some cell organelles; every membrane shown is a trilaminar unit membrane, as indicated at the top of the figure. Ribosomes are shown attached to parts of the endoplasmic reticulum and outer nuclear membranes. The membranes of the Golgi body mature from simple vesicles which coalesce forming cisternae to 'coated vesicles' budded off from the cisternae at the extreme left.

ER cavity, and thence isolated into a vesicle for secretion. The ER is also the site of enzymes responsible for lipid synthesis.

The nuclear envelope is connected with this system of ER membranes and is probably best regarded as a special region of ER; it often bears ribosomes on its outer, cytoplasmic surface, and may be reformed from elements of ER following nuclear division. The two unit membranes which comprise the nuclear envelope are separated by a space of about 20–30 nm, which is seen in some electron micrographs to be continuous with the space within ER channels. Pores which occur in the nuclear envelope (p. 43) provide a route for chemical interactions between the cytoplasm and components within the nucleus.

Another membranous structure associated with the ER is the cluster of flattened vesicles or cisternae called the Golgi body or dictyosome (Fig. 2.2). These are very numerous in some protists, notably flagellates, but appear to be absent from metamonad flagellates and some amoebae, and are weakly developed in some other amoebae, microspora and most ciliates. They are concerned with membrane maturation and with the elaboration or storage of products of cell synthesis (Alberts *et al.* 1983). It is frequently possible to distinguish between one (forming) face of a dictyosome at which vesicles derived from

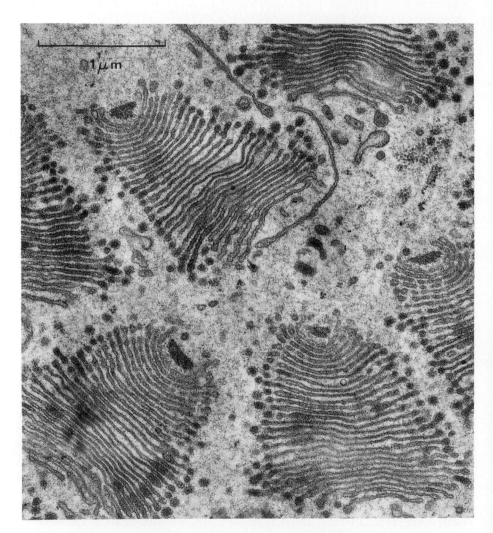

Fig. 2.2.   Electron micrograph of sections of dictyosomes that form part of the Golgi complex or parabasal body of the flagellate *Joenia annectens*. In each dictyosome the forming face with less distinct membranes lies against a dense body that is a section of a parabasal fibre; towards the opposite 'mature' face of the dictyosome the membranes become more distinct and characteristic dense vesicles are budded off from the edges of the cisternae. (Micrograph by Hollande and Valentin, 1969).

the ER or nuclear envelope aggregate to form the cisternae, and the opposite (mature) face at which vesicles (usually 'coated vesicles') are formed from dictyosome cisternae and emigrate, often to the plasma membrane. The character of the vesicles at the two faces is often different, and the cisternal membranes can often be seen to increase in thickness between the forming face and the mature face of the dictyosome. Vesicles containing polypeptides synthesised at ribosomes on the ER are transported to the forming face of the dictyosome, and within the cisternae the membrane is modified and the enclosed materials are elaborated into their final form before being transported away from the mature face. One of the best examples of this is the formation of complex scales within the Golgi or ER vesicles of certain flagellates and amoebae (pp. 129, 167), before the vesicles move to the plasma membrane and the scales are incorporated into the surface covering of the organism (Manton and Leedale, 1961, 1969). Even more spectacular structures formed within Golgi-type vesicles are the 'cell wall' valves of diatoms (p. 134), formed within a silicalemma vacuole (Volcani, 1981), and elements of siliceous skeletons of radiolarians formed within a cytokalymma (Anderson, 1983).

A class of intracellular organelles synthesised in Golgi or ER vesicles are the extrusomes, found in some amoeboid, ciliate and flagellate protists (Hausmann, 1978). These organelles, which are enclosed in a single membrane, take up a position beneath the cell membrane specified by a cluster of IMPs within the membrane. On receipt of a suitable stimulus, membrane fusion takes place between the organelle and cell membranes and the contents of the extrusome are released (Satir, B. *et al.*, 1973); in the extrusion of trichocysts of *Paramecium*, both the membrane fusion and the expansion of the trichocyst matrix are calcium-dependent events. Many extrusomes function in food capture, e.g. kinetocysts of centrohelids (p. 178), haptocysts of suctorians (p. 64) or toxicysts of gymnostome ciliates (p. 200), but others like the trichocysts of *Paramecium* (p. 208), the mucocysts of *Tetrahymena* (p. 191) or ejectisomes of cryptophytes (p. 129) have no known function, although they could be defensive (Hausmann, 1978).

A more elaborate membrane arrangement is found in mitochondria and plastids, both of which are concerned with energy transformations. Mitochondria are usually about 1 μm across and may be one to several μm long; some cells contain many mitochondria, but often only one simple or elaborate mitochondrion may be present. They are constructed from two membranes, the outer of which forms the surface of the structure, while the inner, which is separated from the outer by a distance of about 10 nm in most places, is continuous with projections to the interior of the mitochondrion (Fig. 2.1) which form lamellar, vesicular or tubular cristae. If mitochondria had a symbiotic origin (p. 3), the inner membrane would be derived from the bacterial symbiont and the outer one from the vacuole of the host cell. The rather impermeable inner membrane is specialised, with not only relevant transport proteins and respiratory chain (ATP-generating) enzymes (p. 60), but also an abnormal lipid composition; the outer membrane is freely permeable to many small molecules (Alberts *et al.* 1983). The membrane system forms two compartments; the inner membrane encloses a mitochondrial matrix compartment, with citric acid cycle (p. 58) and other enzymes, mitochondrial DNA and ribosomes, and the space between the two membranes and within the projecting cristae forms a second compartment containing other enzymes. Tubular cristae are found in ciliates and some flagellates (e.g. Fig. 2.3a), but lamellar or discoidal cristae occur in some other flagellates and amoebae (Figs. 2.3c,d), and the cristae of some other amoebae and sporozoa appear as vesicles rather than lamellae (Fig. 2.3b). Some protozoa which live in environments devoid of oxygen, such as parts of the vertebrate gut, have been reported as having mitochondria without cristae, but these structures are probably microbodies, since normally anaerobic protozoa lack mitochondria. At least some of the proteins required by mitochondria are coded for by genes within the mitochondrion, while others depend on nuclear genes. The genes, ribosomes and protein synthesis mechanism within mitochondria show prokaryotic features consistent with the supposed symbiotic origin of these organelles (Margulis, 1981). It may be speculated whether different symbiotic events, involving different bacteria, may have given rise to mitochondria with different forms of cristae, or biochemical differences (Yang *et al.*, 1985), but it is known that different forms of cristae occur in mitochondria of different cells of the mammalian body.

Microbodies form a class of organelles around 0.5 to 1.0 μm in diameter and bounded by a single membrane. Their biochemical functions vary in detail, but involve participation in various oxidation-reduction reactions. The best known are peroxysomes, characterised by the presence of oxidase enzymes which release hydrogen peroxide, and catalase which oxidises the peroxide. The presence of these respiratory organelles is widespread in eukaryotes; including protists like *Acanthamoeba*, *Ochromonas* and *Tetrahymena* (Muller, 1975). In many cases, including *Tetrahymena*, enzymes of the glyoxylate cycle (p. 58) are also present in the same organelles, which are sometimes given the alternative name of glyoxysomes. The primary function of the glyoxylate cycle is normally the synthesis of carbohydate from fat; in *Tetrahymena* mitochondria and peroxysomes must work together, since in this ciliate some enzymes of the glyoxylate cycle have been found only in the mitochondria. Parabasalian flagellates like *Trichomonas*, which have no mitochondria, have microbodies called hydrogenosomes, which are also respiratory organelles; they contain enzymes which oxidise pyruvate to acetate and CO_2, resulting under anaerobic conditions in the release of hydrogen, though water is formed if oxygen is available (Muller, 1980, 1985). Symbiotic methanogenic bacteria often accompany the hydrogenosomes in anaerobic protists; the H_2 and CO_2 released by the hydrogenosomes being used by methanogens in production of methane (p. 281). Glycosomes, found in trypanosomatid flagellates, contain a set of glycolytic enzymes which normally occur free in the cytoplasm (Opperdoes and Borst, 1977). Some authors believe that microbodies were derived from symbionts, possibly from mitochondria (see Cavalier–Smith, 1987).

The plastids of autotrophic protists are also limited by at least two unit membranes, but these enclose numerous flattened sacs (thylakoids), within whose membranes the photosynthetic pigments and associated enzymes are incorporated (Fig. 2.3e). The arrangement of the thylakoids and the pigments they contain varies widely in protists; this variation is discussed in Chapter 5. Many plastids contain a pyrenoid region, rich in protein, and often associated with deposits of stored carbohydrate. Although nuclear genes and cytoplasmic ribosomes produce many of the plastid proteins, all plastids contain DNA and prokaryote-type ribosomes responsible for independent synthesis of some

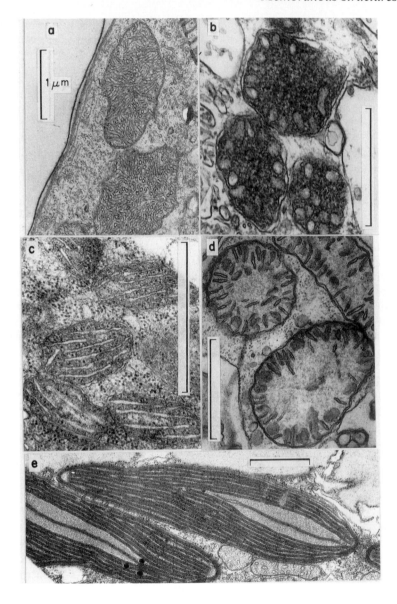

Fig. 2.3. Electron micrographs of mitochondria and chloroplasts. **a** Mitochondria of the ciliate *Euplotes* with tubular cristae; **b** mitochondria of the heliozoan *Actinosphaerium* with vesicular cristae; **c**, mitochondria of the centrohelid amoeba *Acanthocystis* with lamellate cristae; **d**, mitochondria of the euglenoid flagellate *Peranema* with discoidal cristae; and **e** chloroplasts of the diatom *Melosira*. (Micrographs by R. Gliddon, A.C. Macdonald, D.J. Patterson, B. Nisbet and R.M. Crawford, respectively.) All scale bars 1 μm.

plastid RNAs and proteins. These features are consistent with a symbiotic ancestry of plastids (Margulis, 1981), which would suggest that the outer limiting membrane is of host cell origin and the inner one of symbiont origin.

There are also other vacuole systems with specialised functions, namely the contractile and food vacuoles, which deserve some comprehensive treatment here because of their particular importance in protists.

Contractile Vacuoles

Contractile vacuoles are present in most ciliates and in many flagellates and amoebae as well as in freshwater sponges (Kitching, 1967, Patterson, 1980); they are probably most conveniently studied in ciliates, particularly sedentary forms. Cells of some species have several contractile vacuoles, others have only one. Most vacuoles occur close to the plasma membrane, and are seen to swell slowly (diastole) before suddenly contracting (systole) and expelling their fluid contents to the outside. They are filled from systems of canals, small channels or vesicles that collect fluid from a differentiated region of the cytoplasm called the spongiome. In protists with a specialised surface the contractile vacuoles have fixed positions and open to the outside during systole through permanent pores. In many amoeboid forms the contractile vacuole has a less organised structure and may be carried around in the cytoplasm before systole at the cell surface.

In electron micrographs of the simple contractile vacuole of *Amoeba* (Flickinger, 1973), the unit membrane enclosing the clear vacuole is surrounded by a region containing many small tubules, vesicles and mitochondria (Fig. 2.4a). Some of the small vesicles may be seen to have direct continuity with the vacuole, so it appears that small vesicles form in the spongiome cytoplasm and fuse with the main vacuole during diastole. The expulsion of fluid at systole of this type of vacuole is rather feeble; following systole the membrane of the vacuole fragments and many small vacuoles appear in its place and soon coalesce, initiating diastole.

A more elaborate contractile vacuole complex is found in *Paramecium* (Fig. 2.4b) (Patterson, 1977), where the main vacuole is fed by about six collecting canals which extend out into the spongiome. These canals are surrounded by a zone of smooth-surfaced, branching tubules 20–80 nm in diameter, which communicate at their outer ends with more regular straight tubules about 50 nm in diameter and bearing arrays of particles on their outer surfaces. The branching tubules open to the collecting canals during diastole, but are thought to close to reduce backflow during systole of the canals; perhaps there are contractile elements which cause systole of the canals and also constrict these tubules? Diastole of the main vacuole is initiated by systole of the canals (phase 2, Fig. 2.4c). Soon after transfer of fluid from the canals into the main vacuole, the latter ceases to swell (phase 4) and commences its systolic contraction by becoming more distinctly spherical (round-up, phase 5) (Patterson and Sleigh, 1976). Phases 5 and 6 (Fig. 2.4c) are interpreted as a period of increasing pressure on the vacuolar contents before the two membranes at the pore (Fig. 2.4b) fuse together, opening the pore and allowing expulsion of the fluid (phase 7, Fig. 2.4c). During round-up (phase 5) some fluid may be forced back into the collecting canals. It is believed that fibrous contractile elements (presumably made of actin, since they bind fluorescent anti-actin antibodies)

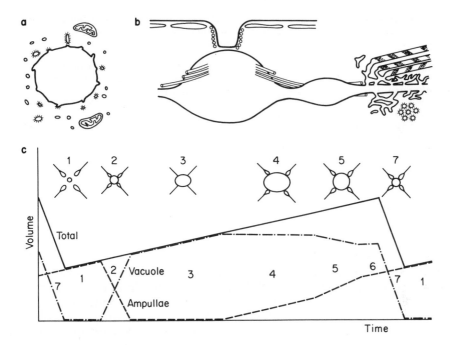

Fig. 2.4 Contractile vacuoles and their activity. **a**, A diagram of a section through a contractile vacuole of *Amoeba* surrounded by vesicles and mitochondria (see text). **b**, A diagram of part of a section through a contractile vacuole complex of *Paramecium*, the main vacuole underlies a permanent pore in the pellicle and connects to several canals which collect fluid from a system of tubules in the surrounding spongiome (*see* text); both the canals and the pore are supported by microtubules. **c**, Changes in volume of the vacuole, the canal ampullae and the total volume of the main chambers of the vacuole complex of *Paramecium* during the cycle of activity shown in the sequence of vacuole outlines and described in the text. (Information from Patterson 1977, 1980.)

surround the vacuole and exert pressure upon it to initiate expulsion, though the completion of the expulsion process probably depends upon a combination of contractile pressure and cytoplasmic hydrostatic pressure. Contractile activity of the vacuole is rhythmic; if the pore of the vacuole is blocked, contraction will cause rounding-up which persists for a while and then relaxation allows renewed diastole before a further period of systolic contraction; this may be repeated through several cycles of alternating diastole and round-up before either the blockage is released or the cell bursts. Although the frequency of contraction can be experimentally modified, e.g. by changing cation concentrations, the mechanism of timing is not understood.

The activity of contractile vacuoles is normally much higher in freshwater forms than in marine protists or endosymbionts; this observation, together with experimental evidence, suggests that the primary function of the contractile vacuole is osmotic regulation. The osmotic pressure of the cytoplasm of freshwater protists (50–150 mOsm l^{-1}) is well above that of the surrounding medium (< 10 mOsm l^{-1}), and some marine protozoa have been shown to be hypertonic to seawater. As a result water will flow by osmosis into the bodies of protists; it

Table 2.1   Comparative data on the contractile vacuole activity of several Protozoa.

Species	Duration of one cycle seconds	Rate of output μm^3/sec	Time to expel body volume minutes	Temperature °C	Medium
Amoeba proteus	150–800	54–109	230–800	19–27	Hay infusion
Paramecium caudatum	6–30	54–258	15–49	15–23	Culture medium
Carchesium aselli	6–39	6–20	25	14.5–16	London Tap Water
Discophrya collini	25–100	3–15	25–120	20–25	Bristol Tap Water
Cothurnia curvula	30–1800	0.1–1.7	240	14.5–16	Sea Water
Acanthamoeba castellanii	mean ~50	mean 2.1	25–30	24–25	1% Peptone

will also enter in food vacuoles and may be produced in metabolism. Vacuolar output depends upon the volume of the vacuole and the duration of the vacuolar cycle. The frequency of systole varies widely (Table 2.1), ranging from once every few seconds in smaller freshwater ciliates to only once or twice in an hour in larger endosymbiotic ciliates. The relative rate of removal of water also appears to be greatest in the smallest forms, e.g. a small ciliate may expel a water volume equal to its body volume in a few minutes, while a large ciliate may take several hours to pump out the equivalent of its body volume. Some typical data given in Table 2.1 suggest a relationship between output and surface area; a study of vacuole output of colpodid ciliates of different sizes (Lynn, 1982) confirms this relationship, but output per unit of surface area increases with body size rather than remaining constant.

Changes in the rate of vacuolar output are important as part of the response of the cell to change in osmolarity of the external medium. J. A. Kitching (1951) found that when the suctorian *Discophrya* is placed in a dilute solution of sucrose (to which the body surface has low permeability), the rate of output of the vacuole quickly falls to a new level, and quickly returns to the original rate if the suctorian is soon returned to water. In such experiments the vacuolar output is proportional to sucrose concentration, without obvious body shrinkage, in sucrose solutions up to about 60 mM 1^{-1} (Fig. 2.5a); at higher concentrations the vacuole soon ceases to fill and the body shrinks. This response by the contractile vacuole to the rate of inward water flux is important for the volume regulation of the cell in the short term. In the longer term the cell responds by adjustment of its internal osmotic pressure to restore an osmotic differential between inside and outside (Fig. 2.5b) (Stoner and Dunham, 1970). Thus a rise in external osmolarity is followed by an increase in internal osmolarity through mobilisation of free amino acids, or more rarely of polyhydric alcohols or free cations; similarly, a fall in osmolarity of the external medium is followed by a reduction in cytoplasmic concentrations of osmolytes by rebinding of amino acids etc. (Kaneshiro *et al.*, 1969). In the long term the freshwater protist maintains a cytoplasmic osmotic pressure above that of the external medium, and, where contractile vacuoles are present, the water flux through the cell is

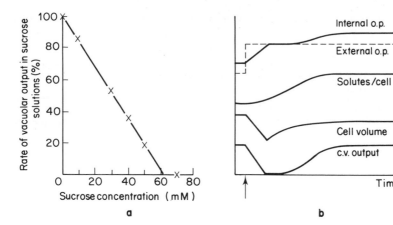

Fig. 2.5 a, The effect on the rate of contractile vacuole output of **Discophrya** of various sucrose solutions, compared with the rate in water (output is calculated from frequency of contraction and maximum diameter attained before contraction). **b,** The longer term effects of a change in environmental osmotic pressure on a protist; a rise in external osmotic pressure causes a reduced output and temporary stoppage of the contractile vacuole, associated with a reduction in cell volume and a concomitant rise in internal osmotic pressure, and followed by an increase in concentration of the free osmolytes in the cell that restores internal hypertonicity, so that the contractile vacuole fills once more and the cell volume increases towards its original size. A series of reverse changes would accompany a reduction in external ospmotic pressure.

maintained with continued activity of the contractile vacuole. Protists with cell walls, which normally lack contractile vacuoles, respond in the same long-term manner by adjusting internal osmolarity; in the short-term, swelling in dilute media is prevented by the cell wall, but in hyperosmotic media the cells show plasmolysis until regulation of the internal osmotic pressure has been achieved. Responses are necessarily more complex when changes in external osmolarity are caused by substances to which the membrane is permeable, particularly ions.

In order to exert osmoregulatory control, the contractile vacuole should expel a fluid of low osmotic pressure. The osmolarity of the contractile vacuole fluid of large amoebae was found by direct measurement to be less than half that of the cytoplasm, but much higher than the osmolarity of the medium (Schmidt-Nielsen and Schrauger, 1963; Riddick, 1968). However, expulsion of this fluid still results in a loss of osmotically-active materials, and the constant osmolarity of the cytoplasm can only be maintained if the loss of osmolytes is made good in feeding or active uptake from the medium. How does the contractile vacuole produce a fluid that is hypotonic to the cytoplasm? The concentration of sodium ions is much higher and potassium concentrations much lower in vacuolar fluid than in the cytoplasm, so evidently some ionic regulation occurs. It is thought that probably the segregation of water from cytoplasm in the spongiome occurs by active secretion of solutes (sodium ions?) into the vesicles or tubules, whose contents become hyperosmotic to cytoplasm, so that water follows; selective reabsorbtion of solutes (especially ions?), could then reduce the osmotic pressure of the vacuolar fluid.

Ionic regulation is often closely coupled to the regulation of water content. In addition to the selective extrusion of sodium ions and retention of potassium ions at the contractile vacuole, there is evidence of active uptake of potassium, sodium and phosphate ions and active extrusion of sodium, calcium and probably chloride ions at the surface membrane (Connolly and Kerkut, 1983), as well as uptake from food and passive leakage.

It is often assumed that metabolic products and waste nitrogenous substances are removed from the body in the vacuolar fluid; however, while urea has been found in the contractile vacuole fluid of the large ciliate *Spirostomum*, it is likely that most nitrogenous wastes of protists are liberated as ammonia and diffuse rapidly away.

The dinoflagellates have vacuoles called pusules, sometimes formed into complex systems (p. 116), whose function has been thought to be the same as that of contractile vacuoles (Dodge, 1972). However, pusules only rarely show swelling and shrinkage and do not undergo regular contractions. In complex pusule systems there is a main vacuole surrounded by subsidiary vacuoles in a specialised region of cytoplasm and connected to the surface by one or more tubular canals.

Food Vacuoles

Another important type of vacuole is that which is concerned with the intake and digestion of food (Allen, 1984; Kitching, 1956; Nilsson, 1979). Food vacuoles are conventionally divided into phagocytic vacuoles, which enclose large food particles, and pinocytotic vesicles, which enclose invisible food materials, either in solution or adsorbed on the surface membrane. No absolute distinction can be made between these, but some protozoa clearly form vacuoles of both types while others use only one or the other. For example, phagocytic vacuoles are formed following food capture by various means discussed in Chapter 3, either at permanent cytostomes used repeatedly or at 'food cups' formed at whatever site food is encountered. The development of invaginations, which may be cut off to form pinocytotic vesicles, has been observed frequently in amoebae; the rate of formation of the vesicles may be stimulated by protein, basic dyes and some salts in the medium, but not by carbohydrates. The coated pits formed around parasomal sacs of *Tetrahymena* (p. 191) are also sites of pinocytosis (Nilsson and van Deurs, 1983). Autophagic vacuoles may surround certain organelles prior to their autodigestion during morphogenetic reorganisations of the body.

The organic contents of phagocytic, pinocytotic or autophagic food vacuoles are digested by the activity of lysosomal enzymes during a sequential process that has been best documented for phagocytic vacuoles of ciliates like *Tetrahymena* (Nilsson, 1979) and *Paramecium* (Allen, 1984) (Fig. 2.6). In ciliates membrane for food vacuole formation cannot be obtained from the cell surface, since the pellicle is complex, but is recruited from discoidal vesicles which fuse with the vacuole membrane around the cytostome as captured food is passed into the vacuole. The filled vacuole is pinched off at the cytostome and the vacuole is carried away into the cell interior. Small acidosome vesicles, with few intramembranous particles (IMPs), fuse with the vacuole and the pH of its contents quickly falls to below 3, killing most prey. Within about five minutes after formation (in *Paramecium*) the vacuole has decreased markedly in size by the formation of tubular vesicles at its surface which remove both fluid and

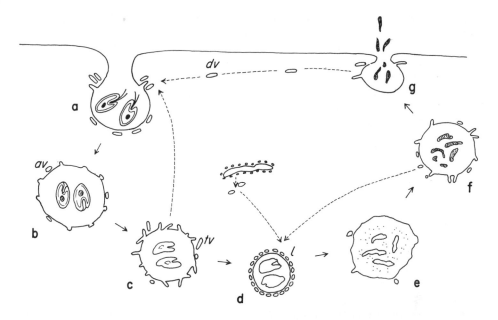

Fig. 2.6   The food vacuole cycle of a ciliate like *Paramecium*. The food vacuole **(a)** receives captured food particles at the cytostome, where it also receives membrane from discoidal vesicles *(dv)*; the food vacuole is pinched off from the surface membrane, and as it moves away **(b)** it fuses with acidosome vesicles *(av)*; fluid is removed from the vacuole **(c)** into tubular vesicles *(tv)*, which may recycle to the cytostome; the shrunken vacuole **(d)** receives enzymes in lysosomes *(l)*, that are either newly produced from ER or recycled from old food vacuoles; following digestion **(e)** micropinocytosis occurs around the vacuole and later **(f)** some lysosomal materials may be retrieved and recycled before the spent vacuole reaches the cytoproct **(g)**, where undigested materials may be released and membrane retrieved to be returned to the cytostome as discoidal vesicles. (Information from Nilsson, 1979 and Allen, 1984).

membrane (probably recycling discoidal vesicle membrane to the cytostome); this shrinkage is accompanied by a loss of IMPs from the vacuolar membrane. In the next few minutes lysosomal vesicles with a distinct glycocalyx first bind to and then fuse with the food vacuole membrane, releasing into the vacuole acid hydrolase enzymes which digest the contents. As a result the vacuole swells, its pH rises once more and more IMPs (from lysosome membranes?) are present in its membrane. The products of digestion may be absorbed through the vacuole membrane or may be removed by micropinocytosis around the vacuole surface. By about 20 minutes acid phosphatase activity is no longer detectable but tubular evaginations occur around the vacuole; these appear to be sites of retrieval of glycocalyx-bearing lysosomal membrane, which may form vesicles and be recycled after fusion with newly synthesised primary lysosomes. The vacuole shrinks again at this time and it is finally carried to the cytoproct (cytopyge) where the membrane of the spent food vacuole fuses with the plasma membrane and releases its contents to the exterior. The residual food vacuole membrane is rapidly pinched off into small vesicles which are recycled by being transported back to the cytostome as discoidal vesicles.

The food vacuoles of other phagotrophic protists seem to behave in a similar manner, although the timing varies widely. Thus, while the life of a food vacuole in **Paramecium** lasts about 30 minutes, it takes about 2 hours in **Tetrahymena** and 12–16 hours in **Actinophrys** (Patterson and Hausmann, 1981). Such active processes as food engulfment in amoeboid forms, pinching off of food vacuoles at the ciliate cytostome, fusion of vesicles at the food vacuole surface and transport of food vacuoles within the cell appear to involve actin filaments, since in at least some cases they are inhibited by cytochalasin (p. 28).

Fibrous structures

The cytoplasm of eukaryote cells contains various structural proteins formed into fibrous elements of three main classes: microtubules, microfilaments and intermediate filaments (Alberts *et al.*, 1983). These are usually formed from globular proteins polymerised together into chains called protofilaments (Fig. 2.7a). Two protofilaments of actin molecules (M. W. 42 kD) are twisted around one another to form microfilaments about 7 nm thick (Fig. 2.7b). Microtubules are cylindrical elements normally about 25 nm in diameter with walls 5 nm thick formed from 13 protofilaments of tubulin (M. W. 55 kD) (Fig. 2.7d). Intermediate filaments (6–12 nm thick) composed of a variety of different proteins, usually with linear rather than globular molecules, have been recently recognised in cells of vertebrates; filaments that may be strictly equivalent have not been adequately studied in protists, although there are many reports of filaments in protists that appear to be neither microfilaments nor microtubules. All three types of element form parts of cytoskeletons; microfilaments and microtubules can also form parts of motile systems, and although contractility has not been reported for intermediate filaments from vertebrates, some comparable filaments in protists are contractile.

Microtubules and microfilaments are labile structures, the polymerised fibrous state being often in dynamic equilibrium with the depolymerised mono-

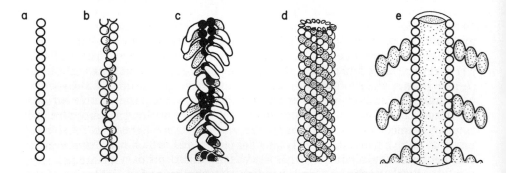

Fig. 2.7   Protein fibres. Globular molecules may form protofilaments **a**; two protofilaments of actin spiral around one another to form a microfilament **b**, and 13 protofilaments of tubulin form the walls of a microtubule **d**. Myosin head fragments bind to an actin microfilament in a polarised 'arrowhead' array **c**, and molecular complexes of dynein bind to a microtubule (seen in section at **e**) in a polarised fashion.

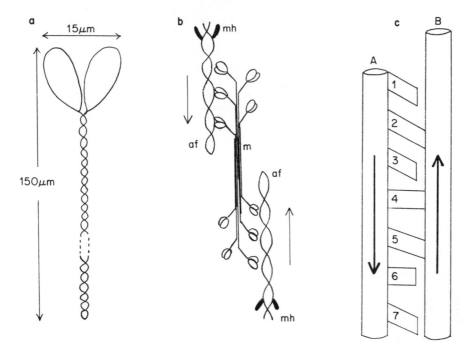

Fig. 2.8 The mechano-chemical proteins myosin and dynein and their role in contraction. The myosin molecule **a** is formed of a long polypeptide backbone with paired heads. Myosin molecules aggregate to form bipolar filaments (*m*) shown in **b**, where heads at one end of the mysoin filament link to an actin filament (*af*) of one polarity, while heads at the other end of the myosin filament link to an actin filament of opposite polarity (the polarity of the actin filaments is indicated by the way they bind isolated myosin heads (*mh*): bending movements of the attached myosin heads will draw the opposite actin filaments towards one another. Dynein molecules (1–7) shown in **c** projecting from microtubule A may link distally to a second microtubule (B), and, once linked (2), they change orientation from a downward tilt to stand perpendicular to the microtubules (5,4), before releasing their distal attachment (6) and returning to the original orientation (7); during the change in shape of attached dynein molecules, the two microtubules are moved in opposite directions.

mers. Both structures form polarised fibres that can exist under conditions in which assembly (and growth) of the polymer is favoured at one end of the fibre, whilst disassembly is favoured at the other end; commonly some stability is given to the structures because one or both ends may be 'capped' by other proteins. Either microfilaments or microtubules can take part in active motility when pairs of similar fibres are linked together by special mechano-chemical proteins. Molecules of the protein myosin (M. W. 500 kD) (Fig. 2.8a) form filaments whose two ends have opposite polarity; the heads of myosin molecules at either end of the filament can link with actin filaments of opposite polarity, and by cyclical attachment, neck-bending, detachment and straightening of the myosin heads, using energy from ATP, the two actin filaments can be drawn together (Fig. 2.8b). Experimental exposure of actin filaments to a preparation of myosin head fragments, which bind to the actin monomers, not only shows that the filaments are made of actin, but also displays their polarity (Fig. 2.7c).

Microtubules bind the protein dynein (M. W. 1.3–1.9 MD) in a similar way (Fig. 2.7e), but dynein molecules do not form separate filaments. Where dynein causes relative sliding between two microtubules, as in cilia, the bases of the dynein molecules are fixed along one side of one microtubule and the dynein tips form temporary attachments with another microtubule of the same polarity; when the dynein molecules undergo cycles of ATP-powered attachment, straightening, detachment and recovery they push one microtubule down and the other up (Fig. 2.8c).

The mechanical nature of the rope-like actin microfilaments makes them useful as tensile elements. Contractile structures involving actin are present in most if not all eukaryote cells – plants and protists as well as animals. Changes in shape of amoeboid cells (p. 159), phagocytosis, cell cleavage and some forms of cytoplasmic flow, among other cell functions, appear to be actin-based. Bundles of actin filaments also form non-motile anchoring fibres, and networks of actin filaments cross-linked by other proteins provide the structural framework of many cytoplasmic gels. Actin filament functions are disrupted by cytochalasins and phalloidin.

Microtubules possess some resistance to longitudinal compression and lateral distortion as well as extension, so they can provide stiffness and can be used for both pulling and pushing. This makes them suitable for structural functions different from those of microfilaments. Microtubules are used to maintain the shape of cells or their components, and bundles of cross-linked microtubules support cellular projections like heliozoan axopods (p. 174), strong internal structures used in feeding (p. 66), or pellicles (pp. 122,190, 227). The bending cycles of cilia require stiffness of the microtubules as well as the pushing and pulling of cross links made of dynein (p. 36), and in the axostyles of oxymonad flagellates microtubules and 'dynein' again interact to produce undulatory movements (p. 121).

Certain forms of cytoplasmic flow and organelle transport and perhaps some movements in mitosis may depend on dynein arms projecting from microtubules. However, active movements can also be caused by the change in length of microtubules through controlled assembly and disassembly (Sleigh and Pitelka, 1974). For example, rapid shortening of axopods (powered by surface tension) following stimulation of microtubule disassembly is used in food capture by centrohelids (p. 177), and slower length changes of axopods occur in locomotion of heliozoans; length changes of microtubules, which may contribute motile force, also occur during mitosis (p. 47). Microtubules only form in the presence of microtubule organising centres (MTOCs), which need not themselves be microtubular, or by elongation of existing lengths of microtubules (Alberts *et al.*, 1983). Colchicine and vinblastine bind to tubulin and disrupt the more labile forms of microtubule (Alberts *et al.*, 1983); heavy metals are also disruptive (Shigenaka, 1976).

Fibrous proteins of other types abound in ciliates, larger flagellates and some amoebae, and may be present in most or all protists. Whilst many types of cytoskeletal filaments may be present in protists, few of them have attracted detailed attention except those that are contractile. One exception is the type of crossstriated rootlet fibre, formed of filament bundles, attached to the basal bodies of many cilia and flagella; although in a few cases, e.g. ***Platymonas*** (p. 141), these also are contractile (Salisbury, 1983). Other filamentous contractile structures include the spasmoneme of peritrich stalks (p. 211) (Amos, 1975), myonemes of certain ciliates (p. 194) (Huang and Mazia, 1975), myonemes of acantharians (p. 183) (Febvre, 1981), filament bundles in the stalk appendage (piston) (Cachon and Cachon, 1981) and also in the flagellum of certain dino-flagellates (p. 110) (Maruyama, 1982). The 3–6 nm thick filaments in these structures all seem to share some common features and may all be formed of similar proteins that change configuration (contract), without a need for ATP, when they bind calcium ions. For example, the spasmoneme of *Vorticella* appears to contain a single fibrous protein (spasmin or spastin, M. W. 20 kD) which binds calcium at micromolar concentrations and contracts, but relaxes if calcium ions are removed by reducing the concentration; this explanation is almost certainly oversimplified (Amos, 1975; Yamada and Asai, 1982). Although protein filaments of this family are widespread, but morphologically diverse, in protists, they have not been reported in other eukaryotes. Other fila-ments that could have a motile function exist in gliding gregarines (p. 226) where two types of membrane-associated and a third type of ectoplasmic fila-ment occur in the surface folds (Walker *et al.*, 1979). The costa of *Trichomonas* (p. 116) is a large cross-striated rod, associated with flagellar bases, which propagates undulations along its length; it is composed of thin lamellae and appears to undergo ATP-dependent folding of the lamellae as it bends (Amos *et al.*, 1979). The component responsible for contractility of membrane stacks in several ciliates is not known (Sleigh, 1984b), though filaments associated with membrane arrays in *Tontonia* may be contractile (Greuet *et al.*, 1986).

The structure of cilia and flagella

Cilia and flagella are characteristic organelles of a great many protists and deserve detailed description(*see* Sleigh, 1974; Gibbons, 1981). The two types of organelle are structurally the same, but have minor functional differences; although it is probably correct to describe cilia as a special class of flagella, the word cilium will be used to refer to the generalised organelle. Cilia are cylin-drical organelles about 0.25 μm in diameter, composed of a longitudinal bundle of microtubules (the axoneme) surrounded by cytoplasm and enclosed within an extension of the cell surface membrane. The axoneme extends into the surface region of the cell as the basal body, sometimes called the blepharoplast in flagel-lates or the kinetosome in ciliates. Although some of our knowledge of ciliary organisation and function has been derived from metazoan material, the cilia of *Tetrahymena* and *Paramecium* and the flagella of *Chlamydomonas* have been used in much of the research described below.

The microtubular fibrils of the axoneme are arranged in a precise pattern, best seen in cross-section, known as the 9 plus 2 array (Fig. 2.9). The 9 peri-pheral doublets are arranged in a circle about 0.2 μm in outside diameter, and each consists of a complete A microtubule which shares part of its wall with an incomplete B microtubule. These and the two separate central microtubules appear to run straight along the cilium without spiralling. A basic 8 nm periodi-city of the microtubules of the axoneme, provided by the dimers of tubulin of which they are composed, is reflected in the arrangement of other protein components attached to the axonemal microtubules. In most cilia each central microtubule bears curved projections attached at 16 nm intervals in two rows;

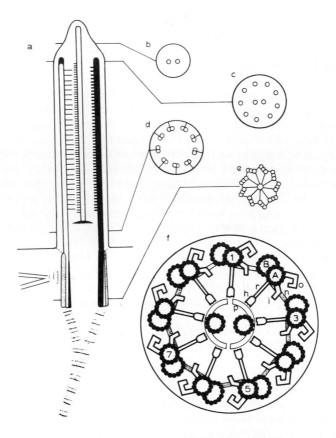

Fig. 2.9   Diagrams of details of ciliary structure described in the text, as seen in longitudinal section at **a** and in transverse sections at various levels in **b-f**. At the left in **a** the arrangement of radial spokes and central projections is shown as seen in a median longitudinal section and at the right the arrangement of dynein arms seen in a more tangential section is shown. The structures seen in a transverse section of the main part of the ciliary shaft are shown in detail at **f**; they include the outer doublets numbered 1 to 9 and formed of A and B microtubules, the A microtubules bearing outer (*o*) and inner (*i*) dynein arms, radial spokes (*r*) with dilated spoke heads (*h*), and connected to adjacent doublets by nexin links (*n*), as well as the central singlet microtubules with curved projections (*p*). The outer doublets become triplets in the basal body (*see* **a** and **e**), where microtubular and/or striated roots may arise from associated dense material.

these were formerly regarded as a central sheath, but the central microtubules and projections together are now called the central complex. Paired links connect the two central microtubules at intervals of 16 nm in some species. At intervals of about 32 nm each A microtubule bears centripetal projections (spokes), formed of a spoke shaft and a spoke head, the latter apparently being able to link with the central complex. Connections between adjacent doublets occur at 96 nm intervals in the form of inter-doublet (nexin) links. Finally, each A microtubule bears lateral arms that project at intervals of 24 nm in inner and outer rows towards the B microtubule of the adjacent doublet (i.e. in a clockwise direction when looking from the base towards the ciliary tip); these dynein arms are complex structures with 2 or 3 major subunits and are about 9 nm thick, 20–25 nm long and tilted downwards at an angle of about 32° (Warner and Mitchell, 1978; Gibbons, 1983).

It has proved possible using two dimensional gel electrophoresis to separate about 200 polypeptides present in the flagellar axoneme of *Chlamydomonas*, and the isolation of mutants that lack specific structural features allows the identification of polypeptides which correspond with certain axonemal structures (Luck, 1984). Thus mutants from which the outer arms are missing lack 13 polypeptides while 9 different polypeptides are absent from mutants lacking the shorter inner arms; mutants lacking radial spokes lack 17 polypeptides, only five of which are missing when only spoke heads are absent. Such 'genetic dissection' of the axoneme is being coupled with biochemical and physiological studies to understand the functioning of axonemal components.

At the ciliary tip the B microtubule normally terminates before the A microtubule, but there are no lateral structures on the A microtubule after it becomes single. Usually the central microtubules are longer than the peripheral ones and may extend into a prolonged narrow tip. In at least some cases the central microtubules terminate in a cap structure which is attached to the membrane of the ciliary tip, and the ends of the A microtubules may also be linked to the ciliary membrane near the tip (Dentler, 1980). The arrangement of the axonemal microtubules is also modified in the transition zone between the base of the shaft and the basal body. The base of the shaft is marked by the termination of the central microtubules, often at a plate or granule, and no dynein arms or radial spokes are present below this level. The top of the basal body is marked by the addition of another incomplete microtubule attached to the side of the B tubule of each doublet to make triplets (Fig. 2.9). In the transition zone additional links between the peripheral doublets may occur, e.g. on their outer surface or as a stellate pattern of filaments within the axoneme connecting adjacent doublets. The doublets are frequently connected to the ciliary membrane by short links in the transition zone (and sometimes also elsewhere); the site of attachment of these links to the membrane may be marked by rows of IMPs (Fig. 2.10) but the origins of other particle patterns on ciliary membranes are unknown (Bardele, 1981).

The basal body is usually about 0.5 μm long. The triplets of microtubules show an axial twist so that the A microtubule comes to lie nearer the centre; thin strands interconnect the triplets in the basal body and at its inner end a hub-and-spokes array of filaments is often seen (Fig. 2.9e). Thin filaments may connect the distal end of each triplet to the cell membrane near the ciliary base. Dense granular material often surrounds parts of the basal body, and may provide a site for attachment of rootlet structures, either microtubules or bundles of cross-striated filaments, whose nature and arrangement vary widely in different organisms as described later.

The centriole associated with the mitotic spindle in many organisms has the same detailed structure as a basal body; it serves the function of an MTOC and was probably derived from a basal body in an ancestral flagellate. Both basal bodies and centrioles have been regarded as self-replicating organelles because new ones arise close beside (and normally perpendicular to the basal end of) existing ones. However, a former belief that they contain DNA has now been

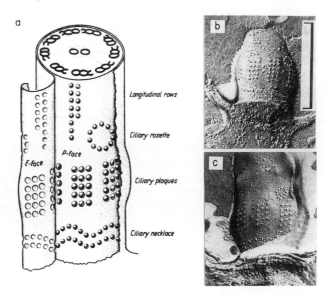

Fig. 2.10 Intra-membranous particles in the ciliary membrane. Diagram **a** shows how particles are revealed on the protoplasmic half (p-face) and imprints of particles are left on the external half (e-face) of the membrane's lipid bilayer by the freeze-fracture technique. A ciliary necklace formed of two rows of particles has been found in all ciliates examined (flagellates often have 3 or 4 rows). Plaques, rosettes and longitudinal rows of particles characterise particular groups; *Frontonia* cilia, for example, have particle plaques and rosettes near their bases, as shown, in **b** and **c**. (Diagram and micrographs by Bardele, 1981). Scale 0.5 μm.

abandoned, the possible presence of RNA has no established relevance, and in some cases basal bodies or centrioles appear to arise *de novo*, from some other, unidentified, MTOC. Basal bodies are essential for ciliary development, but detached cilia continue to swim *in vitro* (Goldstein, 1974), and *Chlorogonium* flagella still beat normally in forward and reverse mode after detachment of basal bodies at division (Hoops and Witman, 1985).

Cilia are frequently combined together to form compound organelles called cirri, membranelles or ciliary undulating membranes; no membranes or formed structures seem to be present to hold the component organelles together, although adhesion by secreted mucous material is a possibility in some cases. An additional outer membrane, the perilemma, is present around the cilia of some ciliates, and may cover the whole body (Bardele, 1981). With these exceptions, the structure of motile cilia is usually very stereotyped, although length varies widely with function. By contrast, there are several interesting variants of the flagellar pattern. They include intraflagellar material lying alongside the axoneme, hairs, scales or other structures attached outside the flagellar membrane, and extensions of the flagellar membrane which may form a flagellar undulating membrane, one edge of which is attached to the surface of the cell while the other contains the axoneme. These variants characterise particular groups of flagellates and are described in Chapter 5. While additional internal material may be important in providing additional stiffness within the flagellum, the external structures may have an effect on the hydrodynamics of flagellar motion. A few structures of protists are derived from modified ciliary organelles, e.g. the outer region of the stalk of **Vorticella**, which develops from a scopula (p. 215).

Certain flagellates carry an organelle called a haptonema (Parke, Manton and Clarke, 1955), which is similar in dimensions to a flagellum but has a different internal structure and moves in a different manner. The lengths of haptonemata vary in different species (*see* Fig. 5.31), and they usually show slow uncoiling movements and more rapid coiling-up (Leadbeater, 1971); there is no information on the mechanism of movement. The name haptonema indicates that they are adhesive and may attach to foreign structures, perhaps in feeding or for anchorage. The outer membrane of the haptonema, continuous with the cell membrane, is mostly underlain by a tube of endoplasmic reticulum, formed of two inner membranes around a vesicular cavity. Within this is a bundle of 5–9 single longitudinal microtubules. Normally the haptonema arises between two flagella (Fig. 5.31), and its microtubules extend down into the cell to terminate close to the flagellar basal bodies; it is suggested that it arose from a bundle of flagellar rootlet microtubules (see p. 150).

Movement of cilia and flagella

Protists have been used frequently in studies of the movement of flagella and cilia and of the coordination of cilia; they provide a rich variety of examples (*see* Sleigh, 1974; 1981). Since these organelles beat very quickly (often more than 20 times per second), it is necessary to employ special methods for studying them. The simplest apparatus is a stroboscope, which may be a mirror or perforated disc rotated by an electric motor, or an electronic flash tube driven by a timing circuit; in either case the flashes of light are made to occur at the same instant in each cycle of beat of the organelle, which therefore appears to stand still. This technique allows the study of the metachronal waves of cilia, and if the stroboscope is set at a slightly lower frequency of beat than the cilium the form of beat of the slowly moving cilium can be studied. High-speed cine photography and video filming are also most valuable tools in the study of the movement and coordination of cilia, allowing detailed analysis of the form of beat, of the construction of metachronal waves and of the currents they produce. Photographs taken with electronic flash may be used to supplement these techniques.

The movement of external motile organelles may be slowed down considerably by the use of such viscous agents as methyl cellulose; in these media flagella and cilia show an abnormal beat, and cilia show modified coordination patterns, so that observations must be interpreted with extreme caution (Machemer, 1972b). The shape of cilia at various stages of their cycle of beat and the form of metachronal waves have been studied on 'instantaneously-fixed' organisms by light and scanning electron microscopy, using either chemical (Parducz, 1967; Tamm and Horridge, 1970) or rapid freezing techniques (Barlow and Sleigh, 1979). In this case also it is necessary to insist on cautious interpretation of the pictures obtained, although H. Machemer (1972a) has shown that profiles of cilia seen in flash photographs of living **Paramecium** appear the same as those of cilia on instantaneously fixed specimens. For stationary organisms with a regular ciliary beat the stroboscope remains the most

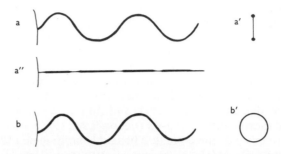

Fig. 2.11 The movement of flagella. In some flagella the propagated waves are planar and the flagellum appears different when viewed in different directions (**a** in side view and **a''** with the flagellum viewed in the plane of the waves), while in other flagella the waves are helical and appear the same when viewed from different angles around the flagellum, **b**. The appearance of the two types of waves as seen from the flagellar tip is shown at **a'** and **b'**.

convenient aid in observation, but for actively motile organisms or those with irregularly beating organelles the use of cine or video methods is necessary, though flash photography also helps.

Flagella move in an undulating manner (Fig. 2.11). The movement may take place approximately in a single plane, so that the envelope of movement is almost a flat rectangle, or the flagellum may move out of the plane to a greater or lesser extent, so that the envelope of movement becomes more nearly cylindrical as the beat tends towards the helical form. Such movements are normally regular and symmetrical in a healthy flagellum, and several complete waves of movement of the organelle may frequently be seen within the length of the flagellar shaft. In some cases the movements towards either side become asymmetrical, and an extreme form of such a unilateral tendency is the ciliary beat described below. Helical or planar undulations may travel along the flagellar shaft in either direction (Fig. 5.6, p. 107), but waves of movement originate at the flagellar base much more commonly than at the tip (Holwill, 1964, 1966). The movement of waves along the flagellum produces a resulant force on the water acting along the long axis of the organelle in the direction of the waves, provided the flagellum has a smooth surface. In considering the effective propulsive force of a flagellum it is important to realise that inertial forces are negligible in comparison with viscous forces, in structures as thin as a flagellum moving in water at relatively low speeds (although the frequency of beat may be high). Part of the available energy must be used up in changing the shape of the flagellum (i.e. in overcoming internal elastic resistance) and part is used in doing work against the viscous resistance of the water around the flagellum (i.e. in propulsion). Under optimal conditions the viscous work may be about three times the elastic work, and C. Brokaw calculated that the total energy dissipation of a 40 μm long flagellum beating at 30 Hz with a wavelength of about 30 μm is about 3×10^{-7} erg s^{-1} at 16°C (Brokaw, 1965). The rate of propulsion achieved by a given flagellar waveform depends principally on flagellar length, the amplitude and wavelength of the waves and the speed at which these waves move along the flagellum; the thickness of the structure is of rather less importance. The viscosity of the medium is only important in that it may affect the shape of the waves and their frequency (Holwill, 1966, 1974).

Lateral hairs on the flagellum modify the movement of water that is produced by the flagellar undulations. These hairs are of at least two types: long slender hairs about 5 nm thick, frequently in one row, as in *Euglena* (Leedale, 1971), and thicker, stiffer projections about 20 nm thick and 1 μm or so long (Figs. 2.12 and 5.5) (Bouck, 1971). It is common to refer to all types of flagellar hairs as flimmer filaments and to reserve the term mastigoneme for the thicker, usually tripartite, projections. When mastigonemes project rigidly from the flagellum in two rows arranged in the plane of the flagellar undulation, they cause a reversal of the flow of water produced by the undulation of the flagellum, so that the propagation of bending waves along the flagellar axis from base

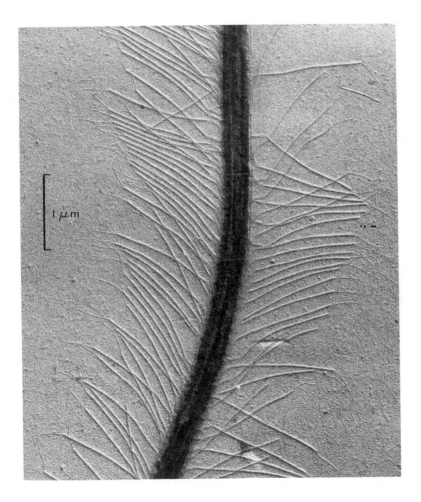

Fig. 2.12 Part of a flagellum of **Paraphysomonas** showing the form of the thick lateral hairs (mastigonemes), which have three regions, a basal part, a main shaft and fine bristles at the tip (*see also* Fig. 5.5). (Electron micrograph from Leadbeater, 1972).

to tip results in a flow of water towards the flagellar base (Fig. 5.6h, p. 107). The observed rate of water flow produced by an ***Ochromonas*** flagellum bearing mastigonemes was of the same order as the rate calculated by M. E. J. Holwill, using hydrodynamic equations involving surface coefficients of resistance to movement of the flagellum and its mastigonemes (Holwill and Sleigh, 1967).

The use of the techniques of observation mentioned above has also shown a rich diversity of patterns of ciliary beating (Sleigh, 1968, 1974). The ciliary beat involves two phases, one in which the cilium moves towards one side while bending only in the basal region, and a second in which the region of bending is propagated up the ciliary shaft to the tip and returns the cilium to the starting position (Fig. 2.13a). The latter phase usually occupies much more than half of the beat cycle and is referred to as the recovery stroke, while the rigid swing is referred to as the effective stroke since this stroke produces the main water flow. Occasionally the beat takes place more or less in one plane, and where the beat is not planar it appears that the more rapid movement of the effective stroke is approximately in one plane, while the movement of the recovery stroke involves a swing of the cilium to one side. In protozoa this swing appears to be consistently to the left of the direction of the effective stroke, so that viewed from the ciliary tip the recovery stroke moves anticlockwise (Machemer, 1972a; Parducz, 1967) (Fig. 2.13b), but a swing to the right occurs in some Metazoa (Aiello and Sleigh, 1972). Beat patterns show many adaptations for specialised functions.

The bending of the ciliary axoneme that produces these beat patterns results from active sliding between adjacent doublets generated by unidirectional activity of the dynein arms. During these cycles of change in configuration of the dynein arms, coupled with attachment to and detachment from the adjacent doublet, each dynein arm splits ATP and propels the adjacent doublet tip-wards (Sale and Satir, 1977). Normally the effective stroke occurs towards doublets 5 and 6 (Fig. 2.9), so activity of dynein arms on doublets 1–4 produces the effective stroke and activity of arms on doublets 6–9 produces the recovery stroke, reverse sliding being passive on other doublets in each stroke. The sliding of doublets is thought to be converted to bending of the axoneme through the

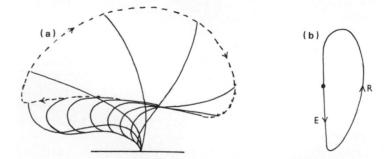

Fig. 2.13   The movement of a cilium. A series of profiles showing an unspecialised example of a cycle of ciliary beating as viewed from the side at **a** may be subdivided into the effective stroke, during which the ciliary shaft swings from left to right through a large arc, and the recovery stroke, during which the bent cilium unrolls towards the left. The path followed by the ciliary tip sometimes follows an anticlockwise loop, as shown at **b**, where the cilium is viewed from the tip, with part *E* representing the effective stroke and part *R* the recovery stroke.

presence of such transverse elements as radial spokes, and variations in beat pattern could result from differences in the pattern of timing of the active sliding of doublets at the two sides (Sleigh and Barlow, 1982b).

When the ciliary or flagellar beat is planar, the plane of bending is at right angles to a line drawn through the two central microtubules. This relationship between the plane of bending and the orientation of central microtubules has also been shown to hold locally within regions of cilia beating in three dimensions (Tamm and Horridge, 1970). A systematic change in the orientation of the central microtubules at the base of **Paramecium** cilia during the beat cycle has led to the suggestion that the central pair rotate through 360° during each beat cycle, and that this rotation is concerned with the production of three-dimensional beating (Omoto and Kung, 1980). There is also evidence for central-pair rotation in the flagellate **Micromonas** (Omoto and Witman, 1981). It is difficult to evaluate this evidence and further corroboration is awaited; Machemer has suggested that it may be mechanically necessary for the central pair to rotate if the three-dimensional swing of the cilium is extensive.

The analysis of high-speed cine films has provided data for calculations of the work performed against viscous resistance during the ciliary beat (Sleigh and Holwill, 1969), and of the stiffness of the ciliary shaft at different parts of the cycle (Rikmenspoel and Sleigh, 1970). The viscous work done during the effective stroke, and thus the propulsive effect on the water, is dependent on ciliary length^3 and angular velocity^2; the viscosity of the medium and the duration of the stroke are also important, but the ciliary diameter is of less importance. In one example (Holwill and Sleigh, 1969), the work done against viscosity in one effective stroke lasting 24 ms by a cilium 32 μm long was calculated to be about 4×10^{-9} ergs. In the recovery stroke of the same cilium, lasting 36 ms, the viscous work was calculated to be about 1×10^{-9} ergs. The work done against the stiffness of the cilium depends on the amount of bending (the radius of curvature) and on structural features. There is some evidence that the stiffness of the cilium is some 10 times less during the recovery stroke than during the effective stroke, the additional stiffness of the latter being provided by active forces within the cilium (Rikmenspoel and Sleigh, 1970). On this basis it is estimated that the cilium mentioned above, which performed about 5×10^{-9} ergs of viscous work per cycle, would perform between 10^{-9} and 10^{-8} ergs of elastic work against the stiffness of the cilium in each cycle. This requires a total power dissipation by the cilium of between 10^{-7} and 10^{-6} ergs s^{-1}, which is closely comparable with the power output of a flagellum (Hiramoto, 1974).

The propulsion of water by a cilium depends on the fact that a structure moving in water carries with it a surrounding layer of water which forms a no-slip layer in contact with the surface, and is progressively less influenced by the moving structure at greater distances from the surface (Blake and Sleigh, 1974). The presence of a cell surface also influences the flow of water over it; as a result the moving cilium carries around itself a zone of water whose diameter increases with distance from the cell surface (Fig. 2.14a). In the effective stroke the zone of water carried by the cilium is large and conical because the cilium is extended straight out from the cell surface, while in the recovery stroke the zone is small and cylindrical because the cilium moves parallel and close to the cell surface; during the beat cycle shown in Fig. 2.14b, therefore, more water is carried to the right in the effective stroke than is carried back to the left in the recovery stroke.

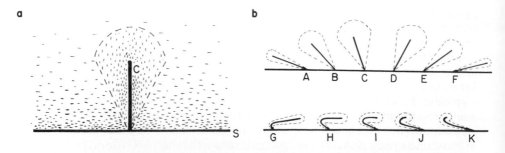

Fig. 2.14 The propulsion of water by a cilium is a process dominated by viscous forces. When a cilium (*c*) moves in water (part **a**) it carries with it a surrounding zone of water, whose extent varies with the distance from the cell surface, while movement of water further away from the cilium is restrained by the presence of the cell surface (*s*). During the beat cycle **b** the cilium influences a larger volume of water (dotted outlines) in the effective stroke (profiles A–F) than in the recovery stroke (profiles G–K) and therefore achieves a net transport of water towards the right. (From Sleigh, 1984a).

The longer the cilium and the more frequently it beats, the greater the amount of water it can propel. But, if a cilium is too long, or moves close to a solid surface, it will bend backwards and lose propulsive efficiency, so cilia have an optimal length which is influenced by their function and by whether they provide mutual assistance for one another through the formation of compound structures or metachronal waves (see below) (Sleigh, 1984a). The activity of cilia is integrated to produce locomotion or feeding currents (Sleigh and Barlow, 1982a), using mechanisms discussed in the next section.

The coordination and control of cilia and flagella

These phenomena and the mechanisms that underlie them have attracted much interest; metachronal coordination of cilia has been discussed by Machemer (1974) and Sleigh (1974) and ciliary control by Naitoh and Eckert (1974) and Sleigh and Barlow (1982b).

When two cilia lie close enough for their transported water layers to overlap, interference will occur between the movements of the two cilia, and the activity of the two cilia will probably become hydrodynamically linked (Sleigh, 1984a). Such hydrodynamic linkage will also occur between two moving flagella which lie close together; this accounts for the synchrony of beat seen in tufts of flagella and the coordination between the flagella of separate spermatozoa seen in packed suspensions. It also explains the waves of coordination of the flagella which densely cover the bodies of some of the larger flagellates, and even the coordinated movement of spirochaetes that live attached to the large flagellate *Mixotricha* (Cleveland and Grimstone, 1964).

In most cases the coordinated activity of cilia is held to result from hydrodynamic linkage of this type. The form of interaction between two adjacent cilia will depend on the positional relationship between them (distance and position with respect to plane of beat), on their length and on the pattern of ciliary beat that they perform. The compound ciliary organelles of protozoa contain cilia which are structurally separate; they appear to beat as a single compound unit in life, but fray apart quickly on death. The bases of the component cilia in cirri and membranelles lie so close together that there is tight viscous-mechanical coupling (i.e. hydrodynamic linkage) between adjacent cilia, and the beat of all of the units is closely coupled unless the compound structure is artificially split apart. In a ciliary undulating membrane (p. 196) the basal bodies of the cilia lie very close together, but are arranged in a long line; if a cilium at one end of the line beats, the motion of this cilium is coupled to the next and is passed to all members of the row in sequence, so that the cilia beat metachronally and waves of movement pass along the membrane. In many arrays of cilia the bases are further apart, but the cilia still beat in metachronal waves if there is sufficient hydrodynamic coupling between them.

The patterns of metachronal coordination which result are diverse (Fig. 2.15). In linear arrays of cilia the plane of the effective stroke is usually approximately perpendicular to the row along which the metachronal waves pass, and the metachronism is referred to as diaplectic, conforming either to the dexioplectic pattern when the beat is towards the right or the laeoplectic pattern when the beat is towards the left of the line of propagation of the waves. Most fields of cilia show diaplectic metachrony; in these cases the strongest coupling occurs during the effective stroke and produces synchronous beating in this plane, while the metachronal waves move towards the direction in which the

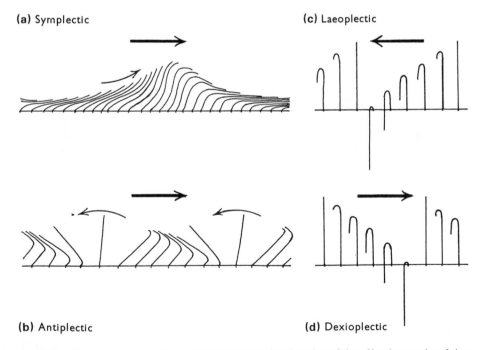

(a) Symplectic

(c) Laeoplectic

(b) Antiplectic

(d) Dexioplectic

Fig. 2.15 Diagrams to show the relation between the direction of the effective stroke of the ciliary beat and the direction of propagation of metachronal waves in four main patterns recognised and named by Knight–Jones (1954). In **c** and **d** the effective stroke of the ciliary beat is directed towards the observer.

cilia bend in their recovery stroke; thus, if the ciliary tip follows an anticlockwise path in the recovery stroke (seen from above), as it does in ciliates, the metachrony is diaplectic. Laeoplectic metachrony occurs where the tip follows a clockwise path (as in molluscs). In a few cases fields of cilia show symplectic metachrony in which the waves pass in the same direction as the effective stroke; this pattern seems to depend upon continuous coupling of moderate intensity produced by close spacing or high viscosity at lower frequencies. Finally, some widely-spaced metazoan cilia show antiplectic metachrony where waves pass in the direction opposite to the effective stroke and the beat of successive cilia of the row is individually triggered (Sleigh, 1974). There is no evidence in any protist that the coordination of cilia to form metachronal waves ever depends upon any internal conduction process; all current evidence suggests that hydrodynamic interaction provides an adequate explanation for this metachronism.

Many ciliates as well as some flagellates and *Opalina* can show a reversal response in which a change in the pattern of beating alters the direction of swimming (Naitoh and Eckert, 1974). In all cases so far studied normal beating occurs at a low intraciliary calcium concentration ($< 10^{-7}$ M) and a rise in intraciliary calcium concentration (to 10^{-6} M or above) causes modification of beating. The form of response varies; *Crithidia* normally swims with the flagellum held out in front while waves pass from tip to base, but upon stimulation the direction of wave propagation is reversed and the flagellum pushes the cell (Holwill and McGregor, 1976): *Chlamydomonas* swims forward by moving its two flagella in a synchronised ciliary beat, breast-stroke fashion, but on stimulation the two flagella are thrown forward and proceed to propagate undulations from base to tip which propel the cell backwards (Hyams and Borisy, 1978): the cilia of *Paramecium* reverse the orientation of the effective stroke upon stimulation so that the ciliate is propelled in reverse (Naitoh and Kaneko, 1973). In each case studies using demembranated models have shown that all that is required to change the beat pattern is a change in calcium ion concentration, and that this appears to act upon the axonemes themselves. The calcium binding protein calmodulin has been shown to be associated with the axoneme (Gitelman and Witman, 1980; Jamieson, Vanaman and Blum, 1979). Progressive increase of calcium concentration apparently induces the following sequence of changes in beating of various organelles: normal ciliary beat, symmetrical flagellar beat, reversed ciliary beat, active ciliary arrest (Sleigh and Barlow, 1982b).

The control of ciliary beating in *Paramecium* has been studied in some detail, using microelectrodes to monitor or control membrane potentials and induce current flows, and demembranated models to control ionic concentrations (Machemer, 1977; Naitoh and Eckert, 1974). A typical forward-swimming *Paramecium* has a resting membrane potential of -30 mV, based largely upon the electrochemical gradient of K^+ ions, to which the surface membrane is fairly permeable; under these conditions there is a steep concentration gradient of 'free' calcium ions (10^{-7} M inside and about 10^{-3} M outside), but the calcium gates in the membranes (particularly ciliary membranes) are closed, so calcium contributes little to the resting potential. If the ciliate is stimulated, e.g. by an anterior collision, calcium gates in the cell membrane at the site of stimulation are opened allowing inflow of calcium ions that causes depolarisation of the membrane, leading to further calcium influx and often positive overshoot of membrane potential. This electrical change is rapidly conducted over the whole cell surface, including ciliary membranes, so that calcium gates in all cilia open for a short time, calcium enters and all cilia beat in reverse. The calcium gates quickly close again, potassium ions flow out of the cell in a period of raised potassium conductance to restore the normal resting potential, and, once the excess calcium ions have been pumped out or absorbed into cellular organelles, the normal beat direction is resumed. The external concentrations of these ions, or other ions which interact with them, will clearly affect the resting potential or the depolarisation response or both. Mutants with altered responses have been used to study this behaviour (Kung *et al.* 1975). Internal ionic levels, probably principally Ca^{2+}, also affect the rate of ciliary beat (Machemer, 1976), and relate to gradients of membrane specialisations so that, for example, a posterior stimulus causes outflow of K^+ ions, membrane hyperpolarisation and faster forward swimming (probably by changing local calcium levels (Ogura and Machemer, 1980)). Comparable responses to membrane potential changes have been reported for the cirri of *Stylonychia*, but the membranelles respond differently and appear to have a different form of control of calcium gates (p. 217) (Deitmer, Machemer and Martinac, 1984). There is no evidence that internal fibre bundles associated with the cirri of such complex ciliates conduct impulses (Naitoh and Eckert, 1969), although they were once thought to do so. Perhaps it should be stressed that the 'normal' direction of beat of a cilium is that specified by its implantation at the cell surface; if an area of cell cortex is reversed, the cilia continue to beat in a normal direction in relation to their attached rootlets, i.e. opposite to that of the cortex into which they are grafted (Tamm *et al.*, 1975).

If ciliates like *Paramecium* or *Tetrahymena* are exposed to an electric field they exhibit galvanotactic migration towards the cathode. This is believed to be due to activation of voltage-sensitive channels in the cell membranes, depolarising membranes at the cathodal side and allowing Ca^{2+} entry and ciliary reversal, while membranes at the anodal side are hyperpolarised with exit of K^+ ions and show faster beating; as a result the ciliate turns its anterior end towards the cathode and swims towards it, the reversal response in this case being weaker than the acceleration at the posterior end (Brown *et al.*, 1981; Machemer and de Peyer, 1977). Other ciliates may adopt other orientations, e.g. oblique galvanotaxis in *Stylonychia* (Dryl, 1963) and transverse orientation in *Spirostomum*. *Opalina* migrates towards the anode in an electric field; the same ciliary changes occur as in *Paramecium*, but the reversal at the cathodal side exerts a dominating influence on locomotion (Naitoh and Eckert, 1974).

The rate and direction of swimming of protists may be influenced by light (Halldall, 1964). In particular, photosynthetic flagellates and algal zoospores demonstrate effective phototaxis (Greuet, 1982; Piccinni and Omodeo, 1975). The essential principle is similar in all cases, namely that the cell rotates continuously on its longitudinal axis as it swims, and the light sensor, orientated perpendicular to the long axis, is subjected to maximal fluctuations of light intensity if the flagellate (by alternate full exposure and shading) swims at right angles to the light path, and minimal fluctuations of intensity when it swims towards or away from the light. Signals from the light sensor to the flagellum allow successive corrections to the swimming direction until it is optimal. Although this principle is uniform, the physics of the light sensor mechanism and the means of communication with the flagellum are variable (Foster and Smyth, 1980; Song and Walker, 1981; Melkonian, 1984a). The most intensively studied example is probably **Euglena** which is positively phototactic at lower light intensities and negatively phototactic at very high light intensities. Negative phototaxis has been reported for some amoebae and ciliates (Grebecki and Klopocka, 1981; Halldall, 1964), but other ciliates and chytrid zoospores show positive phototaxis (Kazama, 1972).

Many protists show gravitational responses. In **Loxodes** the Muller's bodies (p. 199) act as statocyst-like gravity receptors (Fenchel and Finlay, 1984) in directing vertical migrations (p. 269). Density differences, hydrodynamic effects of swimming and responsive changes in ciliary propulsion have been proposed, inconclusively, as mechanisms determining positive (e.g. **Blepharisma**) or negative (e.g. **Paramecium**) geotaxis in flagellates and ciliates without gravity receptors (Bean, 1984; Roberts, 1981). **Amoeba proteus** shows a weak negative geotaxis (Klopocka, 1983).

Chemosensitivity is associated with ciliary motility in various ways in protists. Noxious stimuli cause an escape response either by reversed swimming or by reorientation followed by forward swimming away from the chemical. Some amoebae, flagellates and ciliates show a 'chemotactic' response to food and other attractant chemicals (van Houten et al., 1981). Positive 'chemotaxis' is known for male gametes of various types, and among protists and related forms is best known from brown algae (Geller and Muller, 1981) and oomycetes (Pommerville, 1978); in these cases the locomotory behaviour of the flagellate gamete changes in response to change in concentration of sex attractants released by female cells. These and other examples of responses of protists to chemical stimuli are more correctly regarded as kinetic reponses rather than tactic ones, since they depend on changes in the rate of swimming, or rate of change of direction, rather than being orientation responses to the stimulus. Many protists that show no evidence of 'chemotaxis' are stimulated to aggregate by pheromones and may adhere together by reactions involving specific membrane components (p. 86).

The nucleus

The structure of protistan nuclei

All protists have at least one nucleus, and in many species several or even many nuclei may be enclosed within the same body of cytoplasm (*see* Raikov, 1982). Ciliates characteristically have nuclei of two types (macronuclei and micronuclei, p. 186) and at a certain stage in the life cycle of some foraminifera nuclei of two types are present within the same cell (p. 78); the special features of these forms are considered later. All nuclei contain the genetic material deoxyribonucleic acid (DNA) in the chromosomes within the nucleoplasm. The forms of nuclei and their manner of multiplication are more diverse in protists than in higher animals and plants. They vary in diameter from 1 μm or less to over 100 μm. Most commonly protistan nuclei are vesicular and spherical or oval, the best known exceptions being the macronuclei of ciliates which are more dense and often have complex shapes.

The nuclear envelope consists of two unit membranes (p. 15), the inner and outer components of which come together to form the margins of pores that provide a connection between the cytoplasm and the interior of the nucleus. These pores are commonly 80–100 nm in diameter, and appear in surface view as dense rings with eight large protein granules around their margins, forming a nuclear pore complex. It is estimated that the open channel through the pore between nucleus and cytoplasm is only about 9 nm in diameter, though apparently molecules larger than this can be transported through them, e.g. proteins required by the nucleus but synthesised at cytoplasmic ribosomes pass inwards and ribosome subunits pass outwards.

The nuclear envelope is often thickened because of the presence of a fibrous lamina, usually at the inner side of the inner membrane, but it may occur outside. For example, a curious honeycomb structure is associated with the inside

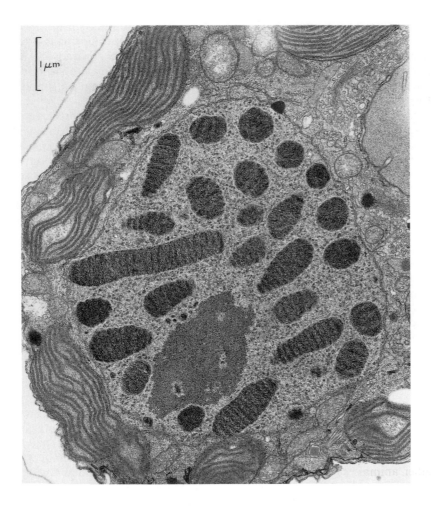

Fig. 2.16 A section through the dinoflagellate *Heterocapsa triquetra* showing the appearance of interphase chromosomes in members of this group; note the fine coiled filaments within the chromosomes. (Electron micrograph from Dodge, 1971).

or outside of the nuclear envelope in some species. In *Amoeba proteus* this consists of a layer of hexagonal membranous tubes about 160 nm across and 300 nm long extending into the nucleus from the inner membrane, where each tube surrounds a pore in the envelope (Daniels, 1973). In other species the pattern is generally less well developed. The function of the honeycomb layer is not known.

In electron micrographs the interior of most interphase (non-dividing) nuclei has a characteristic finely granular appearance, with occasional aggregations of denser material which represent chromatin (DNA + histone protein) bodies (Fig. 2.18). Dense chromatin strands shorten and thicken as visible chromosomes before nuclear division. In the nuclei of some protists, notably dinoflagellates, euglenids and hypermastigid flagellates, the chromosomes are visible throughout interphase in a more or less condensed condition in the configuration normally seen during mitotic prophase (Fig. 2.16). Protists of some groups have a single (haploid) set of chromosomes in their normal cells, while in nuclei of other groups a double (diploid) set of chromosomes is present (see p. 73). Large nuclei are often polyploid, with several chromosome sets in each nucleus, or may have undergone selective gene multiplication (p. 187). In ciliates the micronucleus is diploid and the macronucleus is normally polyploid (186).

The chromatin of eukaryotes is formed into strings of nucleosomes, each unit being composed of a length of DNA of about 200 base pairs (Alberts *et al.*, 1983). Typically, one section of about 140 base pairs of the DNA is looped twice around a cluster of histone molecules, two molecules of each of four types (H2A, H2B, H3 and H4), to form the nucleosome itself, whilst a linking section of about 60 base pairs of DNA accompanied by a molecule of H1 histone joins adjacent nucleosomes. Nucleosomes have not been found in dinoflagellates, which apparently contain only small amounts of histone-like protein (Sigee, 1986). Present evidence indicates that individual histones vary slightly in different eukaryote groups, and that in some cases there may be fewer than five types of histone.

Nuclei usually contain one or several nucleoli associated with one or more of the chromosomes. These bodies are rich in RNA, principally as ribonucleoprotein, clustered around loops of nucleolar organiser DNA. They are associated with the synthesis of ribosomes, whose ribosomal RNA (rRNA) is coupled to proteins imported from the cytoplasm to make the large and small ribosomal subunits that are separately exported to the cytoplasm, and there join as functional units. Nucleoli usually, but not always, disappear during nuclear division (p.48). Ciliate micronuclei and some gametic nuclei lack nucleoli.

The nucleus during the cell cycle

The nucleus has two major functions, the replication of the genetic material of the cell and the release of genetic information to the synthetic machinery of the cell. The genetic information content of the nucleus encoded in the base sequence of the chromosomal DNA is involved in both of these functions. Replication involves the synthesis of new DNA to duplicate the chromosomes, and subsequently the separation of chromosomes into daughter nuclei, normally immediately before cell division. Unlike DNA synthesis in prokar-

1. *Trypanosoma mega* (18.9h)

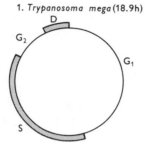

2. *Tetrahymena pyriformis* (3.7h, 29°C)

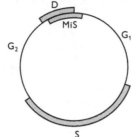

3. *Amoeba proteus* (36h, 23°C)

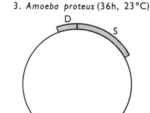

4. *Euplotes eurystomus* (12h)

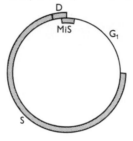

Fig. 2.17 Diagrams to show the proportion of time occupied by the main phases of the cell cycle in four protozoa. In the ciliates the micronuclear S phase (*MiS*) is shown within the cirle for comparison with the timing of events in the macronuclear cycle shown around the outside of the circle. An average figure for the duration of the cycle is indicated in each case. (Information mainly from Prescott and Stone, 1967).

yotes, which is normally continuous, the sythesis of DNA in eukaryotic nuclei is confined to only part of the cell cycle, and this is usually at a different time in the cycle from the division of the nucleus. The cell cycle may be divided into four sections: the division period (D), the time from the end of division to the beginning of DNA synthesis (G_1), the DNA synthesis period (S) and the time from the end of DNA synthesis to the beginning of division (G_2) (*see* Prescott and Stone, 1967). The timing of these events in some protozoan cycles is shown in Fig. 2.17. In some ciliates the S periods of the macronuclei and micronuclei occur at different times.

There is evidence in some cells that once the transition from G_1 to S has taken place the events leading up to nuclear division and the subsequent cell division have been set in motion. G_1 may thus be the most variable phase. Prescott and Stone (1967) found that if **Tetrahymena** is deprived of the amino acids tryptophan and histidine during G_1, then there is only partial DNA synthesis and no cell division; if the ciliates are deprived of these amino acids after the G_1-S transition, then DNA synthesis is completed and cell division follows. It is concluded that before the beginning of the S phase these amino acids are required

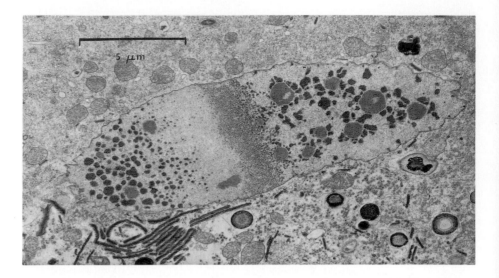

Fig. 2.18   A section through part of the macronucleus of ***Euplotes eurystomus*** during the passage of a 'replication band' from left to right. Both before synthesis of DNA (at the right) and after synthesis (at the left) the nucleus contains smaller, more dense chromatin masses and larger, less dense nucleoli; both of these structures disappear during the passage of the replication band. (Electron micrograph by R. Gliddon.)

for the synthesis of enzymes required for DNA synthesis, that these enzymes are broken down after DNA synthesis and that they are made anew in each cell cycle; if the amino acids are not available at the appropriate moment the enzymes are not made and DNA synthesis cannot take place.

Simultaneous with the synthesis of new DNA there is active synthesis of nuclear histone proteins, which also double in quantity. This has been studied in the macronucleus of *Euplotes*, in which, as in many other ciliates, the new synthesis of DNA occurs as 'replication bands' travel along the nucleus (Fig. 2.18). In *Euplotes* zones of reorganisation are seen to originate at either end of the elongate nucleus and to move towards the centre. The passage of a replication band over a section of the nucleus is associated with a doubling in the quantity of both histones and DNA. It is interesting that the replication bands indicate that synthesis of DNA begins simultaneously at both ends of the macronucleus. The signal for commencement of macronuclear DNA synthesis is not known: it appears to be specific, for micronuclear DNA does not replicate at this time, but during and immediately after micronuclear telophase (Fig. 2.17d). Other proteins are also accumulated by nuclei; in *Amoeba* they may be released on the breakdown of the nuclear membrane during mitosis, but are reaccumulated within 30 minutes or so after reformation of the nuclear envelope. The importance of these types of protein in nuclear functioning is not established, but many of them must be enzymes concerned in the mechanisms controlling release of genetic information to the cell.

Nuclear division

Division of the protistan cell is usually immediately preceded by division of the nucleus, except in some multinucleate forms where it may be possible for the organism to separate into nucleated parts without an associated phase of nuclear division. The genetic function of the nucleus demands that at division each daughter nucleus should receive a complete set of chromosomes identical with that possessed by the parent cell; this is achieved by the complex processes associated with mitosis (karyokinesis). The events in mitosis vary in many details among protists, but its essential features remain constant (Heath, 1980; Raikov, 1982). Haploid or diploid nuclei undergo a mitotic process in which the replicate chromatids resulting from DNA synthesis in the preceding S period separate longitudinally and move with products of separation of other chromosomes so that identical complete sets of chromosomes come together in each of the two daughter nuclei. The pattern of division of ciliate polyploid nuclei is less clear, but it is thought that normally several complete chromosome sets (genomes) separate into each daughter nucleus and that a balance between

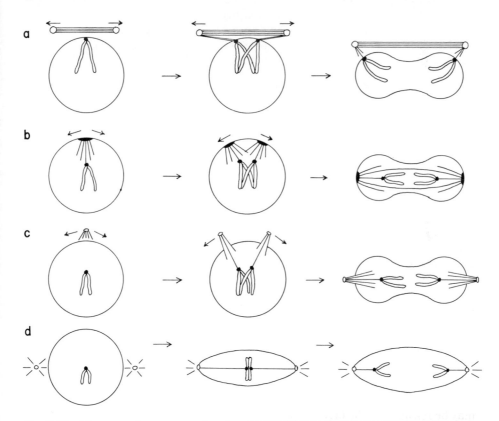

Fig. 2.19 Diagrams of some examples of nuclear division in protists described in the text. **a.** Closed mitosis with extranuclear spindle, as found in parabasalian flagellates; **b**, closed pleuro-mitosis with intranuclear spindle found in foraminiferans and many other groups; **c**, semi-open pleuromitosis found in many sporozoans; and **d**, open orthomitosis found in metazoa, land plants and many groups of protists.

chromosomes of various types is maintained. The macronuclei of hypotrich ciliates appear to contain multiple copies of free genes which are thought to segregate randomly at division (Raikov, 1982).

The first sign of an approaching nuclear division is usually the multiplication of the centrioles or other form of spindle pole body (Fig. 2.19a–d). This is usually accompanied or followed by condensation and shortening of the chromosomes. Microtubular fibrils of the mitotic spindle grow towards the chromosomes from the spindle pole bodies, and often also outwards from the centromere regions of the chromosomes. Banded kinetochore structures appear at either side of each centromere, one associated with each chromatid. Spindle microtubules associated with each kinetochore interact with polar microtubules so that one chromatid of each chromosome becomes connected to one pole of the spindle and the other chromatid is linked to the other pole. Equal tensions on the microtubules associated with each pole presumably cause the chromosomes to line up in metaphase at the equator of the spindle before the centromere region replicates and allows the chromatids to separate (e.g. Fig. 2.19d). As the spindle poles move apart, or the spindle fibres attached to kinetochores shorten (or both), two identical sets of chromatids are guided to form two separate groups that make up the chromosome complements of the two daughter nuclei.

The nuclear division of different protists varies in the persistence or breakdown of the nuclear envelope, the nature and position of the spindle pole body (if any), whether the spindle microtubules form as a full, axially symmetrical spindle (orthomitosis, Fig. 2.19d) or as separate half spindles at an angle to one another (pleuromitosis, Fig. 2.19c), whether the spindles of closed nuclei (with persistent nuclear envelopes) are entirely extranuclear or intranuclear (Fig. 2.19a,b), and whether the chromosomes remain attached to the persistent nuclear membrane (Fig. 2.19a). In addition, in a few groups (euglenids, kinetoplastids, dinoflagellates, some amoebae and some coccidians) the nucleolus remains visible throughout mitosis and divides into two, whilst in the majority of groups it disperses early in mitosis and reforms at telophase; no explanation for this difference has much support, and it has no clear relation to any other feature of mitosis. The diversity of mitotic features suggests that evolution of mitosis has taken place within the protists, and so it is relevant to outline here the possible course of that evolution. Further details concerning the variants and their distribution have been given by I. B. Heath (1980) and I. B. Raikov (1982).

The single chromosome of prokaryotes remains attached to the cell membrane throughout the cell cycle; replication of the chromosome begins with division of its membrane-attachment site, and the two new attachment sites move apart by 'membrane growth' or some other form of interpolation of membrane material between them. In parabasalia (trichomonad and hypermastigid flagellates) and dinoflagellates, which appear to have primitive nuclei, the nuclear envelope remains closed throughout mitosis and the chromosomes remain permanently anchored to the nuclear membrane by the kinetochores (Fig. 2.19a). In examples of each group the division of the centromere region of the chromosomes to form two daughter kinetochores is followed by separation of the kinetochores, apparently initially by membrane growth (*cf.* chromosome separation in prokaryotes). These kinetochores are all located at one end of the nucleus in parabasalia, and across this area a bundle of microtubules forms a short central spindle in the cytoplasm between the spindle pole bodies. Chromo-

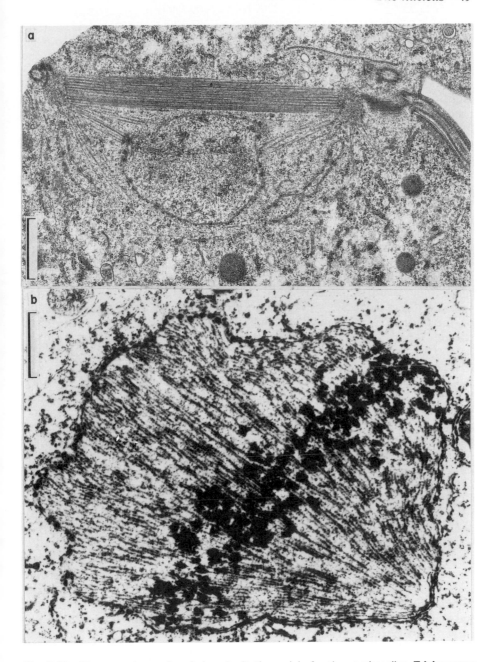

Fig. 2.20 Electron microgrphs of closed mitotic nuclei of **a** the parabasalian *Trichomonas vaginalis*, with an extranuclear spindle, and **b** the micronucleus of the ciliate *Nassula* with an intranuclear spindle. (Micrograph **a** from Brugerolle, 1975, and **b** from Tucker, 1967.) Scales both 1 μm.

somal microtubules from the poles become attached to the outside of the nuclear envelope where kinetochores attach at the inside or penetrate through it (Fig. 2.20a). These chromosomal microtubules do not change in length, but the poles move apart as the central spindle lengthens, pulling the kinetochores apart as two groups and thereby segregating the chromosomes (Fig. 2.19a). In some dinoflagellates the central spindle runs in a furrow in the nuclear surface, and in other cases this furrow sinks into the nucleus to form one or more tunnels lined by nuclear envelope, the central spindle running right through the nucleus whilst the chromosomal microtubules terminate at kinetochores embedded in the walls of the tunnels. The spindle pole bodies in these groups are sometimes centrioles, sometimes fibrous structures connected to flagellar bases, sometimes amorphous bodies and sometimes no structure can be seen at the poles. Opinion is divided about whether this or the second type is the most primitive mitotic pattern.

A more common mitotic pattern is that where the nuclear envelope remains closed but the spindle is entirely intranuclear. At the start of this mitosis (in its basic form) microtubules extend from a granular plaque (the MTOC) lying at the inner side of the nuclear envelope to the centromere regions of the chromosomes. The centriolar plaque divides into two and the daughter plaques separate (Fig. 2.19b), presumably by membrane growth between them. As the plaques move apart the kinetochores of daughter chromatids of each chromosome separate because in each case one kinetochore is attached to one plaque and its partner to the other plaque. Initially a single half spindle appears inside the nucleus and this becomes split into two half spindles whose poles move apart until they lie at opposite sides of the nucleus. Elongation of the nucleus, carrying the poles further apart, with or without shortening of the microtubules attached to the kinetochores, draws the chromatids into two groups, forming daughter nuclei by pinching off of the persistent nuclear envelope. This closed intranuclear pleuromitosis occurs in kinetoplastid and oxymonad flagellates, oomycetes, some chytrids, many higher fungi, foraminiferans, radiolarians, microsporans and ascetosporans. In some cases centrioles lie outside the nucleus opposite the polar plaques.

The closed intranuclear mitosis of some other groups is probably derived from this. These are cases where the half spindles develop from plaques that have already separated to opposite poles of the nucleus, and are therefore orthomitoses. Some amoebae, some gregarines, some green algae and xanthophytes show this pattern. An essentially similar pattern, but without any obvious plaque or other MTOC is found in other amoebae, euglenoid flagellates, the micronuclei of ciliates (Fig. 2.20b) and probably in opalinids. The acentric spindle of these forms consists of many nearly parallel microtubules ending near the inner nuclear membrane. Although the nuclear envelope remains intact in these forms throughout mitosis, in some ciliates and a few other forms new nuclear envelopes form around daughter nuclei within the existing nuclear envelope before the old membranes break up at the end of the process. In the case of the testacean amoeba *Euglypha* the intranuclear orthomitotic spindle formed from plaques inside the nuclear envelope is accompanied by microtubules that extend around the outside of the nuclear envelope between centrosphere structures (darkly staining bodies without centrioles) in the cytoplasm opposite the polar plaques.

In semiopen pleuromitosis (Fig. 2.19c) the development of the spindle is essentially the same as in closed intranuclear pleuromitosis except that the spindle pole bodies lie outside the nucleus and spindle microtubules enter the nucleus through small gaps (fenestrae) in the nuclear envelope. A single half spindle appears, divides into two and the two half spindles, which already make contact with the kinetochores, move apart to opposite sides of the nucleus before the chromosomes are separated into two groups. This type of mitosis is found in sporozoans, with cytoplasmic pole bodies near pairs of centrioles in coccidians and plaques embedded in a porous area of the nuclear membrane in haemosporidians and piroplasms.

The spindle pole bodies may take up positions at either side of the nucleus before spindle microtubules enter the nucleus through polar fenestrae in semi-open orthomitoses. Diffuse or compact pole bodies may or may not be associated with flagellar basal bodies or centrioles in this case. Many volvocid and chloromonad flagellates, as well as actinophryid heliozoans, some gregarines, some chytrids, some myxomycetes, green, brown and red algae show this pattern. The open orthomitotic pattern (Fig. 2.19d), in which the nuclear envelope breaks down completely and spindle poles lie at either side of the nucleus from the start, is the most familiar because it occurs in higher animals and green land plants. It also occurs in chrysomonad, cryptomonad, prymnesiid and some prasinophyte flagellates, diatoms, some myxomycetes, some amoebae, helioflagellates, labyrinthulids and some gregarines. Centrioles may form the spindle pole bodies in some cases, other spindles are acentric, as in higher plants, or may have other types of centrosphere or rhizoplast (flagellar root) at the spindle poles.

Macronuclear division in ciliates is very variable, although the nuclear envelope always persists (Raikov, 1982). Normally it is not possible to see any traces of the normal mitosis in the division of polyploid macronuclei; the large, sometimes multiple, macronuclei usually condense to a single compact body which extends, constricts and separates in a process of 'amitosis'. The constriction of the macronucleus may be preceded by a splitting of shortened chromosomes without participation of a microtubular spindle of fibres (endomitosis), although some nuclei have internal microtubules which appear to push the daughter nuclei apart, and extra-nuclear microtubules may also be present. Sometimes macronuclear division is accompanied by the extrusion of some DNA material into the cytoplasm, and its later reabsorption; this process may have some role in regulating the DNA balance of macronuclei.

Nucleic acids and protein synthesis

The genetic information of the nucleus is expressed in the synthesis of proteins, both structural molecules and enzymes (*see* Alberts *et al.*, 1983). While protein synthesis occurs at the cytoplasmic ribosomes, it is dependent on the continued presence of the nucleus, for protein synthesis ceases soon after removal of the nucleus. This is because any protein that is synthesised must be specified by a short-lived messenger RNA (mRNA) which becomes temporarily associated with the ribosome (Fig. 2.21). The mRNA carries a transcript of the DNA genetic code in the form of triplets of nucleotide bases whose linear sequence is determined by base pairing during the synthesis of mRNA against a DNA strand

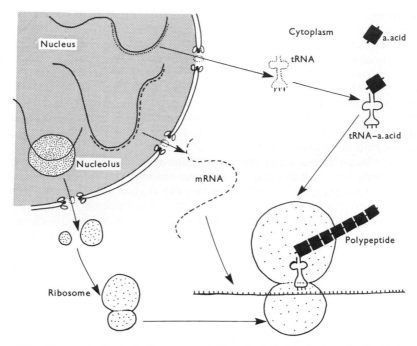

Fig. 2.21   Diagram to illustrate the sequence of events during protein synthesis, as described in the text. *mRNA* is messenger *RNA*, *tRNA* is transfer RNA and *a.acid* is an amino acid.

in the nucleus. A ribosome attaches to one end of the mRNA molecule and moves along to the other end; as the ribosome passes each triplet of bases of the mRNA a temporary linkage is made by base pairing with a molecule of transfer RNA (tRNA) coupled to a particular amino acid of the type specified by the triplet of bases on the mRNA. There is a species of tRNA for every triplet of bases that occurs as a code for an amino acid on mRNAs. In the cytoplasm the tRNAs form temporary compounds with the appropriate amino acid under the influence of specific activating enzymes, so that a pool of aminoacyl-tRNA molecules is available in the cytoplasm for use at the ribosomes. Each successive tRNA which is lined up against the mRNA strand at the ribosome brings its amino acid into close contact with amino acids specified by previous triplets and brought by previous tRNAs; this contact leads to the formation of a peptide bond which adds the amino acid to a linear polypeptide chain. The polypeptide is completed when all triplets of the mRNA have been 'read' by the ribosome, and it may then fold up in a specific manner to form the complete protein, with or without combination with other polypeptides. Cytoplasmic proteins may be synthesised at ribosomes that lie free in the cytoplasm, but membrane proteins and proteins to be exported from the cell are formed at ribosomes attached to the ER (p. 14). Several ribosomes may read the same mRNA stand, and polysomes are groups of ribosomes moving in succession along the same mRNA strand. The same ribosome may associate successively with different mRNA strands and be concerned with the synthesis of proteins of different types.

While the synthesis of proteins at ribosomes in the cytoplasm only needs amino acids and energy from cytoplasmic sources, it requires RNA of three types which must be supplied by the nucleus. The synthesis of all proteins requires continuous supplies if rRNA (from the nucleolus) and tRNA, but specific mRNA strands are produced only when required for the production of specific proteins. Following the formation of mRNA strands by transcription from DNA, the mRNA is 'processed' in the nucleus by addition of specific terminal groups and by removal of 'intron' sequences that do not form part of the functional message that is to be translated at the ribosome. The nuclear envelope thus separates transcription and processing in the nucleus from translation and many other enzymic processes in the cytoplasm.

Many proteins, including numerous cell enzymes, must be continuously required by the cell, and the mRNA for such proteins is probably continuously produced. Marked changes in the production of other enzyme proteins occur during the life of a protistan cell, e.g. during the DNA synthesis period, and during the phase of the cycle when new cell organelles are being made; the formation of each of these special proteins results from the induced synthesis of the appropriate mRNA as a result of the receipt of a specific chemical signal originating within or outside the cell. Cells which show more complexity in their life cycle, such as the development of a spore or cyst stage, show many characteristics of the differentiation of cells of multicellular organisms. In the soil amoeba *Acanthamoeba*, for example, the encystment process has been shown to involve new macromolecular syntheses which result in the production of first an outer protein coat and later an inner cellulose layer (Neff and Neff, 1969). Formation of a cyst wall only occurs in the presence of a nucleus; it involves the synthesis of new RNA and new protein, and is blocked by inhibitors of these syntheses. Selective transcription of the genetic information content of the nucleus occurs in protists just as it does in other organisms, but is less extensive than in highly differentiated animals and plants where the characteristics of a large diversity of cell types within an individual are specified by the same set of chromosomes.

Extracellular products

Materials manufactured within protistan cells may be passed to the outside of the cell membrane for some function in relation to the life of the organism. The best-known forms of this secreted material are permanent or temporary shells, tests or cyst walls. Many protists produce quantities of mucoid material, apparently as an aid to locomotion or feeding, and the production of such extra-cellular structures as trichocysts and similar threads is widespread. Waste products of metabolism, which may move by diffusion or with fluid expelled by the contractile vacuole, and undigested remains released from spent food vacuoles are also passed out of the cell, but are not included under this heading.

The pellicles of ciliates and such flagellates as *Euglena*, and the pseudopods of some amoebae, may be strengthened or stiffened with protein fibres or skeletal rods or plates of calcareous or siliceous material, but these structures are within the cell membrane. By contrast, many protists have a cell wall, usually of cellu-lose or a similar polysaccharide, outside the cell membrane; cellulose is laid down largely by polymerisation outside the cell of carbohydrate secreted through the membrane, but other materials are exocytosed from Golgi-derived vesicles. The formation of the silica 'cell wall' of diatoms is different, since it is secreted within a vacuole and becomes exposed by loss or retraction of the superficial membrane layers (p. 135). On the outer surface of the membrane of many flagellates and some amoebae are adherent scales or spines that are also clearly extracellular products.

Incomplete envelopes found around various protists are variously described as shells or tests when they fit closely, e.g. around some amoebae (p. 162), or as loricas when they are chambers within which flagellates or ciliates move more freely (pp. 133, 211). These envelopes are often composed of a proteinaceous or mucopolysaccharide secretion, usually described as 'chitin', 'pseudochitin' or 'tectin'. This may compose the entire structure of the shell of the amoeba *Arcella* (p. 163) or the lorica of the peritrich ciliate *Cothurnia* (p. 211), or it may be reinforced by foreign materials, such as sand grains or diatom frustules, in testaceous amoebae like *Difflugia*, tintinnid ciliates like *Codonella* (p. 217) or foraminifera like *Haliphysema* (p. 171). Further secreted material, in the form of internally prefabricated siliceous scales, is incorporated in the shell of the testaceous amoeba *Euglypha* (p. 167), and in most foraminifera the organic layer is supported by a heavy secretion of calcareous material. Gelatinous or mucous secretions frequently aggregate debris from the environment to form a temporary loose lorica around flagellates or ciliates, and may also provide a means of temporary or more permanent colony formation. Aggregates of faecal pellets, each enclosed in a membrane, are collected with foreign debris within the secreted tubular test of xenophyophorea (p. 174). Secreted loricas or naked protists may be supported by secreted stalks; the extracellular sheath of the stalk of such peritrich ciliates as *Vorticella* (p. 211) is composed of proteinaceous material of the keratin type.

Many protists form resistant spore or cyst walls at certain times in the life cycle, particularly in such parasites as coccidia (p. 229), or in response to certain environmental conditions, as discussed in Chapter 9. These walls, which are secreted as close-fitting extracellular coats, may involve a variety of materials; e.g. in *Acanthamoeba* a phosphoprotein layer is secreted first, followed by an inner cellulose layer, while in the coccidian *Eimeria* the very resistant wall of the oocyst is formed of an outer layer of quinone-tanned protein and an inner lipo-protein layer.

Materials secreted to form extracellular structures, including scales and spines as well as shells and cyst walls, are known in some cases to appear first in vesicles of the Golgi system before they are carried to the surface and extruded (Manton and Leedale, 1961, 1969). The production of mucus in membrane-bound mucocysts is probably a comparable phenomenon (Tokuyasu and Scherbaum, 1965). Such materials may be released in quantity to form a complete new test or a new chamber of a foraminiferan shell, or they may be secreted more gradually in the continuous extension of a lorica or stalk, e.g. the growth of the peritrich stalk from the scopula (p. 215).

3
Nutrition, Metabolism and Growth

Nutrition of Protista

All organisms require a supply of materials to build their body substance and a supply of energy for the performance of synthetic reactions and such necessary activities as movement and transport. The necessary materials must provide a metabolic pool of small organic molecules which can be used as 'building blocks' for growth or the laying down of food reserves, and as a source of chemical energy to be made available by the process of respiration (Fig. 3.1). While the immediate source of energy for cell processes is normally the breakdown of the smaller organic molecules, the intake of energy into the cell may be from either or both of two main sources. Many organisms obtain energy from the breakdown of organic materials derived from other organisms, and these organic molecules also provide many of the small molecules of the metabolic pool; such organisms practise heterotrophic nutrition. Some organisms practise autotrophic nutrition in which energy derived from light (in photoautotrophs) or from the oxidation of inorganic substances (in chemoautotrophs) is used to build up organic molecules from inorganic materials. Photoautotrophic nutrition is found in those protists which possess chlorophyll pigments, but chemoautotrophic nutrition is not known to occur in eukaryotes – the important chemoautotrophs are bacteria. Examples from many phyla of autotrophic protists have been shown to be capable of heterotrophic nutrition, at least to a limited extent.

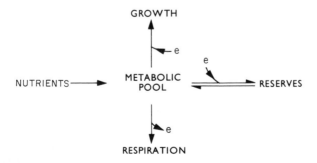

Fig. 3.1 Diagram showing the central position of the metabolic pool of organic molecules in relation to the pathways of metabolism and the utilisation and release of energy (*e*).

The organic molecules of which an organism is composed contain a number of important elements which must be present in their intake of nutrients. Carbon, hydrogen and oxygen are present in most organic compounds, proteins contain also nitrogen and often sulphur, lipids often contain phosphorus and nucleic acids contain nitrogen and phosphorus. An autotrophic organism may be able to synthesise all the complex molecules it requires from simple inorganic compounds containing these elements, together with traces of such elements as copper, iron and magnesium. A heterotroph which feeds on the tissues of other animals or plants is likely to take in complex molecules of many of the types used in metabolism, and its synthetic powers are often limited. Such a heterotroph may require not only an organic energy source, but also one or more classes of organic molecules as organic carbon sources and organic nitrogen compounds; it may have a limited ability to interconvert amino acids, so that it may require many or most of about 20 amino acids as well as purines and pyrimidines. Many organisms, including some that are autotrophs, require a range of growth factors (vitamins) which are generally complex molecules needed to produce coenzymes.

Nutritional requirements

The nutritional requirements of a few representative protists are shown in Table 3.1 (*See also* Droop, 1974; Hall, 1967). Such flagellates as *Chlamydomonas moewusii* are obligate autotrophs, requiring the intake only of inorganic materials and light energy for survival and growth. This complete dependence on autotrophic nutrition is probably rather rare, for many organisms which can perform photosynthesis are known also to be able to grow heterotrophically in the dark, including *Euglena gracilis*, flagellates from many other groups, desmids and diatoms (Droop, 1974). The forms of organic molecules which can be used, and the range of growth factors required, vary widely in different examples; even the different strains of *Euglena gracilis* differ in that some will grow on carbohydrate and others will not – all will grow on acetate, and *Prymnesium parvum* will grow on glycerol. Many of these forms have few organic needs, even for heterotrophic nutrition, and have extensive synthetic abilities, for which an intake of inorganic nutrients is still necessary. While numerous species can survive adequately on either autotrophic or heterotrophic nutrition, it is thought that *Ochromonas malhamensis* is an example which practises both forms of nutrition but is unable to fix enough energy by photosynthesis for growth to occur; its autotrophic nutrition must always be supplemented by a heterotrophic source of energy. The range of organic molecules required by heterotrophic forms varies widely, particularly as regards amino acids, purines, pyrimidines and growth factors. Generally the flagellates related to autotrophic forms are less demanding than amoebae, and these in turn are less demanding than ciliates and parasitic flagellates.

It is only possible to assess the nutritional requirements of a species if it can be grown in pure (axenic) culture in the absence of any other living organism. It is normally easier to maintain axenic cultures of autotrophic species than heterotrophs, but axenic cultures of protists from many groups have been established and have proved most valuable, not only in nutrition studies but also in research on many other aspects of protistan physiology and biochemistry. Several species

Table 3.1 *Some examples of Nutritional needs in the Protista.* Data mainly from R. P. Hall (1967).

	Group	Energy source	Carbon source	Nitrogen source	Other known needs
Chlamydomonas moewusii	Chlorophyceae (green)	Light	CO_2	NO_3^- (NH_4^+)	None
Polytoma uvella	Chlorophyceae (colourless)	Organic	Acetate (some organic acids)	NH_4^+ $(NO_3^- ?)$	None
Euglena gracilis	Euglenophyceae (green)	Light or organic	CO_2, Acetate Ethanol (not carbohydrate)	NH_4^+ $(NO_3^-$ NO_2^- amino acids)	Vitamin B_{12} Thiamine
Ochromonas malhamensis	Chrysophyceae (pigmented)	Light plus organic	CO_2, Starch Saccharides Glycerol	NH_4^+	Thiamine Vitamin B_{12} Biotin
Chilomonas paramecium	Cryptophyceae (colourless)	Organic	CO_2, Acetate, Lactate, Ethanol, some fatty acids (not carbohydrate or amino acids)	NH_4^+ (amino acids amides)	Thiamine
Crithidia fasciculata	Kinetoplastida (flagellate, insect parasite)	Organic	Carbohydrate	10 amino acids Adenine or guanine	Thiamine Riboflavin Pyridoxine Pantothenate Biopterin Folic acid Biotin Nicotinic acid
Tetrahymena pyriformis	Ciliophora	Organic	CO_2 Acetate Lactate Carbohydrate Amino acids	10 amino acids Guanine Uracil or cytidine	Thiamine Riboflavin Pyridoxine Pantothenate Folic acid Biotin Nicotinic acid Thioctic acid
Acanthamoeba castellanii	Sarcodina	Organic	Carbohydrate Glycerol Acetate, Lactate Amino acids	?	Thiamine Vitamin B_{12}

of protist have been used in bioassay procedures for the detection and measurement of very small quantities of vitamins and growth factors; axenic cultures are essential for this work, and precisely defined media with high-purity components must be used. Two protists that require a specific vitamin for growth and that have been found useful in such assays are *Euglena gracilis* and *Ochromonas malhamensis* which can be used in estimation of vitamin B_{12} at concentrations down to about 1 ng 1^{-1}.

Biochemical pathways of metabolism

Studies on axenic cultures have been important in the elucidation and confirmation of biochemical pathways used in various protists for the synthesis and breakdown of molecules used in the cell and for the provision of energy. The interrelations of some of the more important biochemical pathways are summarised in Fig. 3.2, and further details are given by Danforth (1967) and Ragan and Chapman (1978).

The catabolic (breakdown) reactions shown here are principally concerned with the production of ATP. In aerobic cells the catabolism of carbohydrate, fat and protein converges on the citric acid (Krebs or tricarboxylic acid) cycle, whose enzymes are located in mitochondria and whose oxidative activity is coupled to the reduction of coenzymes (NAD and FAD). The citric acid cycle requires an input of a 2C group from acetyl-CoA (acetyl-coenzyme A complex) to combine with oxalacetate (4C). Since several intermediates of the citric acid cycle may be used as components in synthetic reactions, and therefore taken out of circulation, an alternative source of oxalacetate is required; this is provided by the glyoxylate cycle, which has been shown to function in various protists (Muller, 1975). Two acetyl-CoA molecules are used in the production of each molecule of oxalacetate in this cycle, the enzymes for which are found in peroxysomes (glyoxysomes) (p. 18). The glyoxylate cycle is particularly important in protists which use acetate, ethanol or fatty acids as energy sources (and often also as carbon sources); some of these forms can assimilate CO_2 into organic molecules using energy derived from catabolism of nutrients obtained heterotrophically. Thus some euglenoids, volvocids and cryptophytes use photosynthesis and heterotrophic oxidative assimilation as interchangeable and essentially equivalent sources of larger organic molecules.

The breakdown of carbohydrate by the glycolysis pathway is better known than that via the hexose monophosphate pathway, but the latter is important because it provides a source of pentoses for nucleotide synthesis. Most reactions of both of these pathways are reversible, and for those steps which cannot be used reversibly for carbohydrate synthesis there are alternative anabolic by-pass reactions.

The photosynthesis reactions impinge on these two pathways. Light energy absorbed by the chlorophyll and carotenoid pigments (p. 103) of the antenna complexes in the thylakoid membranes of plastids is passed to the reaction centre chlorophyll molecule; here the energy may be used either in the removal of electrons from water and energising them to reduce cytochromes (in photosystem II) or in taking electrons from oxidised cytochromes and energising them to reduce NADP (in photosystem I). Oxidation of the cytochromes is coupled to proton transport across the thylakoid membranes and the resulting proton gradient is used to synthesise ATP (Alberts *et al.*, 1983). The NADPH and ATP are used in the assimilation of CO_2 in the dark reaction cycle of photosynthesis, but ATP has many other uses so the NADPH may be used to reduce NAD and produce ATP in mitochondria (see below) or photosynthesis may be operated cyclically, generating ATP but not NADPH.

The synthesis of fatty acids from acetyl-CoA involves a different route from the catabolic process of fatty acid oxidation. Some amino acids are formed from pyruvic acid and various organic acids of the citric acid cycle; nitrogen

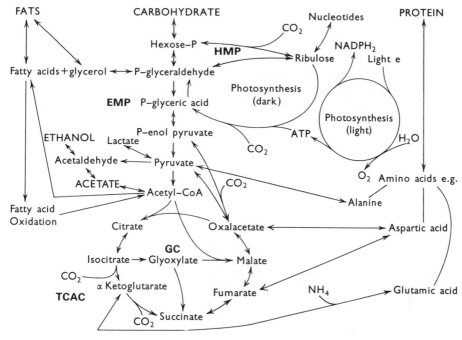

Fig. 3.2 A brief outline of some of the more important biochemical pathways and their interrelationships. Substances named in capitals are common starting points for nutrition in protists. HMP, hexose monophosphate pathway; EMP, Embden-Meyerhof pathway; TCAC, tricarboxylic acid (citric acid) cycle and GC, glyoxylate cycle.

compounds are required for this, since no eukaryote is known to fix atmospheric nitrogen, and while NH_4^+ is most commonly used, some forms can use NO_3^-. These amino acids are used to produce other acids required for synthesis of proteins.

The reduced coenzymes (NADH and $FADH_2$) are reoxidised in mitochondria where reduction and oxidation of membrane-bound cytochromes is associated with the pumping of protons into the intermembrane space of the mitochondrion and ATP synthesis driven by the proton gradient. Two or three ATP molecules and one molecule of water are produced for each pair of electrons provided by the coenzymes. Final oxidation of the cytochrome requires molecular oxygen and does not take place in anaerobic organisms, where the citric acid cycle and oxidation of fatty acids are of no value in energy production; indeed, alternative hydrogen acceptors must be found if oxidation reactions are to proceed, so that anaerobic organisms may often produce lactic acid or ethanol in fermentation reactions, and usually lack mitochondria. One exception is the ciliate *Loxodes*, found in anoxic water of freshwater lakes, which appears to use NO_3^- as the final electron acceptor in the mitochondria and contains a nitrate reductase enzyme typically found in bacteria (Finlay, 1985).

Feeding mechanisms

Heterotrophic protists require to take in inorganic food that may be soluble (in saprotrophs) and/or particulate (in phagotrophs), and a variety of mechanisms of food intake may be used depending on the nature of the food and its concentration. This section is concerned therefore with ways in which organic compounds are taken into the cell, either into food vacuoles, or sometimes directly into the cytoplasm. Processes of digestion are mentioned in Chapter 2 (p. 24).

Certain soluble organic molecules may pass through the cell membrane from the environment to the cytoplasm. Simple diffusion may be responsible, especially where a concentration gradient can be maintained by the rapid binding or reaction of molecules that enter. Many other molecules may be actively transported into the cell by permease enzymes located in the surface membrane, which may bind a specific molecule at the outer surface and release it at the inner surface of the membrane, or which provide a specific conductance channel for a molecular species. There are many protists, with and without cell walls, parasitic and free-living, which have not been seen to form surface vacuoles of the types described below, yet which absorb organic molecules ranging from acetate or ethanol to complex vitamins (Droop, 1974; Hall, 1967). Some of these organisms, like intracellular or blood-dwelling parasites, live in an environment where the concentrations of food molecules are high and energy requirements for food uptake may be low, whereas a free-living flagellate, e.g. *Chilomonas*, may live among decaying vegetation at much lower concentrations of available food. Such protists may use locomotory organelles to locate higher food concentrations in much the same manner as bacteria do. Indeed, many protists depend on similar means of food uptake to those used by bacteria, but have a much lower surface-to-volume ratio than bacteria and so require higher food concentrations for survival. Protists capable of forming normal phagocytic food vacuoles may in addition use such carrier-mediated methods for intake of organic molecules, e.g. amino acid and vitamin uptake by *Tetrahymena* (Hoffmann and Rasmussen, 1972; Orias and Rasmussen, 1977).

Soluble molecules may also be actively transported into the cell by pinocytosis ('cell drinking') in vacuoles or vesicles that are much smaller than phagocytic vacuoles and lack particulate contents. The intake of organic molecules may be non-selective, and without evidence of binding of solutes to the cell membrane, or highly selective, and dependent upon binding of specific solutes to membrane receptors before vacuole closure in a process of receptor-mediated endocytosis. Pinocytosis of the former type is typical of *Acanthamoeba* (Bowers and Olszewski, 1972), which takes up glucose, leucine, albumin or insulin at the same rate in a non-saturable manner within pinosomes; these vesicles are too small to be seen with the light microscope, but are formed all over the cell surface, resulting in vigorous membrane cycling estimated to involve turnover of the complete membrane once per minute. The same organism also performs phagocytosis, and, interestingly, the phagocytosed particles are bound to the membrane before enclosure in a vacuole (Thompson and Pauls, 1980).

Selective pinocytosis appears to depend on the specific binding of substances to the glycocalyx and/or to the cell membrane, and the subsequent invagination of this membrane to form vesicles which undergo a normal lysosomal digestive process. In the case of *Amoeba proteus* there is an extensive glycocalyx 0.1–0.2 μm thick to which proteins bind; food substances like albumin, as well as amino acids and numerous dyes, inorganic salts and other substances, induce the production of pinocytic channels which pinch off internally to form small vesicles to whose membranes the inducing substances are bound (Chapman–Andresen, 1973; Stockem, 1977). Many ciliates, flagellates and probably sporozoans also exhibit pinocytosis; e.g. some observations on *Tetrahymena* which could not perform phagocytosis, either because they were mutants with defective cytostomes or because particles for induction of phagocytosis were absent, suggest that macromolecules are taken up in small (1 μm diameter) vesicles formed in the oral region, though intake at parasomal sacs remains a possibility (Nilsson, 1979). Some endoparasitic protists may take in body fluids or components of host cells by pinocytosis, e.g. uptake of blood fluids by *Trypanosoma*, tissue fluids by myxospora, erythrocyte cytoplasm by *Babesia* (Langreth, 1976), or gut fluids by *Opalina* (Munch, 1980).

The intake of particulate food by phagocytosis may take place at one specialised site, the cytostome, as is usual in protists with pellicles, e.g. ciliates or euglenoids, or food vacuoles may be formed over much or all of the body surface, as in naked protists of many amoeboid and flagellate groups. The food particles are captured and brought into the food vacuole in several different ways, which can be divided into three principal categories of active filtration, interception of passing particles and search-and-capture feeding. The food organisms in any of these categories can vary from bacteria to small metazoans, according to the protist involved, though clearly larger protists can deal with larger prey and may find difficulty catching enough bacteria to survive (p. 254).

Filtration of food particles from water currents created by cilia or flagella is familiar in metazoan animals like bivalve molluscs or sponges as well as in ciliate and flagellate protists. Two essentials for this system are the presence of a filter and a means of propelling water through the filter. Flagella or cilia may be used to create water flows through filtering structures; the velocity at which water passes through the filter must be influenced by its mesh characteristics, but flagella may propel water at a few hundred μm s^{-1} and cilia may create flows of 1 mm s^{-1} or more. If the filter is a purely mechanical sieve, its mesh size will rather strictly limit the minimum size of particles that can be captured, but a filter composed of sticky bars, say mucous strands or pseudopodia armed with extrusomes, may intercept small particles that adhere to the filter as well as larger particles retained by the mesh. For example, the slender pseudopodia that make up the collar around the flagellar base of collar flagellates (Fig. 3.3b) acts as a filter to catch particles in the size range between 0.2 and 1 μm in diameter; these particles may be carried down the pseudopodia by membrane flow and incorporated into food vacuoles near the base of the collar (Lapage, 1925; Fenchel, 1987). Purely mechanical filters are also found in ciliates; in *Vorticella* for example, cilia of the polykinety create a feeding current whilst the almost stationary cilia of the haplokinety, standing 'downstream' behind the propulsive cilia, act as a filter that retains bacteria-sized particles; here the ciliary beating also propels the particles around the peristome between the two bands of cilia into the buccal cavity and down to the cytostome (Sleigh and Barlow, 1976). In *Euplotes*, *Stentor* and *Blepharisma* the same cilia (the membranelles) both create the water current and filter out the food particles

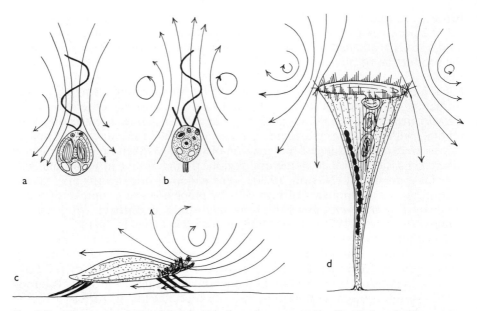

Fig. 3.3 Water currents that bring food to the cell are created by flagella in **a** *Ochromonas*, a member of the Chrysophyta, and **b** *Codosiga*, a choanoflagellate, and by the coordinated activity of cilia in **c** *Euplotes*, **d**, *Stentor*.

'upstream' behind the fanned-out compound cilia (Fig. 3.3c,d), but other cilia may help to carry food to the cytostome (cilia of the peristomial field of *Stentor*) or guide the water flow (undulating membrane of *Blepharisma*) (Fig. 3.4) (Sleigh and Barlow, 1982a). The mesh of the filters of different ciliates varies widely; thus, hymenostome ciliates whose feeding mechanism operates rather like that of *Vorticella* have a fine filter and may show maximal clearance rates for particles around 0.5 μm in diameter, while *Blepharisma* showed maximal clearance of 5 μm particles (Fenchel, 1980a,b). Since such filters are purely mechanical they exhibit no selectivity of food capture. Sticky mucoid filters are found in several invertebrate groups and 'upstream' or 'downstream' mechanical filters in others. It will be noted that the majority of these filter feeding protozoans are stalked or thigmotactic forms; this is because attached forms can maintain higher flows through their filters than free-swimming forms (Lighthill, 1976). Fenchel (1987) has pointed out that forms producing flows parallel to surfaces, e.g. *Euplotes* (Fig. 3.3c), are more efficient than forms producing perpendicular flows when operating close to surfaces or in crevices.

Some flagellates which either swim or create currents around attached cells have the means to intercept particles either on the body (e.g. *Ochromonas*, Fig. 3.3a) or on body projections (e.g. *Actinomonas*, Fig. 5.6). These forms use extrusive organelles or adhesive surfaces in food capture in the same way as other protists that intercept prey swimming or drifting in the surrounding water. These predators include various heliozoans, radiolarians, planktonic foramini-

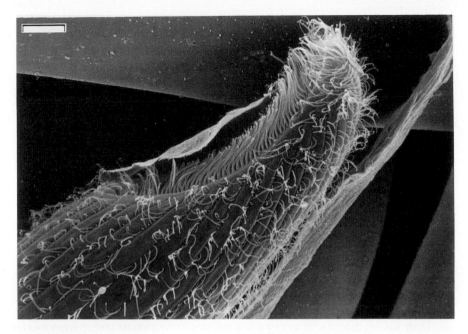

Fig. 3.4 At the anterior end of the ciliate *Blepharisma* an undulating membrane (*UM*) curves over the buccal cavity from the far side, while the membranelles of the adoral zone (*AZM*) draw water into the buccal cavity from the front and filter it as it is propelled out at the near side and rear of the buccal area. (Scanning electron micrograph by D. Barlow.) Scale 10 μm.

ferans and suctorians that extend appendages bearing sticky surfaces or extrusomes (p. 17) which discharge to capture prey that come in contact with the pseudopodia or tentacles. The haptocysts of suctorian tentacles tend to respond only to contact with specific prey, often ciliates, while the sticky rhizopodia of radiolarians, for example, may be used to capture ciliates, diatoms, and even small crustaceans (Anderson, 1983). This method of prey capture is only nutritionally adequate where the size of prey is a reasonable proportion (say at least 10%) of predator size, and so it is not suitable for bactivory by any protists except the smallest flagellates (Fenchel, 1987).

The fate of prey captured by predators of these different classes varies. In the amoeboid forms the prey caught by secretory material of the extrusome is enclosed in a food vacuole, formed either on the pseudopodium or by the combined action of several pseudopodia, according to the relative sizes of prey and predator; in the feeding of the heliozoan *Actinophrys* on the similar sized ciliate *Colpidium*, for example, the cilia of the prey adhere to the pseudopodia of the predator and a thin-walled, funnel-shaped pseudopodium extends to enclose the prey and form a food vacuole in which digestion proceeds (Patterson and Hausmann, 1981). Comparable events occur in radiolaria and foraminiferans (Anderson, 1983).

The suctorian tentacle is a specialised feeding organelle containing an axial

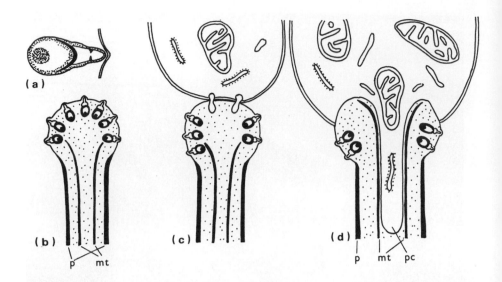

Fig. 3.5 Diagrams to illustrate the appearance **a** and position **b** of haptocysts in a suctorian tentacle, and their role in attachment of the prey to the tentacle tip **c**. In **d**, the prey cytoplasm (*pc*) is seen passing down the tentacle within a cylinder of microtubules (*mt*). The pellicle of the suctorian (*p*) is less well developed at the tip of the tentacle than at the sides. (Information from Rudzinska, 1965, and Bardele and Grell, 1967).

bundle of longitudinal microtubules, seen in cross-sections to be formed of one or two outer rows of microtubules that form a cylinder surrounding longitudinal lamellae formed of short rows of microtubules (Fig. 7.13, p. 207). This microtubule canal extends from the tentacle tip to the main body of the suctorian. After capture of a prey ciliate by the haptocyst at the tentacle tip (Fig. 3.5), the membranes of prey and predator appear to fuse, the pellicle of the prey is penetrated and cytoplasmic contents of the prey are drawn down the tentacle canal within a long cylindrical food vacuole. A supply of membrane material for food vacuole formation appears to come from cytoplasmic dense bodies which pass up the outer region of the tentacle from the cell body. Although it seems certain that activity by the suctorian, rather than pressure within the prey, produces the motive force for food vacuole intake, the precise mechanism is uncertain. The most promising hypothesis suggests that short arm-like projections (possibly homologues of the dynein arms of cilia), directed inwards from the inner lamellae of microtubules, actively propel the food vacuole membrane down the tentacle; the speed of transport is a few μm s^{-1}, which is similar to that of other dynein-based movements (Sleigh and Pitelka, 1974).

Protists that actively search for the prey that they capture or graze upon, rather than waiting for passing prey, tend to use a clearly defined range of mechanisms for food capture and intake. Predatory gymnostome ciliates (p. 199) possess a variety of toxic and harpoon-like trichocysts at the cytostome by which forms like ***Didinium*** or ***Homalozoon*** can capture relatively large prey (Fig. 3.6). Prey ciliates captured in this way are swallowed whole by dilation of the proboscis region and invagination of cytostomal membrane, drawing the

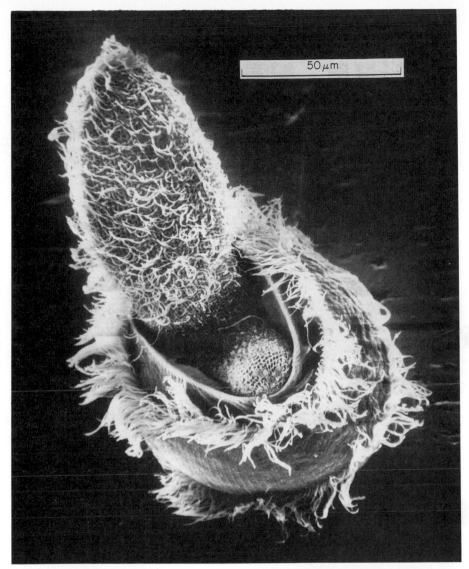

Fig. 3.6 The ciliate *Didinium* in the process of ingesting a *Paramecium*. The proboscis of the predator has been greatly dilated to allow the prey to be drawn in. Compare with a non-feeding *Didinium* in Fig. 7.8. (Scanning electron micrograph by Small and Marszalek, 1969.)

trichocysts and prey into a single large food vacuole (Wessenberg and Antipa, 1970).

The intake of filamentous blue-green algae by hypostome ciliates (p. 202) is, if anything, still more spectacular. Studies on *Nassula*, *Phascolodon* and *Pseudomicrothorax* reveal a broadly similar process of food intake through a cylindrical cytopharyngeal basket whose walls are formed of longitudinal rod-shaped bundles (nemadesmata) of cross-linked microtubules with some accessory sheets and lamellae of microtubules (Fig. 3.7) (Hausmann and Peck, 1978; Tucker, 1968, 1972). When the ventral cytostome area of the ciliate makes

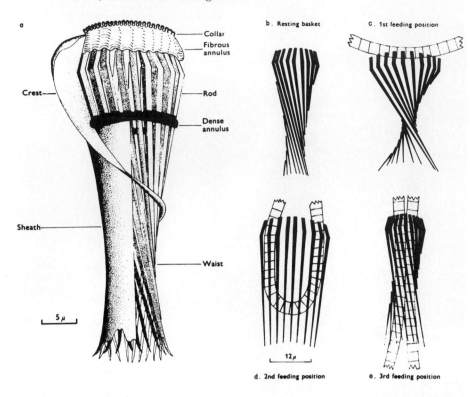

Fig. 3.7 Diagrams of the structure and function of the cytopharyngeal basket of the ciliate *Nassula*. The main structural elements in **a** are formed of microtubules that comprise the rods, sheath and crests shown in the figure as well as two types of lamellae within the basket; the collar is covered by the cell membrane and forms the base of a small depression of the body surface, while the lower end of the basket projects deep into the cytoplasm. During feeding the shape of the basket changes from rest (**b**) by dilating to grasp an algal filament (**c**) and then narrowing whilst the filament is drawn into the cell (**d** and **e**). (Illustrations by Tucker, 1968.)

contact with the tip (or occasionally the side) of an appropriate algal filament, the filament is grasped by the corrugated margins of the cytostome and drawn into a forming food vacuole (Fig. 3.8). In *Pseudomicrothorax* the algal filament passes inwards through the cytopharyngeal basket at about 15 μm s^{-1} and may form a coil within the ciliate which deforms the cell (Hausmann and Peck, 1979). The formation of new membrane to extend the food vacuole at the necessary rate (~ 270 μm^2 s^{-1}) is explained by the fusion of numerous vesicles with the forming vacuole just beneath the cell membrane. These vesicles are brought to this site by a flow of cytoplasm towards the cell surface around the outside of the basket, and they pass between the nemadesmata to the interior of the basket. Microtubular lamellae associated with the nemadesmata bear projecting arms which are seen to be linked to the moving food vacuole membrane by filamentous material, and it is thought that these structures are concerned in transport of the food vacuole membrane, which is presumably linked to the algal fila-

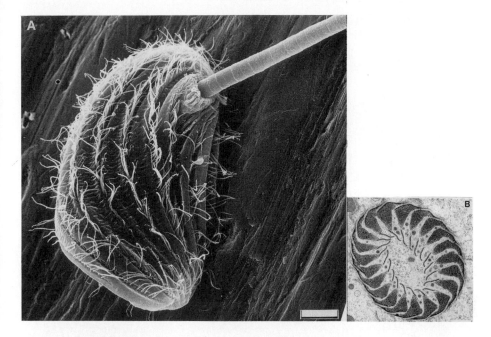

Fig. 3.8 The ciliate *Pseudomicrothorax dubius* feeds on blue-green algae which are drawn into the cell **a** through a cytopharyngeal basket, shown in transverse section **b** to be formed of about 20 rod-shaped bundles of microtubules and associated microtubular lamellae. (Micrographs by Hausmann and Peck, 1979 and 1978, respectively.) Scale bar 10 μ.

ment in some way (Hausmann and Peck, 1979). ATPase activity is associated with this food intake (Hauser and Hausmann, 1982) and actin has been shown to be associated with the basket by the indirect immunofluorescence technique (Hauser *et al.*, 1980). In *Nassula* the inward movement of the food vacuole and its enclosed filament appear to be associated with mass movement of highly gelated cytoplasm which, it is suggested, is being propelled by arms on microtubules of the basket; the basket is somewhat more complex in this case (Tucker, 1978). In these ciliates the algal filaments fragment during digestion and several smaller food vacuoles are formed. Similar methods are used by ciliates for ingestion of diatoms or, in the case of ciliates in the stomachs of ruminant mammals (p. 279), fragments of chewed grass may be ingested as well as starch grains.

Specialised mechanisms of food capture and intake are not limited to ciliates. Flagellates, notably euglenoids, dinoflagellates and kinetoplastids generally have a cytostome and examples have been studied which show a diversity of examples. Amongst euglenids there are various surface-dwelling forms like *Entosiphon* (p. 127), with microtubular feeding organelles which they use to ingest diatoms, and *Peranema* (p. 125), which can swallow other flagellates nearly as large as itself as well as smaller prey (Fig. 3.9) (Nisbet, 1974). On a different size scale, *Rhynchomonas* is a small bodonid (p. 121) which picks up individual bacteria from surfaces with its proboscis-cytostome (Burzell, 1973).

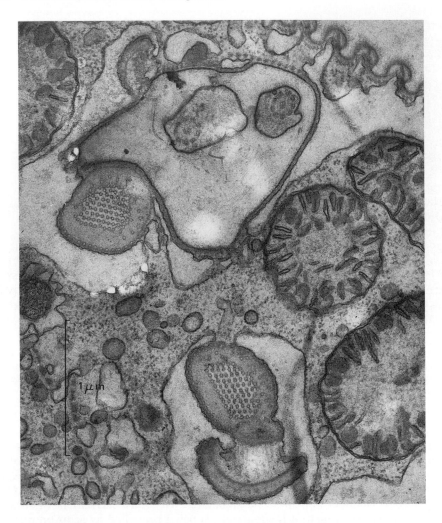

Fig. 3.9 A cross section through the anterior region of the euglenoid **Peranema**. The two flagella are sectioned within the flagellar pocket (top centre), and below are two groups of microtubules enclosed in a homogeneous matrix that are the pharyngeal rods used in food capture. (Electron micrograph by B. Nisbet.)

Several dinoflagellates have been reported to capture and engulf diatoms, cyanobacterial filaments and other flagellates using a pseudopod that emerges from the sulcus region to surround the prey (Gaines and Taylor, 1984); some non-photosynthetic dinoflagellates have quite bizarre means of food capture (p. 112). Hypermastigote flagellates found in the guts of termites (p. 280) ingest wood fragments into a naked posterior region, and even the simple cytostomes of parasitic sporozoans may be used in ingestion of host tissues, e.g. the uptake of erythrocyte cytoplasm by **Plasmodium** (Aikawa, 1971). The active amoebae that creep over surfaces lack specialised cytostomes but are capable of taking in diatoms and algal filaments or even quite active ciliates and flagellates which are somehow 'quietened' whilst they are enclosed in a food vacuole. Benthic foraminiferans move around over their substratum, 'sweeping up' diatoms and other protists as well as bacteria by formation of food vacuoles using their reticulopodia (p. 168), and some even trap crustaceans and ciliates (Lee, 1980).

Growth

In the presence of adequate food most protists go through regular cycles of growth and division (Chapter 4). Since the growth phase usually involves a doubling in volume and the division usually produces only two offspring, the size of the individual organisms in most species does not vary greatly, although starvation causes marked shrinkage. The characteristics of the growth of protists in culture is discussed in Chapter 9. More extended growth followed by multiple division to produce many offspring occurs in members of most groups, and is especially common in parasites. In these forms the life cycle is usually a complex one involving different stages with specialised morphological features. Growth in all of these forms involves a continuous synthesis of materials for cell structures, although certain organelles and such materials as DNA may only be produced at restricted times during the life cycle; such discontinuous activities are associated with reproduction, which forms the subject of the next chapter.

4

Reproduction and Sex

Increase in numbers of higher animals and many plants is generally associated with sexual processes of rather stereotyped forms. In protists a variety of forms of reproduction is seen, many of which involve multiplication by processes having none of the features of sex, but sexual processes of a wide range of types also occur, and in some cases these processes are not directly correlated with an increase in numbers. The essential features of sexual processes are the differentiation of sex cells (gametes), or at least of gametic nuclei, and the fusion of two nuclei at fertilisation; a halving of the chromosome number must occur at some time in the life of organisms which show nuclear fusion. An understanding of the role of the various aspects of the sexual process will require some comment on their genetic significance.

Nuclear and genetic aspects of reproduction

Division of the nucleus in mitosis and meiosis

In the majority of species of protista every individual is capable of reproduction, which may be a complex process, especially in those species with an elaborate organisation which must be replicated. The division of one individual to produce two daughters is the commonest mode of reproduction, but in many species, particularly parasitic ones, this binary fission may be replaced by multiple fission. Division of a cell is normally preceded by nuclear division, by either mitosis or meiosis (Raikov, 1982), although in some multinucleate forms fission and nuclear division are not linked, so that for example, new individuals may be formed by plasmotomy in which the body is simply separated into multinucleate masses – at any time some nuclei may be found in mitosis in such organisms, e.g. *Opalina*. In other multinucleate forms, e.g. *Chaos*, *Physarum*, many ciliates, mitosis may be synchronised. A doubling of the amount of DNA normally occurs during the S phase (p. 45) in each cycle of an organism reproducing by binary fission. In the rapid series of nuclear divisions that may precede multiple fission, or during a succession of fissions without cell growth, such as may occur in the formation of spores or gametes, the situation is more complex; it may require the availability of a pool of nucleotides for rapid DNA replication, or may merely result in depolyploidisation of a polyploid nucleus.

Some protists have diploid nuclei with two chromosomes of each type, such as one finds in the somatic cells of most higher animals and plants, and other protists have haploid nuclei with unpaired chromosomes, such as one finds in the gametes of higher animals and plants; polyploid nuclei with several sets of chromosomes also occur in protists. Diploid nuclei may undergo a process of meiosis to produce haploid nuclei (i.e. a reduction division), but more commonly both haploid and diploid nuclei divide by mitosis to produce two daughter nuclei like themselves.

Before the start of mitosis every gene has been replicated, and each chromosome exists as two duplicate chromatids lying together and joined at the centromere; during mitosis the two chromatids of each chromosome separate, and following division of the centromeres one chromatid of each chromosome passes to one daughter nucleus and the other goes to the other daughter nucleus. Meiosis differs in that the centromeres do not divide and the members of each pair of homologous chromosomes in the diploid nucleus are separated, so that one member of each pair goes to one daughter nucleus, the other member of each pair migrates to the other daughter nucleus and the chromosome number of the nuclei is halved. Most commonly meiosis involves two nuclear divisions, with a separation of chromosomes in the first division and of chromatids in the second, but one-division meiosis occurs in sporozoans, some complex flagellates and probably in dinoflagellates (Beam and Himes, 1980). In mitotic

(a) Mitosis

b Meiosis (2 division)

c Meiosis (1 division)

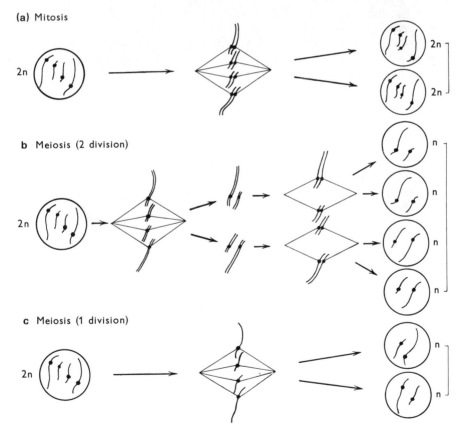

Fig. 4.1 Comparison of the behaviour of chromosomes in **a** mitosis, **b** two-division meiosis and **c** one-division meiosis.

division both chromosomes and centromeres divide, and Cleveland (1950) pointed out that in a one-division meiosis division of both chromosomes and centromeres is suppressed, while in two-division meiosis a division of centromeres is suppressed in the first division and a division of chromosomes does not occur in the second (merely separation of chromosomes duplicated before the first division). The three processes are compared diagrammatically in Fig. 4.1. One division meiosis is probably the primitive meiotic process.

The complete two-division meiosis is normally a more complex process than mitosis, although it has many comparable features. The two divisions usually occur in quick succession. During the extended prophase of the first division the chromosomes (which exist as paired chromatids) associate together in homologous pairs (Fig. 4.2), members of a pair becoming so closely coiled together in 'synaptinemal complexes' that sections of chromatids may be exchanged between them; a process known as crossing over occurs as a result of chiasma

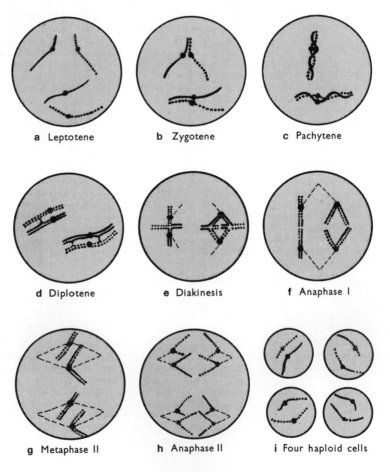

a Leptotene   b Zygotene   c Pachytene

d Diplotene   e Diakinesis   f Anaphase I

g Metaphase II   h Anaphase II   i Four haploid cells

Fig. 4.2 A selection of stages of a two-division meiosis, illustrating the behaviour of chromosomes described in the text.

formation between chromatids. Following further shortening of the chromosomes, the centromeres become attached to spindle microtubules, one chromosome of a pair becoming associated with one pole of the spindle and the other member of the pair with the other pole. The anaphase migrations of the chromosomes separate the two chromosomes of each pair to opposite poles, and crossed-over portions of chromatids separate with the centromeres to which they are now connected, so that two haploid groups of chromosomes are formed. Shortly after this each set of chromosomes usually becomes arranged on a new spindle for metaphase of the second division. This time the centromeres do divide and the two chromatids of each chromosome separate in the subsequent anaphase to form haploid nuclei with chromosomes in the G_1 form. The final telophase of such a two-division meiosis therefore results in the formation of four haploid nuclei from one original diploid nucleus. Cell division may accompany each nuclear division or may not occur till later. In one-division meiosis the chromosomes only aggregate in loose pairs in a simplified meiotic prophase; chiasma formation and crossing over appear not to occur, apparently because either replication of chromosomes or separation of chromatids has not occurred, and all chromosomes maintain the same genetic constitution. Only two haploid nuclei will be formed from each diploid nucleus in one-division meiosis, though daughter nuclei may then proceed to divide mitotically.

Patterns of life cycle and sexual processes in protists

The diversity of these patterns has been well reviewed by Grell (1967, 1973).

Many protists show no trace of any sexual stage in the life cycle; some of these species are believed to have evolved from forms showing meiosis and fertilisation, but others may be primitively asexual (Hawes, 1963); present indications that meiotic processes are rather uniform suggest that meiosis only evolved once in a common ancestor of all sexual forms. It is very difficult to be certain that an organism is entirely asexual, e.g. Amoeba proteus is one of the most intensively studied protozoans and is almost certainly asexual, but one possible report of sexual reproduction in this species raises an unsatisfied doubt. Species that are primitively asexual might be expected to be haploid, but many asexual species are probably diploid or polyploid – in the absence of meiosis the ploidy level may be difficult to determine. Sex has not been confirmed among euglenoid, cryptomonad, eustigmatophyte, choanoflagellate and trichomonad flagellates and schizopyrenid and arcellinid amoebae among others.

The life cycle of an asexual haploid protozoan is illustrated in Fig. 4.3a. The earliest appearance of sexual processes probably involved the fusion of two haploid individuals followed by meiosis of the zygote to produce haploid individuals which proceeded to grow and divide mitotically (Fig. 4.3b). In some cases the diploid zygote undergoes considerable growth before division, and an extension of this growth phase with mitotic divisions could give a life cycle with intermediary meiosis and an alternation of haploid and diploid mitotic phases found in many land plants (Fig. 4.3c). Further extension of the diploid phase leads to the pattern of life cycle found in higher animals (and often therefore regarded as 'normal'), with diploid organisms which undergo meiosis in the formation of gametic nuclei (Fig. 4.3d). All of these types of life cycle occur in protists and examples will be described below.

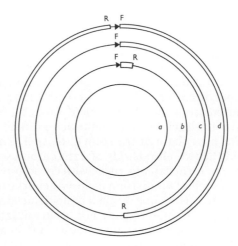

Fig. 4.3   Diagram to illustrate the timing of fertilisation (*F*) and reduction division (*R*) in four different patterns of protozoan life cycle. In **a** the organism is asexual and haploid, **b** is haploid with zygotic meiosis, **c** is haplodiploid with intermediary meiosis and **d** is diploid with gametic meiosis.

Although there are a number of examples of agamic cell fusion in protists, which could give indications of possible origins of sexual processes (Seravin and Goodkov, 1984), known sexual processes in protists all require the occurrence of meiosis in a complete or reduced form at some stage in the life cycle, and all involve the production of haploid gametic nuclei (pronuclei), two of which fuse in fertilisation to form a synkaryon. After these requirements have been met, however, there is still much scope for variation in detail. Gametic nuclei are usually, but not invariably, found in special gamete cells. In some species all gametes look alike (isogametes), but in other protists the gametes may differ in size and structure, so that in many cases it is usual to refer to male and female gametes, by analogy with the anisogametes commonly found in metazoa and higher plants, where the female gamete is an egg cell and oogamy occurs. The mating of gamont cells before these have differentiated to form gametes or gametic nuclei occurs in some groups, and the fusion of gametic nuclei occurs after this differentiation; this form of mating is called gamontogamy to distinguish it from gametogamy where two free gametes fuse. Conjugation is a process in which a cytoplasmic bridge is formed between two cells; in some cases (ciliates) gamete nuclei are exchanged and in others (some algae and fungi) one gametic nucleus or cell migrates through the bridge to fuse with the other. Autogamy is the fusion of gametes or gametic nuclei formed from the same gamont. Many of these variants of the sexual process will be illustrated by examples below.

Life cycles and sexual processes in non-ciliate protists

Haploid protists with zygotic meiosis
The most fully studied of these are probably volvocid flagellates, especially species of **Chlamydomonas** (p. 139). In some species of this genus normal vege-

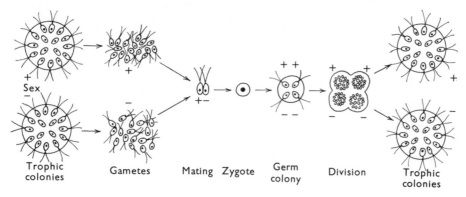

Fig. 4.4 Sexual reproduction in *Gonium pectorale*. (Information from Grell, 1967.)

tative cells transform directly into gametes in response to external conditions, whilst in other species of **Chlamydomonas** gametes are formed after two or more divisions of gamont cells. Compatible (+ and −) isogametes aggregate together by their flagella (see pp. 14 and 82), a papilla from one or both cells extends to make contact with the other cell and total cell fusion, including fusion of the two chloroplasts, follows. The zygote may swim for a while and sooner or later, depending on the intervention of a thick-walled resting stage, it undergoes meiosis liberating haploid flagellates.

Comparable processes occur in other unicellular and colonial volvocids. All cells of some colonial forms may develop into free-swimming isogametes, e.g. **Gonium** (Fig. 4.4), but in larger colonial forms flagellate macrogametes and microgametes may be produced and fuse in pairs, usually heterogametically, but occasionally similar gametes fuse together. **Volvox** species are oogamous, one vegetative cell becoming a single stationary macrogamete (egg cell) while many microgametes are formed by successive mitoses of other vegetative cells. The zygote in these cases undergoes meiosis in the initial divisions that lead to formation of the new colony.

Most dinoflagellates (but not **Noctiluca**) are haploid cells which divide mitotically to produce gametes. In **Peridinium cinctum** fusion of two biflagellate isogametes produces a zygote with two longitudinal flagella that grows for some time before meiosis is completed, motility is lost and a thick-walled resting stage is formed; three nuclei abort and a single haploid flagellate cell emerges from this cyst (Pfiester, 1984). Information on other haploid dinoflagellates indicates that their sexual processes vary in minor details. The sexual processes reported for chrysophytes indicate that these are also haploids, vegetative cells forming flagellate isogametes (Pienaar, 1980). Studies of reproduction of filamentous xanthophytes have revealed the production of isogametes or anisogametes from haploid parents, the common genus **Vaucheria** being oogamous (Hibberd, 1980a).

Some of the large hypermastigid and oxymonad flagellates from the hind guts of termites and the wood-eating roach **Cryptocercus** are haploids (some diploid examples are considered later). These flagellates normally divide mitotically, but sexual stages are found in moulting insects, and it has been shown that the moulting hormone (ecdysone) of the host insect stimulates flagellate sexual

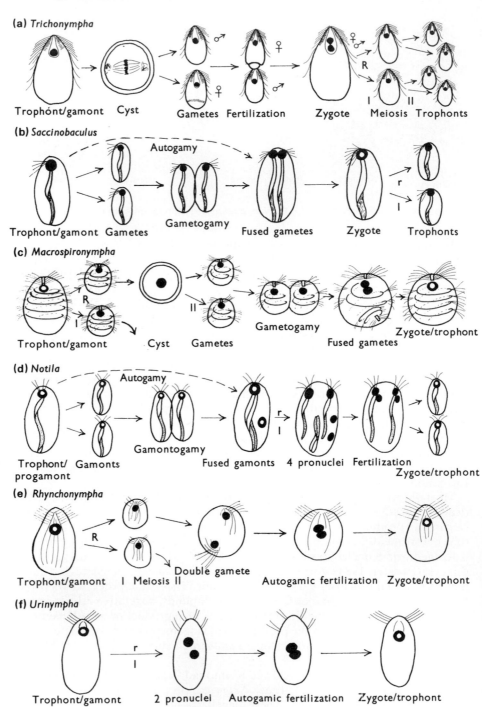

Fig. 4.5 Diagrams summarising events during sexual processes of a selection of parabasalian and metamonad flagellates, as described in the text; nuclei with white centres are diploid, those that are entirely black are haploid; *R* indicates a two-division meiosis and *r* a one-division meiosis. (Information from Cleveland, 1956.)

activity. The rich variety of sexual processes found in these flagellates has been described by L. R. Cleveland (1956). *Trichonympha* is a haploid hypermastigid flagellate whose trophic cells transform into gamonts by encysting and losing most of their extracellular organelles (Fig. 4.5a); these cells then divide to produce two gametes, one of each sex, which develop flagella, escape from the cyst and swim away. Gametes fuse in pairs by the attachment of the anterior of an apparently undifferentiated (male) cell to the posterior clear fertilisation cone of a female cell and the subsequent absorption of the male cell. Nuclear fusion is followed by meiosis within a few hours, and four haploid cells are formed. Sexual processes in other haploid forms vary; for example, in the oxymonad *Saccinobaculus* a cell forms a gamont which divides without encyst-ment to produce two isogametes (Fig. 4.5b). Gametes fuse laterally, with lateral fusion of their axostyles (p. 121), but fusion of the nuclei is delayed for some weeks, and two sets of flagella are temporarily retained. Following nuclear fusion meiosis occurs in a single division and only two haploid cells are formed from each zygote. Autogamy also occurs in this genus when division of the gamont nucleus and replication of cell organelles is not followed by cell division but by fusion of the axostyles and delayed fusion of the two pronuclei within the same cell, which becomes a zygote.

The cellular slime mould *Dictyostelium* is now known to be haploid in the normal amoeboid and spore stages, but the cell which forms the focus of a macrocyst (p. 164) is a zygote formed by fusion of two haploid amoebae; meiosis and mitosis take place within the macrocyst producing many haploid amoebae that are eventually released (O'Day and Lewis, 1981).

Sporozoa display a cytogenetic cycle like that of *Saccinobaculus* (when that flagellate is cross-fertilised). In gregarines two gamonts come together and the pair may remain attached 'in syzygy' for some time before secretion of a gamon-tocyst around the pair of cells (p. 225). Within the common cyst each gamont cell undergoes repeated mitoses to form gametes, usually in large numbers; in *Monocystis* these are amoeboid isogametes, but in some other cases aniso-gametes occur, and sometimes one type of gamete may carry one or more flagella; fertilisation takes place within the cyst and each zygote forms a spore. In most coccidians and haemosporidians gamonts differentiate to form one macrogamete or many microgametes, usually with one to three flagellar axonemes (p. 228); these microgametes swim freely to locate and fertilise a stationary macrogamete. In all sporozoans a one-division meiosis occurs in the zygote, and this may be followed by one or more mitoses to form the haploid sporozoites.

Various non-flagellate green algae are haploids that produce flagellate gametes and oogonia (e.g. *Oedogonium*, *Chara*), or take part in conjugation in which either one non-flagellate cell migrates to fuse with a non-migratory cell, or two amoeboid cells both migrate to meet part-way, and the zygote forms a zygospore (e.g. desmids, *Spirogyra*); in the latter cases nuclear fusion may be delayed and has been seen to occur just before meiosis and subsequent germina-tion of the zygospore.

Haplodiploid protists with intermediary meiosis
Some foraminifera display an alternation of haploid and diploid generations whose individuals may look more or less the same. Mature haploid gamonts produce gametes, and the zygote grows to produce a mature diploid agamont in which meiosis occurs during the formation of numerous small gamonts.

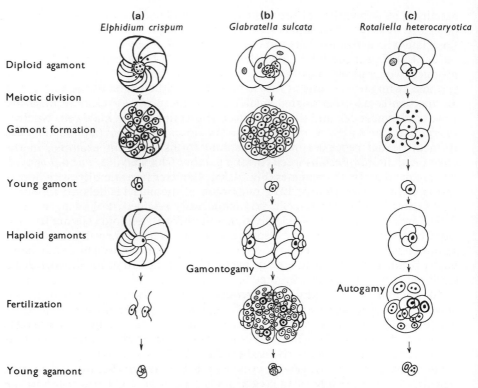

| | **(a)**
Elphidium crispum | **(b)**
Glabratella sulcata | **(c)**
Rotaliella heterocaryotica |

Diploid agamont

Meiotic division

Gamont formation

Young gamont

Haploid gamonts

Gamontogamy

Fertilization

Autogamy

Young agamont

Fig. 4.6 Comparison of three of the life cycle patterns shown by foraminifera, **a** gametogamy of *Elphidium crispum*, **b** gamontogamy of *Glabratella sulcata* and **c** autogamy of *Rotaliella heterocaryotica*. In **b** and **c** the agamont has larger somatic nuclei and smaller generative nuclei. (Information from Grell, 1967.)

Sexual reproduction in foraminifera may involve gametogamy, gamonto-gamy or autogamy, and even closely related forms may differ in this respect; they are usually isogametic, with either flagellate or amoeboid gametes. For example, the mature gamont of *Elphidium crispum* (= *Polystomella crispa*) divides to release numerous biflagellate isogametes into the sea, where fertilisa-tion takes place (Fig. 4.6a). *Glabratella sulcata* is gamontogamous (Fig. 4.6b), and two mature gamonts adhere together so that their shell interiors are continuous; gametes with three flagella are formed by both gamonts, many of them fuse in pairs within the shell and escape as small agamonts after ingesting unpaired gametes. Gamontogamy may involve up to 10 or more gamonts, and in most genera the gametes are amoeboid. Autogamy in *Rotaliella hetero-caryotica* involves the formation and fusion of amoeboid gametes within the shell of a single gamont (Fig. 4.6c).

An alternation of haploid and diploid phases also occurs in at least some haptophytes (p. 143) (Hibberd, 1980b). In *Hymenomonas carterae* a diploid coccolith-bearing flagellate stage reproducing mitotically in the plankton may divide meiotically to produce four haploid swarmers without coccoliths that settle as benthic pseudofilaments; it has been reported that isogametes produced by this benthic stage fuse in pairs and the zygote develops coccoliths and becomes planktonic, but reports of other life forms within the same cycle have yet to be resolved. Other examples of alternate life forms are known in this group, but nuclear details are lacking. Haplodiploid cycles are also found in some filamentous green algae (e.g. *Cladophora*), most brown algae, many red algae and blastocladiid chytrids; in this last case both gametes are uniflagellate.

Diploid protists with gametic meiosis
In addition to the ciliates, which are discussed separately, various flagellates and amoebae come in this category. Myxosporidia are probably diploid, since their sporoblast nuclei are diploid, but the position of meiosis is uncertain (Raikov, 1982), and labyrinthulids (p. 138) are probably diploid since the spindle cells undergo meiosis in the formation of biflagellate zoospores, but the position of nuclear fusion is uncertain; meiosis also occurs in sporogenesis of microsporidia (Loubes, 1979), but again the timing of nuclear fusion is not known.

A simple diploid life cycle is shown by *Opalina* (p. 131), in which diploid gamonts undergo meiosis in the production of multiflagellate gametes; the zygote encysts and after emergence grows as an asexual trophic cell. Multiflagellate gametes are also produced by the hypermastigid flagellate *Macrospironympha*, although after the first meiotic division the two products each encyst before the second meiotic division produces two flagellate gametes, one male and one female; gametes fuse later and male organelles are resorbed (Fig. 4.5c). The first meiotic division of the hypermastigid *Rhynchonympha* produces two cells, but after the second meiotic division the two nuclei remain in the same cell and fuse autogamously (Fig. 4.5e). Gamontogamy of diploid cells occurs in *Notila* (Oxymonadida), and after one division meiosis of both nuclei, the nuclei fuse in pairs before the cell divides (Fig. 4.5d). Autogamy in *Urinympha*

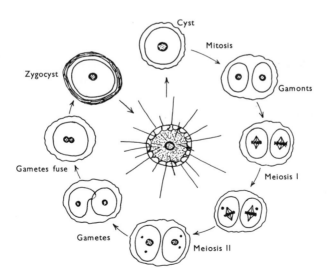

Fig. 4.7   Diagrams illustrating events during the process of autogamy in *Actinophrys* described in the text. (After Belar, 1926.)

involves a one-division meiosis of the diploid nucleus, followed by fusion of the two haploid nuclei (Fig. 4.5f). The dinoflagellate *Noctiluca* divides meiotically and then mitotically to produce numerous uniflagellate isogametes which fuse in pairs, and the zygote grows to form a normal diploid cell (Pfiester, 1984).

Autogamy of diploid cells is also the usual rule in actinophryid heliozoans, e.g. *Actinophrys* (Fig. 4.7). Lack of food following active growth may lead to encystment and one mitotic division; in each cell a two-division meiosis occurs without cell division, one haploid nucleus survives in each cell, the two cells become gametes and fuse together and the zygote secretes a thickened resistant wall from which one diploid heliozoan later emerges. Occasionally two individuals may encyst within a common wall, each forms two gametes, and cross-fertilisation may result.

In the curious life cycle of *Sappinia diploidea* (Fig. 4.8), described many years ago by Hartmann and Nagler (1908), the normal vegetative cell has two nuclei (presumed haploid) which divide each time a cell divides. Periodically two cells encyst together, the nuclei within each cell fuse with one another, the two cells undergo cytoplasmic fusion and each diploid nucleus then undergoes a two-division meiosis from which only one haploid nucleus from each parent survives. A recent study has shown some features consistent with this account, but failed to totally confirm it; the occurrence of occasional uninucleate encysted cells could throw doubt upon its accuracy (Goodfellow, Belcher and Page, 1974).

Many of the larger green algae are oogamous diploids that produce flagellated male gametes. All diatoms are diploid, but while members of the more ancient (centric) group are oogamous with uniflagellate male gametes, pennate diatoms show a form of conjugation (p. 74). Sexual reproduction in the diploid oomycetes involves conjugation between hyphae, and following meiosis male nuclei migrate to fertilise one or more oospheres; biflagellate isogametes are formed by the diploid hyphochytrids.

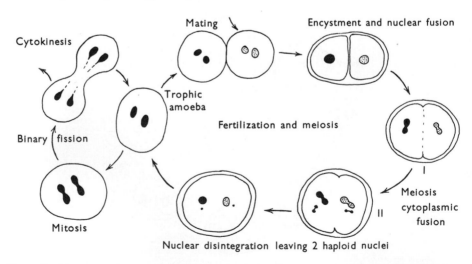

Fig. 4.8 Diagrams of stages in the life cycle of the dikaryotic amoeba *Sappinia diploidea*, showing the behaviour of the nuclei. (Information from Dogiel, 1965.)

Life cycles and sexual processes in ciliate protists

Nuclear events in ciliate sexual processes (Grell, 1973; Raikov, 1982)
Ciliates generally carry nuclei of two types (p. 186). The macronucleus which is typically polyploid and has a somatic function disintegrates and is replaced

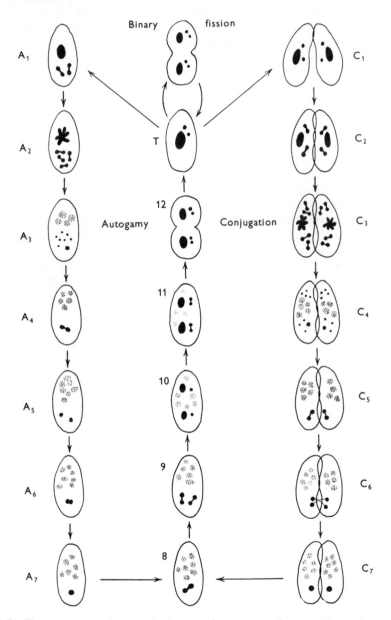

Fig. 4.9 The sequences of events in the sexual processes of conjugation and autogamy in *Paramecium aurelia*; T is the trophic ciliate, A_1–A_7 are stages in autogamy, C_1–C_7 are stages in conjugation and stages 8–12 are common to both processes.

during the sexual process, while the micronucleus is diploid and carries the genetic information through sexual processes. Sexual activity in ciliates commonly involves a special form of gamontogamy called conjugation, in which each gamont is (normally) hermaphrodite and undergoes meiosis in the formation of gametic nuclei; each gamont normally receives a gametic nucleus from the other gamont, and each becomes a zygote. Autogamy is also common. Some variation occurs in the details of these processes, but their main features may be seen in the following example.

Paramecuim aurelia has one macronucleus and two micronuclei. At the beginning of conjugation two ciliates come together (Fig. 4.9), becoming associated first in a clumping reaction due to adhesion of specific cilia and then within half an hour or so they become firmly attached at deciliated areas near the anterior end of their ventral surfaces; later attachments form in the gullet region where membrane fusion leads to the formation of a number of pores providing cytoplasmic bridges between one gamont and the other. Both micronuclei in each conjugant undergo a two-division meiosis, so that eight haploid nuclei are formed in each cell, seven of which disintegrate. The remaining micronuclei undergo mitosis to produce two gamete nuclei in each cell. One gamete nucleus from each gamont migrates across a cytoplasmic bridge into the other cell and fuses with the nucleus which did not migrate, so that synkarya are formed in each cell. Subsequently the cells separate, and by this time the original macronuclei of the two cells have more or less disintegrated. The micronuclear synkaryon divides twice mitotically and two of the four products become polyploid macronuclei (p. 186) and two remain as diploid micronuclei. At the first binary fission of the ex-conjugant one macronucleus passes to each daughter cell and both micronuclei divide mitotically, so that the original complement of nuclei is restored.

In autogamy a single *Paramecium* undergoes the same sequence of nuclear divisions and disintegrations, but following the mitotic division of the haploid nucleus the two gametic nuclei fuse with each other (Fig. 4.9). When two cells come together as if to conjugate, but end up by performing an autogamous self-fertilisation, the process is referred to as selfing or cytogamy. In none of the sexual processes of Paramecium does an increase in number of individuals occur, but this may follow by mitosis and binary fission before the normal nuclear constitution is restored.

The pattern of restoration of the normal nuclear constitution following conjugation of different species of ciliates is particularly diverse (Fig. 4.10) (Grell, 1973). In a simple case, the synkaryon of *Chilodonella uncinata* divides once and one product becomes a macronucleus and one a micronucleus. Two divisions of the synkaryon, as found in *P. aurelia*, are common, but the details vary; in *Euplotes patella* two of the products of such a division degenerate and the other two nuclei form one micronucleus and one macronucleus, while in *Didinium* two of the four products fuse to form a single macronucleus and the other two persist as micronuclei. In *Paramecium caudatum* there are three quick divisions of the synkaryon, three of the products degenerate, one becomes a micronucleus and four become macronuclei; the micronucleus divides mitotically at the next two divisions, but the macronuclei merely separate into the four daughter cells and do not divide, so that four karyonides (see p. 87) are formed from each conjugant in this species.

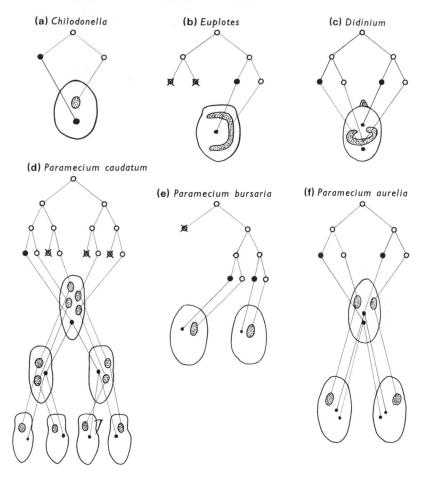

Fig. 4.10 Examples of some patterns of nuclear reorganisation following sexual processes in **a** *Chilodonella uncinata*, **b** *Euplotes patella*, **c** *Didinium nasutum* and **d,e** and **f** three species of *Paramecium*. (Information from Grell, 1956.)

Some ciliates are dioecious, e.g. **Vorticella**, where one gamont contributes only a migratory gamete nucleus and the other contributes only a stationary gamete nucleus. Frequently the cell producing the migratory pronucleus is smaller than the other and undergoes total fusion with the macrogamont.

The karyorelictid ciliates (p. 199) possess diploid macronuclei that do not divide, but are always formed directly from micronuclei (p. 188). Raikov (1969) found that at conjugation in **Trachelocerca phoenicopterus** a number of gametic pronuclei are formed in each conjugant; several of these migrate and fuse with stationary pronuclei in the other conjugant to produce several synkarya, but subsequently all of these become pycnotic except one.

Conditions required for conjugation of ciliates
For some species it is known that two individual ciliates will only conjugate under certain conditions; they must be of the correct mating type, and the age of

Table 4.1   The mating types of the *Paramecium aurelia* complex. Data from Grell (1973) and Sonneborn (1975).

Syngen Number	Mating Type	Other mating types with which conjugation occurs (and %)	Group	+ or − type
1	I	II(95%), X(40%)	A	−
	II	I(95%), IX(40%), XIII(10%), V(1%)		+
2	III	IV(95%)	B	.
	IV	III(95%)		.
3	V	VI(95%), [XVI(40%)],* II(1%)	A	−
	VI	V(95%)		+
4	VII	VIII(95%), [XVI(95%)],	B	−
	VIII	VII(95%), XV(60%)		+
5	IX	X(95%), II(40%)	A	−
	X	IX(95%), I(40%)		+
6	XI	XII(95%)	B	.
	XII	XI(95%)		.
7	XIII	XIV(95%), II(10%)	B	−
	XIV	XIII(95%)		+
8	XV	XVI(95%), VIII(60%)	B	−
	XVI	XV(95%), [VII(95%), V(40%)]		+
9	XVII	XVIII(95%)	A	.
	XVIII	XVII(95%)		.
10	XIX	XX(95%)	B	−
	XX	XX(95%)		+
11	XXI	XXII(95%)	A	.
	XXII	XXI(95%)		.
12	XXIII	XXIV(95%)	B	−
	XXIV	XXIII(95%)		+
13	XXV	XXVI(60%)	C	.
	XXVI	XXV(60%)		.
14	XXVII	XXVIII(95%)	A	.
	XXVIII	XXVII(95%)		.

* Square brackets indicate incomplete mating reactions not resulting in conjugation.

the clone to which they belong must be within certain limits (Sonneborn, 1957, 1975). Ciliates formerly recognised as belonging to the 'species' *Paramecium aurelia* on morphological characters have now been separated into 14 species of the *P. aurelia* complex (Sonneborn, 1975). Originally a number of varieties of *P. aurelia* was recognised in which members of one variety would not normally conjugate with members of another variety, so that each variety was effectively a genetically isolated species (see Table 4.1). Varieties were first called syngens by Sonneborn because they represented groups of ciliates which share the same 'gene pool', but are for this reason now recognised as species although they are not morphologically distinguishable from each other. Within each syngen, or species, of the *P. aurelia* complex there are two mating types, and conjugation will only take place between individuals of different mating type. Mating type specificity in *Paramecium* seems to be expressed in chemical characteristics of the surface membrane (p. 85). The mating types do not represent sexes, since both conjugants are hermaphrodite; the system is better compared with the devices used by flowering plants to ensure cross pollination.

Following separation of the conjugants of *P. aurelia*, each ex-conjugant gives rise to a clone of cells by successive binary fissions. Clonal age influences the ability of individuals to take part in sexual processes; members of a clone are not capable of conjugation until a certain number of fission cycles has been completed since the previous conjugation (Siegel, 1967). Following this period of immaturity, which lasts from 2–10 days, the ciliates are mature for a month or so and will conjugate successfully with ciliates of the complementary mating type. Mating behaviour depends upon the incorporation of complementary mating substances in the ciliary membranes (Kitamura and Hiwatashi, 1978). As the clone grows older and enters a period of senescence, fertility decreases, and as a result conjugations become rare and an increasing proportion of conjugations result in non-viable ex-conjugants; meanwhile, autogamy becomes more common both in separate individuals and in the form of selfing during apparent conjugation. As senescence progresses, autogamy no longer produces viable clones, so that eventually after some months no further sexual processes of any type are successful and the ageing clone reaches a state of genetic death because its members can no longer participate in the foundation of a new clone. Ciliates in the senescent clone may still divide asexually and the clone may persist for some months in a state of reduced vigour before all of the population die and the clone suffers somatic death by ultimate loss of macronuclear function. It is important to recognise that both forms of sexual process, conjugation and autogamy, can result in rejuvenation and the foundation of a new clone; apparently the successful renewal of the macronucleus is the vital requirement in these processes. The clone of ciliates may be compared with a single multicellular animal in respect of ageing and ability to participate in sexual processes. Some ciliates do not show a comparable senescence, since in amicronucleate strains of *Tetrahymena* and *Didinium* conjugation is not possible, yet these strains have been maintained in culture for many years. The extrusion of DNA bodies from ciliate macronuclei has been widely reported, and may be concerned with the maintenance of macronuclear vitality.

Investigations of mating type and sometimes clonal ageing have been made on a few other ciliates, including species of *Euplotes* and *Tetrahymena* as well as other species of *Paramecium*. Some differences are illustrated by *Paramecium bursaria*, where two syngens have four mating types and four syngens have eight mating types (Bomford, 1966). In any syngen a mature member of one mating type will conjugate with a mature member of any of the other mating types of that syngen, but not normally with any member of any other syngen. In *P. bursaria* the immature period lasts several to many weeks and the period of maturity may last for many years, so that the complete life span of the clone may exceed 10 years. At least 50 mating types exist in 17 syngens of the *Tetrahymena pyriformis* complex.

A complex pattern of multiple mating types has been described by Kimball (1942) for *Euplotes patella*. In this species one syngen has six mating types, and mating has been found to be controlled by three substances secreted by the ciliates. A ciliate which produces substance 1 will conjugate in the presence of substances 2 or 3, a ciliate which produces substance 2 will conjugate in the presence of substances 1 and 3, and so on. Ciliates of some mating types produce two of the substances and mate only in the presence of the third. This system normally results in mating between ciliates of different mating types, but experimentally it is found that individuals of the same mating type will conjugate with one another in the presence of an appropriate substance produced by members

Table 4.2 Genotypes of the mating types of *Euplotes patella*. Data derived from tables by R. Kimball (1942).

Mating type	Will mate in filtrate of mating types	Substances produced	Genotype
I	II, III, V	1, 2	mt^1, mt^2
II	I, V, VI	1, 3	mt^1, mt^3
III	I, II, IV, V, VI	3	mt^3, mt^3
IV	I, II, III, V, VI	1	mt^1, mt^1
V	I, II, IV	2, 3	mt^2, mt^3
VI	I, II, III, IV, V	2	mt^2, mt^2

of another mating type. It is concluded that the pattern of mating illustrated in Table 4.2 is controlled by a set of three multiple alleles. This pattern agrees with the proposal of Luporini and Miceli (1986) that mating substances, which they prefer to call pheromones, function in self recognition to prevent self mating, rather than to promote mating between different mating types.

The mating process in **Blepharisma** has been shown to involve mating pheromones (gamones) secreted by individuals of each mating type (Miyake, 1981). In **Blepharisma japonicum** mating types I and II are not completely stable and may change in either direction spontaneously. Mature ciliates (preconjugants) of mating type II placed in cell-free fluid from a culture of preconjugant mating type I cells gain the ability to adhere in pairs among themselves within 1 to 2 hours. Mating type I cells release gamone 1 (blepharmone, an unstable glycoprotein of about 20 kD), which induces an aggregation response in mating type II cells and stimulates them to secrete gamone 2 (blepharismone, a heat-stable, dialysable, tryptophan derivative), which in turn acts on mating type I cells, making them ready to unite, as well as acting as an attractant to type I cells. Each gamone induces protein synthesis in the complementary mating type, stimulating secretion of gamones and preparing the cells for conjugation. In the presence of both gamones, cells of both mating types are 'conditioned' to unite, and upon contact they adhere by their cilia and after a couple of hours membrane fusion and cytoplasmic union follow, with exchange of genetic micronuclei.

Control of the mating type in **P. aurelia** resides in the macronucleus; the mating type of a clone is established at the formation of the new macronucleus following conjugation and is transmitted at macronuclear fission. However, the mating type of a new macronucleus is not necessarily determined by genes derived from the micronuclear synkaryon, since the character of the cytoplasm in which the macronucleus develops can influence the mating type. For example, a newly formed macronucleus of **P. aurelia** has the potentiality of producing either mating type (Sonneborn, 1957). Following conjugation the two ex-conjugants contain the same chromosome complement since each synkaryon received identical haploid sets of chromosomes from the two parental cells. The macronuclei which develop in the two cells will therefore be formed from identical nuclei but will be surrounded by different cytoplasm. In one group of syngens (Group A) mating type seems to be determined randomly during the establishment of each macronucleus, so that the two macronuclei within the same ex-conjugant cell may determine different mating types in the

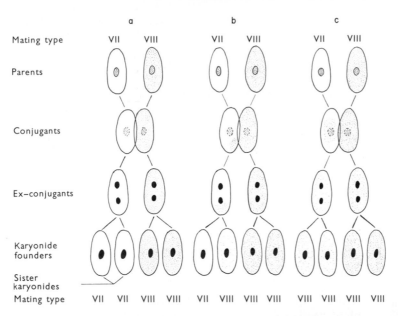

Fig. 4.11 Inheritance of mating type in Group B syngens of *Paramecium aurelia* exemplified by conjugation between individuals of mating types VII and VIII. In **a** the mating type of the two sister karyonides formed from each ex-conjugant is the same and corresponds to that of the parent conjugant, but in **b** and **c**, following a lesser or greater transfer of cytoplasm from the type VIII conjugant to the type VII conjugant, three or all four of the karyonides formed are of the same mating type. (Information from Beale, 1954.)

cells produced at the first fission after conjugation. Within the clone derived from one ex-conjugant there may be two subclones (karyonides – clones with identical macronuclei) of different mating type, one for each of the newly constituted macronuclei; there is also random determination of mating type in the two karyonides established following autogamy. In a second group of syngens (Group B) cytoplasmic factors determine which nuclear character shall be expressed, so that the mating type corresponds to that of the parental cyto-plasm and the mating type of both sister karyonides within a clone will normally be the same (Fig. 4.11a). In exceptional cases of prolonged conjugation a large amount of cytoplasm may leak from one ciliate to the other through the conjugation pores, and the macronucleus which develops in the mixed cytoplasm may have the mating type of the parental cell or of the foreign cyto-plasm (Fig. 4.11b,c). In syngen 13 (Group C) mating type is genetically deter-mined by genes at a single locus, the allele of mating type XXVI being dominant over that of mating type XXV (Butzel, 1974). Karyonidal or genetic control of mating type may also be found in Tetrahymena (see Allen and Gibson, 1973).

Genetic significance of protistan life cycles

The genetic mechanism of all organisms has built into it the possibilities of conservation and variation of information content. Conservation of informa-tion is achieved by accurate copying of the DNA code at both mitosis and meiosis, so that offspring are like their parents. Variation is vital because it provides the raw material for natural selection to work on – the variants of pattern that can be tested against the environment and may lead to an evolutionary change in the population if the variant has a selective value.

The phenomenon of genetic variation has a number of facets that are less widely appreciated than the conservation of genetic information. The source of genetic change is mutation – and alteration in the molecular configuration of the gene which results in a modification of the gene product. Such a change may affect any one nucleotide or any combination of nucleotides within the DNA chain of the gene, so that a vast number of different mutations could affect any gene. A mutated gene may depend for its expression on the other genes in the cell, e.g. it may be a recessive gene masked by a dominant allelic gene, or it may form part of a complex of genes involved in producing enzymes for a chain reaction, and may only find expression if the other genes of the complex are present. In a haploid organism the first of these two could not occur, and the second would be more restricted since only one gene of each type would be present. It is therefore more likely that a gene in a haploid organism will find immediate expression for selection or rejection and so is less likely to remain hidden than in a diploid organism. Mutant genes that are not immediately beneficial may be more likely to survive in forms where the diploid phase predominates, so that genetic variability and the potentiality for genetic change are greater in populations of diploid protists than in haploid populations. On the other hand it is clearly advantageous to employ haploid forms such as **Chlamydomonas** in experimental studies of certain aspects of gene functioning, since mutant genes are more likely to be immediately expressed in the phenotype than in diploids. The occurrence of a mutation in a favourable combination for expression will be very rare in an asexually reproducing organism, especially if it is haploid, so that changes will normally take place slowly unless very large populations of the organisms occur. From the point of view of the conservation of genetic variability, it is particularly advantageous for an asexual organism to be polyploid.

If the organism reproduces sexually the chances of the mutated gene finding itself in a favourable combination are vastly increased by the processes of meiosis and fertilisation. During meiosis, not only do the chromosomes segregate independently, so that the different gametic nuclei produced by an organism contain a full range of different combinations of chromosomes from the set of pairs of homologous chromosomes of the parent, but the occurrence of new combinations of genes within chromosomes is made possible by the exchange of sections of chromosomes during crossing over. Every gametic nucleus may therefore contain a unique set of genes, and at fertilisation two unique sets of genes come together, to make more likely the occurrence of a favourable combination of other genes in the same nucleus as a mutated gene. The spread of genetic variation through a population is greatly enhanced by the occurrence of sexual processes involving assortment of genes at a complete meiosis and recombination of genes from different parents at fertilisation. The more remote the relationship between the parents, the greater the difference between their genetic constitutions, and therefore devices that encourage cross fertilisation, such as the sexes of volvocid flagellates or the mating types of ciliates, will increase the range of new genetic combinations. Similarly, the period of immaturity of the clones of **Paramecium** encourages outbreeding, and the longer the periods of immaturity and maturity the further the clone may spread before mating and the greater the chance that the ciliate will mate with an individual of an unrelated clone.

Reduction and loss of sexual processes

The genetic significance of sexual processes is reduced or lost in many protists, as indicated in Fig. 4.12. The absence of crossing over in one-division meiosis of Sporozoa and some flagellates means a loss of the possibilities of gene reassortment and the production of new patterns of gene linkage. The process of auto-

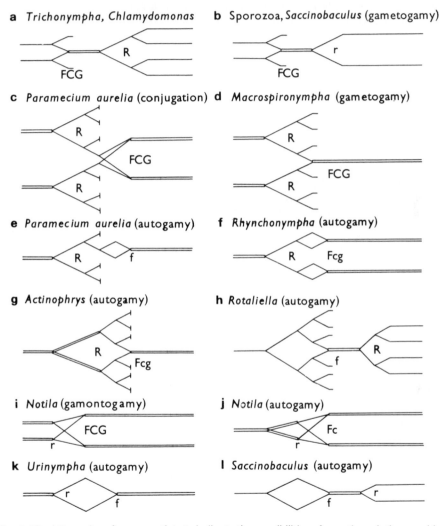

Fig. 4.12 Life cycles of some protists to indicate the possibilities of genetic variation provided by sexual processes. *R* indicates a two-division meiosis involving crossing over as well as independent assortment of chromosomes; *r* indicates a one-division meiosis with assortment of chromosomes but no crossing over; *FCG* indicates fertilisation that brings together new combinations of chromosomes and genes from different parents; *Fcg* indicates fertilisation that brings together new combinations of chromosomes and genes derived from the same parent; *Fc* indicates fertilisation that brings together new combinations of chromosomes from the same parent, but because of the one-division meiosis the gene content of the chromosomes remains unchanged; *f* indicates fertilisation in which the genetic content of the two sets of chromosomes that come together is identical. Single lines indicate haploid stages and double lines indicate diploid stages.

gamy, as seen in the ***Paramecium aurelia*** complex, retains the advantages of meiosis in providing haploid nuclei with reassorted genes but loses the advantage of fertilisation because it brings together two identical sets of genes, so that the nuclei produced by autogamy of this type are completely homozygous. In the heliozoans the situation is somewhat different, since the two pronuclei which fuse in autogamy are products of different meioses, and homozygosity is unlikely to be complete, even after repeated autogamies; because of its special features this form of autogamy is sometimes referred to as paedogamy.

The reduction in genetic significance of. sexual processes in the flagellates studied by Cleveland is particularly striking. The full extent of genetic variation may occur in the haploid ***Trichonympha*** or diploid ***Macrospironympha***, but the variation is reduced because of one-division meiosis in the gametogamy of ***Saccinobaculus*** or gamontogamy of ***Notila***, and is reduced because of autogamy in ***Rhynchonympha***. Although both autogamy and one-division meiosis may occur together in ***Notila***, the process retains some genetic value because the gametic nuclei come from different meioses in which independent segregation of chromosomes may occur. Autogamy of ***Urinympha*** has no genetic significance since the chromosomes which separate at the one-division meiosis are reunited at fertilisation, and similarly autogamous sexual reproduction of ***Saccinobaculus*** results in no genetic change since the zygote must be completely homozygous.

Division of the cell and morphogenesis

Timing of events in the life cycle of *Tetrahymena*

Division of the nucleus is normally followed by division of the cell. These two

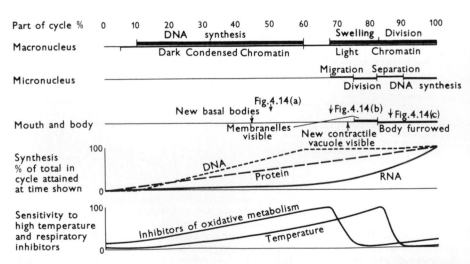

Fig. 4.13   Some events during the life cycle of ***Tetrahymena***. The separation of two daughter cells occurs at time O, and the appearance of the cell at three stages during the cell cycle showing the development of the new mouth structures may be seen in Fig. 4.14. (Information mainly from Holz, 1960.)

events are the most obvious stages of a process which involves many parts of the cell and which commences before any nuclear changes are seen. The most intensively studied protistan life cycle is probably that of members of the ***Tetrahymena pyriformis*** complex, and the timing of some of the events in this life cycle are shown in Fig. 4.13. These organisms have been found particularly suitable for these studies because large numbers of ciliates may be obtained at the same stage in the life cycle by the application of temperature shocks to synchronise cell division (Zeuthen and Scherbaum, 1954; Zeuthen, 1971).

Two transition stages appear to control the timing of this life cycle. There is evidence that once a cell of ***Tetrahymena*** has successfully entered the macronuclear S phase, preparations are set in motion for the next division of the cell (Prescott and Stone, 1967). These preparations involve not only the synthesis of new DNA, but also the production of a variety of proteins, some of which are used in the replication of cell organelles prior to or subsequent to division. The first visible event is the duplication of ciliary basal bodies for the formation of new mouth ciliature. The original mouth of this ciliate is retained by the anterior daughter (proter) and a new mouth for the posterior daughter (opisthe) is formed just posterior to the equatorial fission furrow line and directly behind the existing buccal cavity (Fig. 4.14). These basal bodies first appear as an area of irregularly arranged kinetosomes which later become organised to form the bases of the three membranelles and the undulating membrane (Frankel, 1967), completion of these structures coinciding with the micronuclear mitosis. Late in the formation of the irregular field of kinetosomes the micronucleus migrates peripherally, the macronucleus swells and the ciliate becomes markedly less sensitive to inhibitors of aerobic respiration. A new contractile vacuole develops in the proter and begins to function during the period of micronuclear migration. Micronuclear mitosis is followed almost immediately by synthesis of new micronuclear DNA and by migration of the two micronuclei towards proter and opisthe; at this time macronuclear fission commences, and the appearance of an

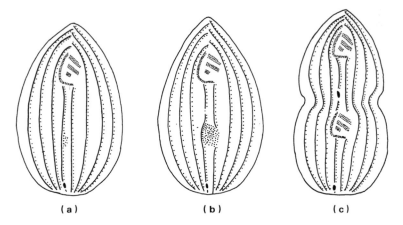

(a)   (b)   (c)

Fig. 4.14   Stages in the development of the mouth of the posterior daughter cell prior to division of ***Tetrahymena***. In **a** the basal bodies that will form the new oral ciliature are beginning to multiply, in **b** they have formed an extensive 'anarchic field' and in **c** the basal bodies have been grouped to form the three membranelles and the undulating membrane. By the third stage the body is markedly furrowed and the new cytoproct of the anterior daughter may be seen.

equatorial fission furrow quickly follows. The furrow is associated with a band of microfibrillar material, whose presence has been reported in many animal cells, flagellates and ciliates (Fig. 4.15) (Jurand and Selman, 1969; Tucker, 1971). This microfibrillar band lies immediately below the pellicle and contracts to constrict the cell until only a narrow neck plugged by fibrillar material remains between the two daughters; the neck finally breaks and the two cells separate.

The second transition point may be demonstrated experimentally by heat treatment (34°C) or by such inhibitors of protein synthesis as puromycin and p-fluorophenylalanine (Zeuthen and Rasmussen, 1972). Development of the oral ciliature is stopped or reversed by these treatments up to a critical time, at about the completion of the membranelles and before the undulating membrane is completed, after which these agents have no effect and the cell is fully committed to the division process. The evidence is consistent with the suggestion of E. Zeuthen that a special heat sensitive 'division protein' synthesised in these cells is required for morphogenetic activities connected with the preparations for division, and that at the second transition point this protein becomes insensitive to heat or is no longer required.

Control of the preparations for division is in fact the basis of the temperature treatment for synchronising cell division of *Tetrahymena*, reviewed by Zeuthen and Rasmussen (1972). Alternating periods of 30 minutes at 34°C and 30 minutes at 29°C permit cell growth, but hold the cells in an early stage of the preparation for division – the heat treatment prevents differentiation and the period at optimum temperature (29°C) is not long enough for morphogenesis to progress as far as the onset of insensitivity to heat; at the end of a number of alternations of this type all of the cells are found to have an irregular field of kinetosomes and the micronuclear division is arrested in anaphase. If the ciliates are then kept at 29°C they will pass through several cycles of division more quickly than usual (Fig. 4.16). Cell cycles can also be synchronised if ciliates are exposed to 30-minute heat shocks spaced one normal cell generation apart (Zeuthen, 1971).

Zeuthen identified two cycles of activity in the life cycle of *Tetrahymena*; a cycle involving alternate periods of DNA synthesis and of no synthesis of DNA, and a second cycle in which morphogenesis and cell division alternate. It would appear that under normal conditions these two cycles are interlinked, events in one cycle serving as signals for events in the other; perhaps the production of a specific protein follows macronuclear DNA synthesis and promotes morphogenesis, while certain stages in morphogenesis exert a blocking effect on macronuclear DNA synthesis. Initiation of the macronuclear S phase may result from an inadequacy of the supply of certain RNAs to the cytoplasm, and the onset of cell division may be triggered when the oral area: cell surface area ratio falls below a certain level. However, the nuclear and cytoplasmic events are to some extent separable, since prolonged synchronising temperature treatment (Fig. 4.16) prevents morphogenesis without preventing the DNA synthesis cycle, so that after the treatment the ciliates have an abnormally large store of DNA and may run through several divisions at a higher rate than usual before a shortage of DNA and need for time for DNA synthesis brings about a normal rephasing of the two cycles. The kinetosomes appear to play a particularly important morphogenetic role, and the state of development of the oral

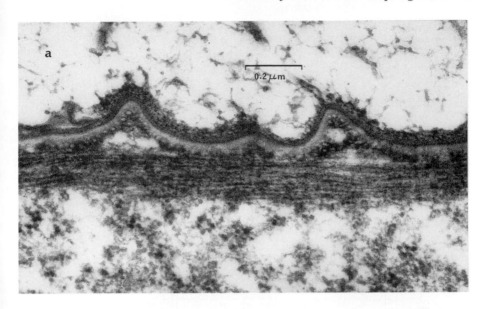

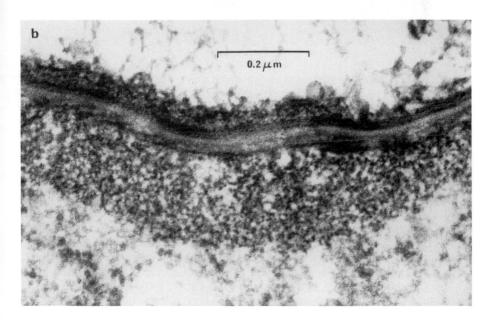

Fig. 4.15 Bundles of microfilaments beneath the cleavage furrow of the ciliate *Nassula*, shown here in **a** longitudinal and **b** transverse section, constitute the contractile ring. (Micrographs by Tucker, 1971.)

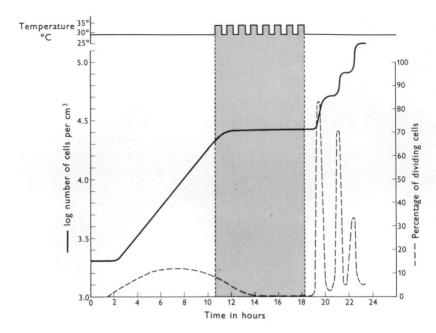

Fig. 4.16   The effect of repeated heat shocks on the population growth and percentage of dividing cells in a culture of ***Tetrahymena***. (From data by Zeuthen and Scherbaum, 1954.)

apparatus may be a key factor in the morphogenesis-cell division cycle; if the development of the oral primordium is blocked by a temperature shock, the existing structures are resorbed and replaced by the development of a new field of basal bodies and their differentiation to form the oral ciliature.

Morphogenesis in *Stentor*

A complete regeneration of mouthparts occurs in many ciliates if the fully formed oral ciliature is damaged or removed. This has been studied in detail by V. Tartar (1961, 1966, 1968) in ***Stentor coeruleus***, where the development of a new mouth primordium occurs in a specific region of the lateral surface at a place where closely spaced kineties meet widely spaced ones (the region of pigment-stripe contrast); grafting experiments showed both the specificity of this site and the nature of the stimulus that leads to primordium formation. It appears from this work and that of N. de Terra (1970, 1977) that an oral primordium will develop when the ratio of oral structures to body size, or oral area to body surface area, falls below a certain level; the primordium development may lead to body reorganisation, regeneration or division, and the newly formed membranelles may be added to the existing membranelles of a short row to extend it, may form a complete new mouth ciliature for a damaged individual or may form the feeding ciliature for the posterior daughter in division. The determination of whether primordium formation in an undamaged cell results in reorganisation or division is a function of cell size – only larger cells divide. Following the appearance of a primordium the macronucleus always undergoes reorganisation by coalescence of its many nodes and subsequent renodulation; these processes are shown by nuclear transplantation experiments to be under cytoplasmic control. If the primordium formation leads to constriction of the cell, the macronucleus is divided by the fission furrow while it is in the condensed state, but if the nucleus is not constricted, either because the cell is regenerating rather than dividing, or because the macronucleus has been artificially displaced to one side of the fission line, it does not divide. The degree of control over nuclear events exerted by the cytoplasm is striking, and the ultimate source of the regulation appears to depend upon the spatial relationships of organised structures in the pellicle. It was found in a hypotrich ciliate that the removal of a single cirrus led to complete reorganisation of the body ciliature. Many aspects of morphogenesis in protists were reviewed by Tartar (1967).

Specialised and modified cell cycles

Although the cell cycle has not been studied in comparable detail in many other protists, it is believed that similar controls and interactions operate in cells of other types, and that these may be modified in a variety of ways in more complex cell cycles, e.g. where these involve encystment (p. 262). Evidence of preparation for division before nuclear changes are seen is available for such amoebae as *Euglypha* where plates for the shell of the new daughter accumulate in the cytoplasm of the parent cell (Fig. 6.8, p. 167), and also in such flagellates as *Euglena* where new pellicle ridges appear between existing ones before division of the nucleus. Where centrioles are present these divide before changes are seen in the nucleus, and in flagellates where the centrioles are normally closely connected with the flagellar apparatus the flagellar basal bodies and associated organelles are usually duplicated before nuclear division starts.

The replacement and renewal of cytoplasmic structures at binary fission is a common event in both flagellates and ciliates. In such flagellates as *Macrotrichomonas* and *Lophomonas* new sets of body organelles are formed in both daughters at binary fission, and the organelles of the parent are resorbed. In the ciliate *Nassula* the original feeding structures are broken down and a new cytopharyngeal complex is formed in each daughter (Tucker, 1970). At binary fission of hypotrich ciliates new locomotory cirri and sometimes feeding organelles develop in both daughters (Tuffrau, 1970, Diller, 1975). In these cases the morphogenetic cycle somehow controls both the breakdown and the synthesis of the same type of structure simultaneously.

The control system of the cell cycle must also be modified in those cases where the protist undergoes multiple fission. A class of division that is sometimes referred to as syntomy occurs when the nucleus divides repeatedly and then a body of cytoplasm gathers around each nucleus to form a daughter cell. Such reproduction is common in sporozoans, e.g. in the formation of gametes in *Monocystis*, in the formation of sporozoites of *Plasmodium*, and in the schizogony of *Plasmodium* to form merozoites. A comparable schizogony occurs in *Trypanosoma lewisi* in the gut cells of the rat flea, and in the multiple division of foraminifera and such parasitic amoebae as *Entamoeba*. This form of multiple division is distinguished from the rapid sequence of binary divisions (called palintomy) shown by such ciliates as *Ichthyophthirius* (p. 209), where a giant ciliate encysts and undergoes many fissions, without growth, to produce numerous small ciliates. In the life cycle of this parasitic ciliate periods of growth, during which the reproductive cycle is inhibited, alternate with periods of repeated reproduction, when no growth takes place. Palintomy is also found in flagellates, where the formation of gametes, e.g. the microgametes of *Volvox*, may involve this pattern of division.

Budding is a form of fission in which the parent cell may be only slightly modified in the process of reproduction. In simple budding of a suctorian, e.g. *Paracineta*, or a chonotrich, e.g. *Spirochona*, the sedentary parent ciliate retains its feeding apparatus during nuclear division and the production of a terminal or lateral bud which develops into a ciliated larva. Multiple budding of 4 to 12 larvae simultaneously is found on the suctorian *Ephelota*, while in other suctorians, e.g. *Acineta*, the larva may be budded off within a brood pouch formed by an invagination of the body surface. The formation of linear chains of ciliates by budding is found among astomes (p. 209).

5
Flagellates and related protists

Unicellular flagellate organisms are frequently encountered in fresh and salt water, in soils and as symbionts. In addition to pigmented and colourless trophic cells, such flagellate unicells may be algal or fungal distributive stages (zoospores) or gametes, or possibly the spermatozoa of higher animals or plants. While the last are likely to be recognisable as gametes because of their reduced structure, the zoospores and gametes of algae are often very similar to flagellate trophic cells in the same group or a closely related group. Within many algal groups there are often coccoid, amoeboid, colonial, filamentous and sometimes thalloid organisms as well as the independent unicellular flagellates, but the red algae and conjugating green algae lack flagellate stages. Information on the distribution of life forms of the groups considered in this chapter is given in Table 5.1, but details of the structure and life cycles of multicellular algae are, in general, beyond the scope of this book, where the aim is to concentrate upon microscopic forms. Various flagellate forms are described, usually as gametes or other transient stages, in other chapters of this book; it is expected that when these forms have been more thoroughly studied they will be brought within the evolutionary scheme discussed at the end of this chapter; these groups include the Heterolobosea (p. 165), sporozoa (p. 222), mastigamoebae (p. 162), foraminiferids (p. 168) and helioflagellates (p. 177).

According to the view of protistan evolution presented on pp. 1–6, some early amoeboid eukaryotes developed flagella to assist their food collection for phagotrophy, providing both locomotion and feeding currents. These flagellates adopted varied modes of life and developed a varied morphology, often associated with their nutrition. Phagotrophy brought symbionts as well as food, so that most (but probably not all) flagellates gained mitochondrial symbionts and members of several groups gained autotrophic symbionts. Those forms whose symbionts brought the ability to photosynthesise sometimes gave up phagotrophy, but phagotrophy is retained in many groups with chloroplasts, and even those which have lost the ability to engulf particulate food often retain osmotrophic heterotrophy.

In many textbooks flagellates are divided into 'phytoflagellates' and 'zooflagellates'; these are convenient terms to distinguish photosynthetic from heterotrophic forms, but, since several groups, such as euglenids and dinoflagellates, contain large numbers of species employing each type of nutrition, these categories are not taxonomically useful; rather, the various groups of flagellates and related protists represent more or less independent evolutionary lines. The differences between many of these groups are regarded by some authors as sufficient to justify the status of separate phyla, although some

Table 5.1

	CHLOROPLAST FEATURES ETC								NUCLEAR FEATURES				
	APLASTIDIC MEMBERS	CHLOROPHYLLS PHYCOBILINS (P)	CAROTENOIDS	MAIN XANTHOPHYLLS	THYLAKOIDS/LAMELLA	GIRDLE LAMELLA	NUMBER OF SURROUNDING MEMBR'S	TYPE OF CRISTAE IN MITOCHONDRIA	GAMETE TYPE	HAPLOID OR DIPLOID	CONDENSED CHROMOSOMES	MITOTIC SPINDLE	POLAR STRUCTURES
Parabasalia	–	✓	–	–	–	–	–	–	F	H,D	(✓)	E	R
Dinophyta	Br	✓	ac_2	$\beta\gamma$	$P\,D_3$	3	×	3	T	F	H	✓	E	–,C
Metamonadea	–	✓	–	–	–	–	–	–	–	F	H,D	(✓)	I	K
Heterolobosea	–	✓	–	–	–	–	–	–	D	–	?		I	–
Kinetoplastidea	–	✓	–	–	–	–	–	–	D	–	?	✓	I	–
Euglenophyta	G	✓	ab	$\alpha\beta$	D, N	3+	×	3	D	–	?	(✓)	I	–
Stephanopogon	–	✓	–	–	–	–	–	–	D	–	?		I	?
Cryptophyta	G,R, B	✓	ac_2p	$\alpha\beta$	$A_2\,D_2$	2	×	4	FT	–	?		O	–
Opalinata	–	✓	–	–	–	–	–	–	T	F	D		I	–
Proteromonadea		✓	–	–	–	–	–	–	T	–	?		?	?
Bicosoecida	–	✓	–	–	–	–	–	–	T	–	?		?	?
Chrysophyceae	Y-Br	✓	ac_1c_2	β	$D_1\,D_2\,F$	3	✓	4	T	F	?		I,O	R
Synurophyceae	Y-Br	–	ac_1	β	F	3	✓	4	T	F	?		?	?
Bacillariophyceae	Y-Br	✓	ac_1c_2	$\alpha\beta$	FD_1D_2	3	✓	4	T	A,F	D		O	P
Phaeophyceae	Br	–	ac_1c_2	$\alpha\beta$	FNV_1	3	✓	4	T	F	H-D		O	C
Xanthophyceae	Y-G	–	ac_1c_2	β	$D_1\,D_2\,V_2$	3	✓	4	T	F	H		I	C
Eustigmatophyceae	Y-G	–	a	β	$V_1\,V_2$	3	×	4	T	–	?		?	?
Raphidiophyceae	Y-G		$ac_?$	β	A, L	3	✓ ×	.4	T	–	?		F	K
Oomycota	–	✓	–	–	–	–	–	–	T	A	D		I	C
Hyphochy-tridiomycota	–	✓	–	–	–	–	–	–	T	F	D		F	C
Haptophyta	Y-Br	–	ac_1c_2	$\alpha\beta\gamma$	$FD_1\,D_2$	3	×	4	T	F	H-D		O	(K)
Chlorophyceae	G	✓	ab	$\alpha\beta\gamma$	LNV, Z	3+	×	2	F	F,A	H,D		IFO	C
Prasinophyceae	G	–	ab	$\alpha\beta\gamma$	LA, NZ	3+	×	2	F	–	?		IO	(R)
Choanoflagellata	–	✓	–	–	–	–	–	–	F	–	?		?	?
Chyridiomycota	–	✓	–	–	–	–	–	–	F	F	H		F	C
Blastocladiales	–	✓	–	–	–	–	–	–	F	F	H-D		I	C
Rhodophyta	R	–	ap	$\alpha\beta$	LNZ	1	×	2	F	A	H-D		F	P

(Left-margin label spanning Chrysophyceae–Hyphochytridiomycota: **HETEROKONTA**; spanning Chlorophyceae–Prasinophyceae: **CHLOROPHYTA**)

Key to abbreviations
– None: ( ) in some only: ✓ present: × absent from organelle
Colour Br, brown: G, green: R, red: B, blue: Y-Br, yellow-brown, Y-G, yellow-green
Xanthophylls   P, peridinin: A, antheroxanthin: A_2, alloxanthin:
D_1, diadinoxanthin: D_2, diatoxanthin: D_3, dinoxanthin:
F, fucoxanthin: N, neoxanthin: L, lutein:
V_1, violaxanthin: V_2 vaucheriaxanthin: Z, zeaxanthin
Cristae   T, tubular: D, discoidal: F, flat
Gametes   F, flagellate: A, amoeboid
Spindle   E, external: I, internal: O, open: F, fenestrate
Polar structures   R, flagellar root: P, plaque: C, centriole: K, kinetosome
Flagella   M, many: H, haptonema
Transition Zone H, helical: S, stellate: L, long
Shaft features   PR, paraxial rod: H, hairs: M, mastigonemes: S, scales:
Sm, smooth: U, undulating membrane

FLAGELLAR FEATURES						FOOD STORES				LIFE FORMS			ALTERNATIVE NAMES
NUMBER OF FLACELLA	TRANSITION ZONE TYPE	HETEROKONT ISOKONT OR ANISOKONT	SHAFT FEATURES	FLAGELLATE STAGE(S)	GOLGI TYPE	CARBOHYDRATE	FAT	EXTRUSOMES	EYESPOT TYPE	UNICELLULAR	MULTICELLULAR	CELL COVERING	
4-M		A	(U)	T	D	G		-	-	Fl	-	N	Trichomonads + Hypermastigids
2		A	PR,H	G,TZ	D	α	✓	T,N	C/?	Fl,A,C	Fil	PP	Dinoflagellata
2,4		A	Sm	T	-				-	Fl	-	N	Retortamonads + Diplo. + Oxymonads
2,4		A,I	Sm	T	S?				-	Fl,A	-	N	
(1)2	L	A	PR(H,U)	T	D			(T)	-	Fl	Col	P	
2(4)		A	PR H	T	D	β		M	C/f	Fl	Col	P	Euglenida
M		I	Sm	T				M	-	Fl	-	·N	Pseudociliata
2		A	M	T	D	α	✓	R	P/?	Fl,C	P	Pe	Cryptomonadida
M	H	I	Sm	T	D				-	Fl	-	N	
2,4	H	A	Sm	T	D				-	Fl	-	N	
2		H	M	T					-	Fl	Col	L	Bicoecida
2	H	H	M	T	D	β	✓	M	P/f	A,C,Fl	Fil,Th	N,W,L	Chrysomonadida
2	H	H	M,S	T	D	β			-/f	Fl	Col	SS	
1			M	(G)	D	β	✓	M	-	C	Fil	SW	diatoms
2		H	M	Z,G	D	β	✓		P/f	-	Fil,Th	MW	brown algae
2	H	H	M	Z,G	D	?	✓		P/f	A,C,Fl	Fil,Th	CW	Tribophyceae, Heterochlorida
1(2)	H	(H)	M	G	D	?			C/f	C	-	OW	
2		H	M	T	D	?	✓	T,M	-/f?	Fl	-	N	Chloromonadophyceae
2		H	M	Z	D				-	-	Fil,Th	CW	Oomycetes
1			M	Z,G	D				-	-	Fil,Th	ChW	
2,H	(S)	I,H	Sm	TG	D	β	✓	(m)	(P/f)	Fl,C,A	Fil	OS,CS	Prymnesiophyta
2(4)M	S	I	(H)	T,Z,G	D	α		-	P/m	C,Fl	Col Fil,Th	OW	Volvocida
(1)2(8)	S	(A)I	S	T,Z	D	α		M,T	P/m	Fl	Col	OS	Prasinomonadida
1			H	T	D				-	Fl	Col	OL,SL	Craspedomonadales
1			Sm	ZG	D				-	-	Th	ChW	
1			Sm	ZG	D				-	-	Th	ChW	
-	-	-	-	-	D	α		-	-	C	Fil,Th	OW	red algae

Flagellate stages   T, trophic: G, gamete: Z, zoospore
Golgi   D, dictyosome: S?, separate vesicles?
Food store   G, glycogen: α, α1–4 glucan: β, β1–3 glucan
Extrusomes   T, trichocysts: N, nematocysts: M, mucocysts: R, 'R-bodies'
Eyespot   C, cytoplasmic stigma: P, plastid stigma: f, flagellar receptor: m, membrane receptor:
Life Forms   Fl, flagellate: C, coccoid: A, amoeboid: Fil, filamentous: Th, thalloid: Col, colonial
Cell Covering   N, naked: P, pellicle: PP, pellicle with plates: Pe, periplast: L, lorica: W, wall: SW, silica wall: MW, mucilage wall: CW, cellulose wall: ChW, chitinous wall: OW, organic wall: CS, calcareous scales: OS, organic scales: SS, silica scales: OL, organic lorica: SL, silica lorica

supraphyletic assemblages have also been recognised (e.g. Corliss, 1984); the phyla recognised by Corliss will in general provide the basis for classification used in this chapter. The acquisition of photosynthetic symbionts seems to have occurred in several branches of an evolutionary radiation of protists that were primarily heterotrophic; considerations of relationships between phyla will occur throughout this chapter and will be reviewed at the end (p. 146).

By bringing together, within the Protista, groups that have been subject to one or other (or both) of two different systems of classification, problems of the rank and naming of groups are inevitable; here botanical names will normally be used for groups that are predominantly plant-like and zoological names for predominantly heterotrophic groups, but mixed groups remain a problem. Heywood and Rothschild (1987) have suggested that protistan groups should be distinguished by the suffix-protista; this seems a valuable proposal for such groups as the euglenoids (Euglenoprotista) and cryptophytes (Cryptoprotista, as also used by Margulis and Sagan, 1985), but it seems unnecessary to modify such names as Ciliophora or Dinoflagellata in this way. A major commission on nomenclature may be required to agree a scheme applicable to protists as a whole.

Cellular structures

The various groups discussed in this chapter differ not only in details of their flagellation, flagellar root patterns and other cytoskeletal structures, but also in their nutrition, the presence and type of eyespot, the pattern of nuclear division (see pp. 47–51), the body form and ploidy of different stages of the life cycle, the shape of mitochondrial cristae and the organisation of the cell surface. Nutrition may be heterotrophic, and conform to one or more of the patterns described on pp. 60–69, and/or may be autotrophic, and depend upon distinctive combinations of photosynthetic pigments located in chloroplasts of varied structure and result in different photosynthetic products stored in varied locations. The fact that protistan evolution took place at the cellular level explains the wealth of cytological features that are used in classification and provide evidence of inter-relationships between the groups. Since most groups have flagellate stages, it is possible to compare the structure of flagellate cells from almost all of the autotrophic and heterotrophic groups, and to compare other cellular features in all of the phyla. A large proportion of the relevant information is at the ultrastructural level, and there is now a very large literature on the fine structure of protists; however, only a few species have been studied in some groups, so that the applicability of some conclusions must await confirmation when more species have been studied. Information on the variation of some principal features is given for the main phyla in Table 5.1, and the general characteristics of these features will be discussed in the following paragraphs before the features of individual groups are described. Since a variety of names has been used for the different groups, both formal and informal names are given in Table 5.1; authorities for the names of phyla are listed by Corliss (1984).

Chloroplasts, pyrenoids, food storage and eyespots

Seven major patterns of chloroplast organisation occur in protists (Bisalputra, 1974; Dodge, 1979). The simplest is the rhodophyte pattern in which chloro-

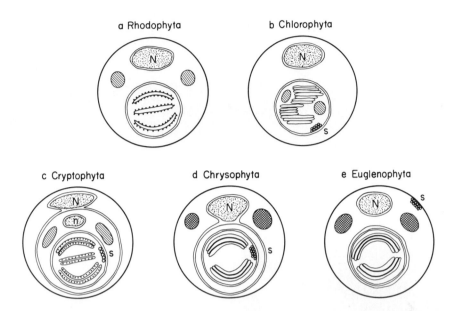

Fig. 5.1 Diagrams showing the arrangement of membranes associated with the Chloroplast, the type and location of stored carbohydrate and the location of the eyespot or stigma (*s*) found in five algal groups, as described in the text. Starch deposits are shaded with diagonal lines and deposits of leucosin or paramylon with cross-hatched shading. Where there are four chloroplast membranes, the outermost of these may be continuous with the outer membrane of the nucleus (*N*), and in cryptophytes the periplastidial space contains both a nucleomorph (*n*) and food storage deposits.

plasts are enclosed by two membranes and contain single separate thylakoids that incorporate chlorophyll a and bear phycobilisome granules containing phycocyanin and phycoerythrin on their outer surfaces (Fig. 5.1a); eyespots are not found, but a pyrenoid may occur in the chloroplast and starch grains occur only in the cytoplasm. Most other groups contain two types of chlorophyll, but cryptophytes are the only other eukaryotes to possess phycobilisomes. In this group again the chloroplasts are distinctive, with loosely paired thylakoids that incorporate chlorophylls a and c and contain internal phycobilisomes (with biliproteins that differ from those in red algae and cyanobacteria) (Fig. 5.1c). Each cryptophyte chloroplast may contain a pyrenoid and a simple eyespot (Fig. 5.2), and is enclosed in two pairs of membranes, the inner two of which form the chloroplast envelope and the outer two enclose a periplastidial cytoplasmic space which contains starch grains and a 'nucleomorph' (see below). The plastids of green algae (chlorophytes and prasinophytes) and higher plants have two bounding membranes and enclose thylakoids that are rather irregularly arranged into bands or grana and contain chlorophylls a and b; a pyrenoid is present and starch is stored within the chloroplast, which may also contain a simple eyespot (Fig. 5.1b). Most of the remaining algae contain chlorophylls *a* and *c* in thylakoids that occur in threes within plastids that are surrounded by two pairs of membranes separated by only a narrow periplastidial space (Fig. 5.1d). Simple eyespots and pyrenoids may be present within the chloroplast but reserve foods are stored as globules of oil or a fluid polysaccharide in

a

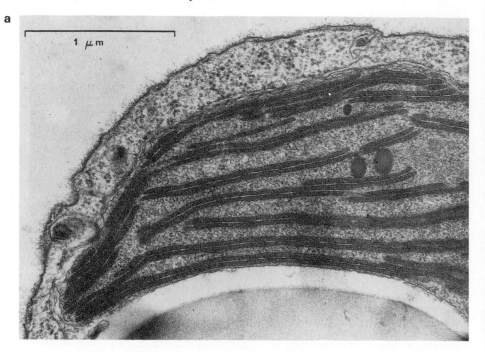

b

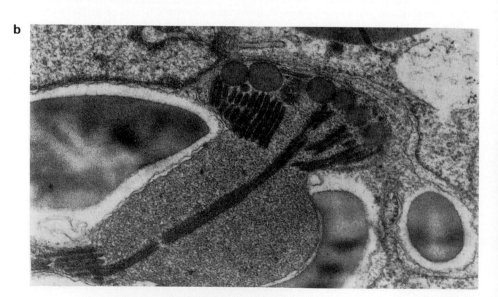

Fig. 5.2 Electron micrographs of sections of chloroplasts of the cryptophyte *Chroomonas mesostigmatica*: **a** shows typical paired thylakoids and **b** shows a projection of the chloroplast containing pigment globules of a simple eyespot, and a pyrenoid region crossed by a single pair of thylakoids and enclosed by starch deposits. Micrographs by J.D. Dodge (1969). Both at the same scale of 1 μm.

the cytoplasm; a girdle lamella of thylakoids runs around the periphery of the chloroplast in several groups (Table 5.1, Fig. 2.3e), but is not present in hapto-phytes, which may also have more complex eyespots. Three membranes have been reported to surround the chloroplasts of euglenids and many dinoflagel-lates, in both of which most thylakoids occur in threes (Fig. 5.1e); here the resemblance ends however, for in euglenoids the plastids contain chlorophylls *a* and *b*, and both eyespots and deposits of the polysaccharide paramylon are found in the cytoplasm, whilst in dinoflagellates the plastids contain chloro-phylls *a* and *c* and sometimes eyespots, with starch grains and sometimes eye-spots in the cytoplasm; the arrangement of pyrenoids is variable in both groups. The chlorophylls are accompanied in the different groups by distinctive combinations of carotenoid and xanthophyll pigments (Table 5.1) (Ragan and Chapman, 1978). Chlorophyll *c* groups were called the Chromophyta.

It is assumed that the rhodophyte plastid is derived from a cyanobacterial symbiont with chlorophyll a and phycobilisomes and the green algal plastid from a chloroxybacterial (***Prochloron***) symbiont with chlorophylls *a* and *b*; the inner and outer membranes presumably representing symbiont and host vacuolar membranes, respectively (Margulis, 1981; Smith and Douglas, 1987). No equivalent brown chloroplast with only two membranes is known to exist today, but it is believed to have existed in the past as a result of symbiosis of a eukaryote with a prokaryote containing chlorophylls *a* and *c* (Sleigh, 1979). Where four membranes are present, it is believed that the plastid is derived from symbiosis with such a brown eukaryote, and that the periplastidial space is the remnant of the cytoplasmic compartment of that symbiont; in cryptophytes this compartment contains a relict of the eukaryote symbiont nucleus (the nucleo-morph), as well as containing ribosomes and stored food, but comparable features are much reduced in other forms. It seems possible that a three-layered envelope results from symbiosis with phagocytosed chloroplasts from other eukaryotes, from a green alga in the case of euglenoids and a brown alga in the case of dinoflagellates (Whatley *et al.*, 1979). Extant groups presumably represent only the most persistent and best integrated of a large diversity of symbioses of protists with photoautotrophs (see p. 283).

Pyrenoids are areas of dense matrix found within chloroplasts; they are some-times associated with the formation of polysaccharides, when they are surrounded or capped by a deposit of starch or some other polysaccharide. Normally the pyrenoid occurs within the body of a chloroplast or in a stalked projection from it (Fig. 5.3); in some cases it may be penetrated by tongues of cytoplasm or even of nuclear material. Thylakoid lamellae usually penetrate the pyrenoid; these may be normal lamellae or they may have fewer thylakoids than usual. The functional implications of the variations in these features are not known.

Clusters of lipid globules containing orange or red carotenoid pigments are referred to as eyespots or stigma structures; they are common in pigmented flagellates and may show varied patterns of relationship with chloroplasts and flagella (Dodge, 1973). In most groups where it is present the eyespot is found within a chloroplast (Fig. 5.2b), and may be associated with a short or back-wardly projecting flagellum, or with flagellar roots; in others it may be unconnected with the flagella. Where the eyespot is situated outside the chloro-plast, it may still be associated with the basal region of a flagellum, often an

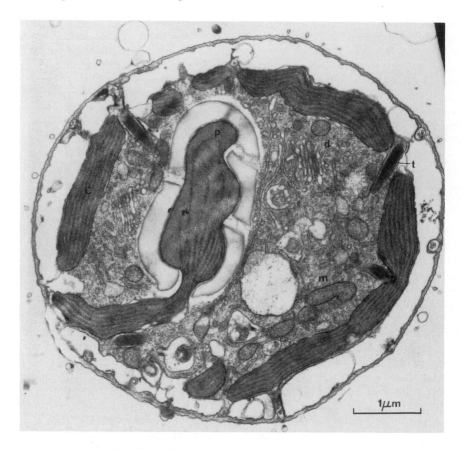

Fig. 5.3. Electron micrograph of a section through the dinoflaglelate *Amphidinium carteri*, showing the chloroplasts (*c*) with lamellae containing three thylakoids and the pyrenoid (*p*) penetrated by lamellae and capped by a starch deposit. Mitochondria (*m*), dictyosomes (*d*) and trichocysts (*t*) are also present. Micrograph by Dodge and Crawford (1968).

active locomotory one. The most complex eyespot structures occur in some dinoflagellates, e.g. *Pouchetia*, where the stigma is not associated with either chloroplast or flagella, and may be a large structure with several components, while in other dinoflagellates the stigma is simpler and lies close to flagellar bases (Fig. 5.4). The eyespot is assumed to be concerned with orientation to light and phototactic behaviour (p. 41), but the reason for the extreme development of the structure in some dinoflagellates is obscure.

Flagella and propulsion

The variations of ultrastructural features of flagella and the way they propel fluids are still imperfectly known for some groups. Variants of extraflagellar and intraflagellar components occur (p. 32) (Fig. 5.5), and tend to have a systematic distribution (Table 5.1). Unilaterally arranged slender hairs on

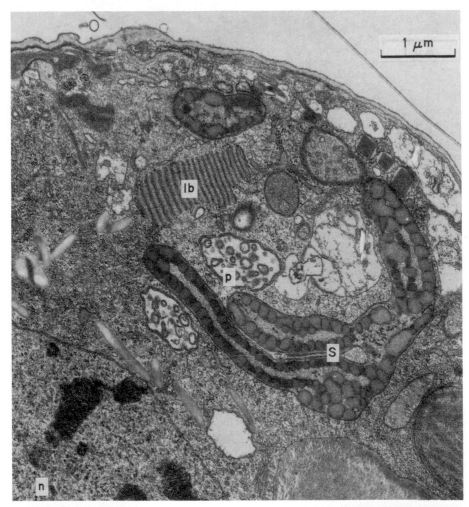

Fig. 5.4 Electron micrograph of a section through the dinoflagellate *Glenodinium foliaceum* showing part of the complex stigma (*s*) and a lamellar body (*lb*), both of which are closely associated with the flagellar bases (not shown here). Parts of the pusule system (*p*) and nucleus (*n*) also appear in the section. Micrograph by Dodge and Crawford (1969).

flagella of some euglenoids and dinoflagellates (sometimes in more complex arrays in the former and in two rows on longitudinal flagella of the latter) are contrasted with the thicker, stiff, tripartite 'mastigonemes' (Fig. 2.12) which occur in two opposite rows on the anterior flagellum of 'heterokonts' (see below). Scales and easily-detached hairs are common on flagella of some types, notably prasinophytes. An intraflagellar rod, varying somewhat in detailed structure, is prominent in dinoflagellates, kinetoplastids (Fig. 5.17) and euglenoids (Fig. 5.19). It may be accompanied by a contractile strand (p. 29). The flagellar axis may extend along the cell surface and its expanded membrane may be attached to the cell surface as an undulating membrane, notably in some

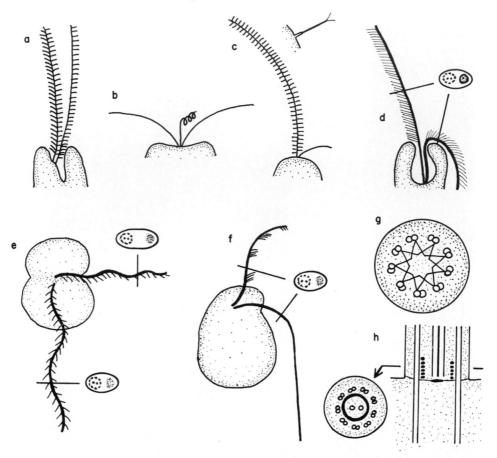

Fig. 5.5 Diagrams showing some features of flagellar diversity mentioned in the text. Cryptophytes (**a**) and dinoflagellates (**e**) tend to have two rows of hairs on one flagellum and one row on the other. Haptophytes (**b**) have two flagella lacking hairs and usually a coiled haptonema. The hairs of chrysophytes (**c**) are thick tripartite structures (inset) borne on one flagellum only (as in other heterokonts). Flagella of euglenoids (**d**) typically carry a unilateral row of slender hairs, as well as other structures; these flagella, as well as those of dinoflagellates (**e**) and kinetoplastids (**f**) contain an intraflagellar paraxial rod alongside the axoneme. The anterior flagellum of bodonid kinetoplastids may carry tufts of hairs. In the transition region of the flagella of chlorophytes the peripheral doublets are interconnected by fine filaments to give a stellate pattern (**g**). Immediately above the transition zone of the flagella of many heterokonts a helical strand is found just inside the peripheral doublets (**h**), but apparently does not contact either central or peripheral microtubules.

kinetoplastids and trichomonads (Fig. 5.14). Interconnections between the peripheral doublets in the basal transition zone of the flagellum (p. 31) form a stellate pattern in some groups and a helical fibre occurs in heterokonts (Hibberd, 1979) (Fig. 5.5; Table 5.1).

Most flagellates have two flagella; where only one is present there are normally two kinetosomes, one of which is barren, while in some groups flagellar multiplication has taken place. The development of the various patterns of flagellar arrangement and the use of flagella in locomotion and food capture is outlined in Fig. 5.6 (Sleigh, 1981). The use of a single flagellum to push a cell is shown by animal sperm, and occurs in chytrids and in choano-flagellates (where it also provides a feeding current). The two flagella on

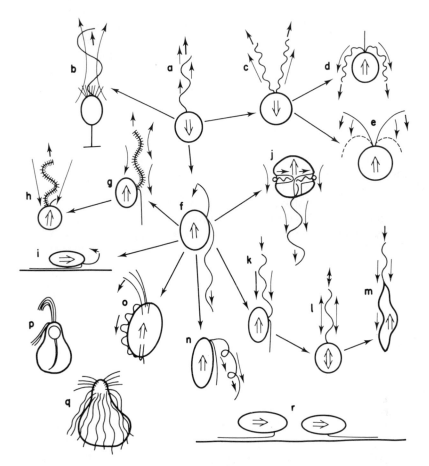

Fig. 5.6 Variants of the arrangement and propulsive function of flagella, arranged to illustrate their possible evolution (Sleigh, 1981). The dotted arrows indicate the direction of water flow, single-shafted arrows the direction of movement of flagellar waves and double-shafted arrows the direction of cell movement. Cells pushed along by single flagella include **a** sperm and chytrids and **b** collar flagellates, e.g. *Codosiga*. Pairs of flagella may move homodynamically to propel *Chlamydomonas* ('primitive' backward mode **c** and 'advanced' forward mode **e**) or the haptophyte *Chrysochromulina* **d**. A hypothetical heterodynamic type **f** could have led to heterokont types **g** (e.g. *Fucus* sperm) and **h** (*Ochromonas*) by development of mastigonemes, to dinoflagellates **j** by development of the girdle grooves in which the transverse flagellum moves, to kinetoplastids **k,l** (*Strigomonas*) and **m** (*Trypanosoma*) by propagation of waves from tip to base along an anterior flagellum, to euglenoids **n** by helical undulations of a back-wardly-directed anterior flagellum, and to the parabasalians (**o**, *Trichomonas,* **p**, *Joenia* and **q**, *Trichonympha*) and metamonads by increase in number of anterior, and sometimes also posterior, flagella. Members of several groups, especially bodonids and euglenoids, use flagella as a means of gliding along surfaces; sometimes these forms supplement gliding with water propulsion (e.g. *Bodo*, **i**), and may be pushed or pulled **r** by the gliding flagellum.

biflagellate cells may move in a similar manner (homodynamic), or may show different patterns of movement (heterodynamic). The homodynamic forms seem less versatile and are generally found on autotrophs; by contrast heterodynamic flagella are very diverse in their appendages and movement. Important heterodynamic variants are the dinoflagellate pattern, with typically a ribbon-shaped transverse flagellum performing a complex three-dimensional beat (Rees and Leedale, 1980) in the equatorial girdle, rotating and propelling the cell, whilst a longitudinal flagellum performs a planar beat; the euglenoid pattern, often with an anterior helically beating flagellum and a posterior trailing flagellum; the kinetoplastid pattern, in at least some of which which the anterior flagellum propagates waves from tip to base to pull the cell along while the posterior one trails; and the heterokont pattern, in which the anterior flagellum with mastigonemes draws water towards the cell and a posterior naked flagellum either trails behind the cell or is used for attachment or food-trapping. An increase in flagellar number, probably involving multiplication of both anterior and posterior flagella of heterodynamic forms, as well as multiplication of homodynamic flagella, has occurred in some groups.

The kinetosomes of the flagellar apparatus may be linked to one another, usually by one or more striated strands formed from fine filaments, and may also be anchored by striated fibres or groups of microtubules which often attach to other organelles after extending out into the endoplasm or beneath the cell membrane. The pattern of fibres varies widely, but is often more consistent within particular groups. Rootlet patterns have been used to support phylogenetic schemes (p. 147), and as more examples are studied, features of rootlet morphology will be more widely used in taxonomy (e.g. Melkonian, 1984). In both simple and complex flagellates the basal bodies or structures attached to them may act as MTOCs for the mitotic spindle (p. 50).

Organisation of the cell surface

While parasitic flagellates and some free-living forms appear naked, with only a more or less developed glycocalyx for protection, many free-living flagellates and their relatives secrete extracellular materials that presumably afford protection or support. Cell walls formed largely of various carbohydrates secreted through the cell membrane characterise some groups, and internally-secreted scales and silica walls take up a superficial position in other groups. Members of several groups secrete loricas which also provide protection. Reinforcement of the surface with cytoplasmic structures that comprise a pellicle (p. 53) consists, in its simplest form, only of a differentiated epiplasm; this may be supplemented by microtubules, and in other forms there may be thickened pellicular strips or membranous sacs (alveoli), which may contain protective plates. Special extracellular coats occur in resting stages, such as cysts, and the form and composition of these varies from group to group.

Features of the main phyla

The phyla considered in this chapter vary greatly in numbers of species and in the information available about their members, so that it is not easy to be certain what features are generally present even in some larger groups. There are also many small groups, or even individual genera, that appear to be unlike any other organisms, and may belong to separate phyla; only more information about a wide range of species will solve these problems of relationships. However, it is possible to recognise a number of groups with distinctive sets of characters, even if the extent of variation within them is not yet known. Features of the nuclei, flagella, chloroplasts and mitochondria have been used to decide the order in

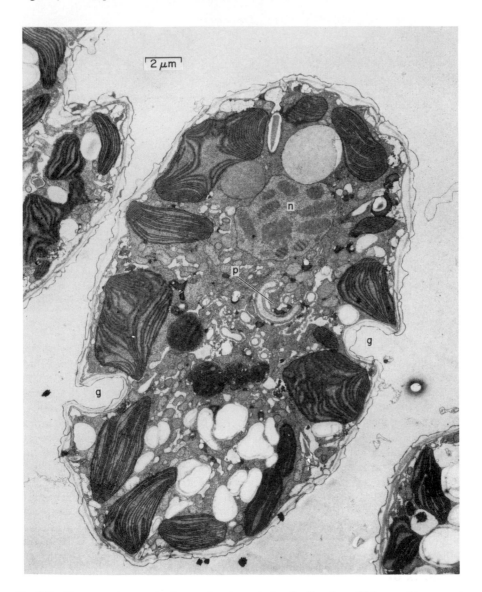

Fig. 5.7 Electron micrograph of a section through the dinoflagellate *Woloszynskia coronata* showing the position and appearance of the pusule system (*p*), the nucleus (*n*), the pellicular alveoli of the theca and their infolding to form the girdle (*g*). Micrograph by J.D. Dodge and R.M. Crawford (unpublished).

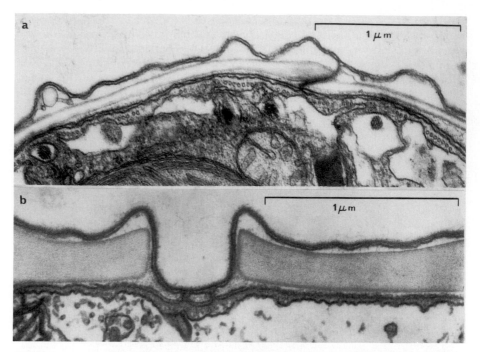

Fig. 5.8   Electron micrographs illustrating the structure of the theca as seen in sections of the dinoflagellates *Glenodinium foliaceum* (**a**) and *Ceratium hirundinella* (**b**). Beneath the cell membrane are membranous alveoli, which in these examples contain thin (**a**) and thick (**b**) skeletal plates; groups of microtubules occur at the inner side of the pellicular layer. Micrographs by Dodge and Crawford (1970a).

which phyla are presented; this order does not indicate a phylogenetic sequence.

The dinoflagellates are an ecologically important and distinctive group, and have been the subject of several books (Dodge, 1985; Spector, 1984; Taylor, 1987), major reviews (e.g. Taylor, 1980), and recent keys (Dodge, 1982, Dodge and Lee, 1985). They have often been regarded as primitive, although Loeblich (1984) challenges that view and suggests that some features may be degenerate. Their large nuclei lack nucleosomes, and have permanently condensed chromosomes (Fig. 2.16) (Sigee, 1986), few histones and a closed mitosis with external spindle, all of which may be primitive features. For this reason the dinoflagellates have sometimes been called 'mesokaryotes' (Dodge, 1965), part-way between prokaryotes and eukaryotes. However, their cytoplasmic features are typically eukaryotic (Fig. 5.7) and their cells are often very complex, with usually two flagella (Fig. 5.5e), mitochondria with tubular cristae and often with brown chloroplasts of a unique type (p. 103). They also have a eukaryotic pellicle (theca or amphiesma) composed of a layer of alveoli beneath the cell membrane, which in turn is underlain by larger or smaller numbers of microtubules (Fig. 5.8) (Dodge and Crawford, 1970; Loeblich, 1983); this structure is reminiscent of the pellicles of ciliates (p. 190) and of sporozoa (p. 223). Frequently there are plate-like deposits of polysaccharide, usually cellulose, within the alveoli; these plates may form a permanent 'armour' (Fig. 5.9), but

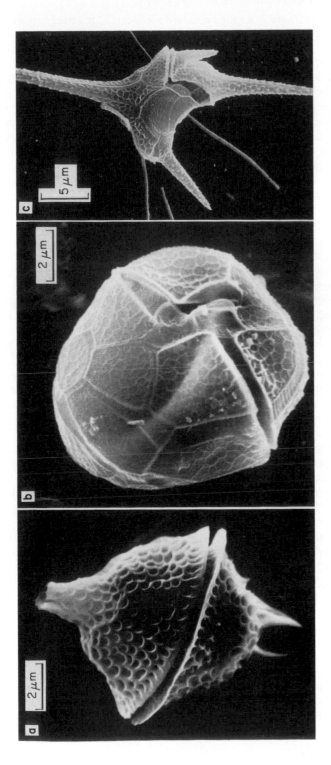

Fig. 5.9 Scanning electron micrographs of the dinoflagellates *Gonyaulax digitale* (**a**), *Peridinium cinctum* (**b**) and *Ceratium hirundinella* (**c**). The equatorial girdle is seen in all three and the outlines of thecal plates are clear in (**b**) and (**c**). Micrographs (**a**) and (**b**) by J.D. Dodge, and (**c**) from Dodge and Crawford (1970b).

many dinoflagellates have a more delicate surface. There are, however, variations in most of these features, and the group shows considerable evolutionary divergence.

Most of the 2000 or so living species of this group are unicellular, free-living flagellates, but some live in protists or metazoa as symbionts (p. 283) or parasites, some form colonies, and some are coccoid or filamentous autotrophs (Dodge and Lee, 1985). Autotrophic species appear brown and the many non-autotrophs are colourless or pinkish; two species have been shown to contain chrysophyte symbionts, for they contain a second nucleus and chrysophyte-type plastids (Dodge, 1975), and a green species with a reduced symbiont containing chlorophylls *a* and *b* has been reported (Watanabe *et al.*, 1987). Heterotrophy is practised by colourless species, and many photosynthetic species can also ingest or extract cytoplasm from other organisms; the true extent of osmotrophic nutrition is not known, but many recent reports show that phagotrophy is widespread. Thecate forms like **Protoperidinium** protrude a veil-like lamellipodium from the sulcus to surround prey and digest them (Gaines and Taylor, 1984), while naked forms (e.g. *Noctiluca*) may capture prey with a prehensile tentacle, or may suck fluids from prey through a tubular, microtubule-lined peduncle (reminiscent of a suctorian tentacle, p. 64) as in **Gymnodinium** (Spero, 1982).

In most orders of planktonic forms the body is grooved equatorially to form a girdle, occupied by the transverse flagellum, and a second groove passes posteriorly from the point of emergence of the flagellar canals as a sulcus, from which the longitudinal flagellum extends (Fig. 5.10a,b). In gymnodiniids, including the compound **Polykrikos** (Fig. 5.10d) as well as planktonic unicells (Fig. 5.10a) and the symbiotic **Symbiodinium** (**Zooxanthella**), the theca is thin and flexible, whilst in peridiniids the theca is thickened and in dinophysiids (Fig. 5.10g) the girdle is often at the anterior end and enclosed in deep flanges (lists). In prorocentrids the two flagella are inserted apically (Fig. 5.10c), and the photosynthetic cell is enclosed in two large thecal plates without grooves. Although their mature forms are quite varied, members of the remaining groups reproduce by motile dinospores, usually with transverse and longitudinal flagella. The noctilucoids are large, colourless, phagotrophic species including some bizarre forms described by J. and M. Cachon (1967, 1969), one like a small medusa (Fig. 5.10h), and others with even more complex shapes (Fig. 5.10i,j), as well as the bioluminescent *Noctiluca* (Fig. 5.10k). The blastodiniids are mostly ectoparasites on marine eggs (e.g. Fig. 5.10l), whilst the syndiniids are endoparasites of other protists or metazoans and may become multicellular (e.g. Fig. 5.10m). The benthic, coccoid or filamentous, phytodiniids are usually found attached to marine plants. Loeblich (1984) considers that the phagotrophic ebriids (Fig. 5.25l) (Deflandre, 1952; Lee, 1985), with a large internal siliceous skeleton, and the ellobiopsids (Grasse, 1952), which parasitise crustaceans or annelids, are also related to dinoflagellates although neither have dinospore larvae; both groups require re-examination.

The position of the colourless genus **Oxyrrhis** is interesting; long considered to be a dinoflagellate allied to gymnodiniids, although not of typical shape (Fig. 5.10f), it has recently been found to contain nucleosome-like structures (Gao, 1980) and possess an intranuclear spindle (Triemer, 1982). If one believes, with Heath (1980) and Loeblich (1981) that the intranuclear spindle is more primitive than the extranuclear one, then **Oxyrrhis** may be an ancestral type and

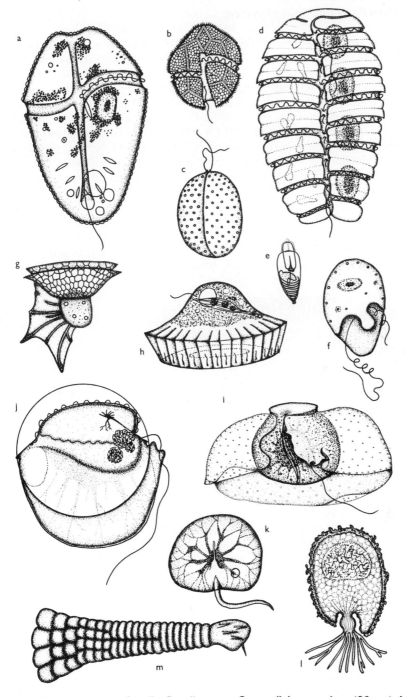

Fig. 5.10 Some representative dinoflagellates; **a**, *Gymnodinium amphora* (30 μm): **b**, *Peridinium tabulatum* (45 μm): **c**, *Exuviella marina* (40 μm): **d**, *Polykrikos schwartzi* (150 μm): **e** nematocyst of *Polykrikos* (12 μm): **f**, *Oxyrrhis marina* (20 μm): **g**, *Ornithocercus splendidus* (120 μm): **h**, *Craspedotella pileolus* (100 μm): **i**, *Cymbodinium elegans* (500 μm): **j**, *Kofoidinium pavillardi* (500 μm): **k**, *Noctiluca scintillans* (800 μm): **l**, *Oodinium ocellatum* (60 μm): **m**, *Haplozoon clymenellae*, from polychaete gut (250 μm).

the lack of nucleosomes in most dinoflagellates may be a degenerate feature; alternatively, the view taken by Triemer and Fritz (1984), and others, is that extranuclear spindles and the lack of nucleosomes are primitive features, so that *Oxyrrhis* may be a derived form, perhaps related to syndiniids. Clearly more details are needed to resolve this problem, but it highlights the problems of trying to infer relationships on the basis of small numbers of described species. The chromosomes of syndiniids and *Oxyrrhis* remain condensed in interphase, but differ in fine structure from the usual dinoflagellate pattern; the syndiniids differ from other dinoflagellates in having centrioles at the poles of the extra-nuclear spindle, and are sometimes considered to represent a separate phylum (e.g. Corliss, 1984).

Reproduction of dinoflagellates is typically by longitudinal fission that commences posteriorly; in some armoured forms the thecal plates are shared between the daughters, and in others the whole armour is shed before division and each produces a complete theca. Several species show sexual reproduction (Pfiester, 1984), and all of them are haploid with zygotic meiosis, except for

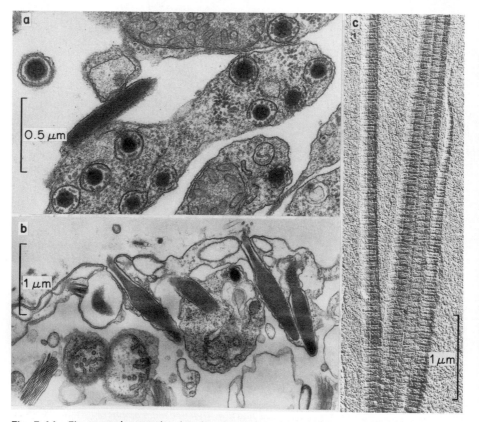

Fig. 5.11 Electron micrographs showing the structure of dinoflagellate trichocysts. Mature trichocysts of *Oxyrrhis marina* are shown in transverse (**a**) and longitudinal (**b**) sections, and extruded trichocyst threads of *Exuviella marie-lebouri* are shown in a shadowed preparation (**c**). Micrographs **a** and **b** from Dodge and Crawford (1971). **c** by the same authors, unpublished.

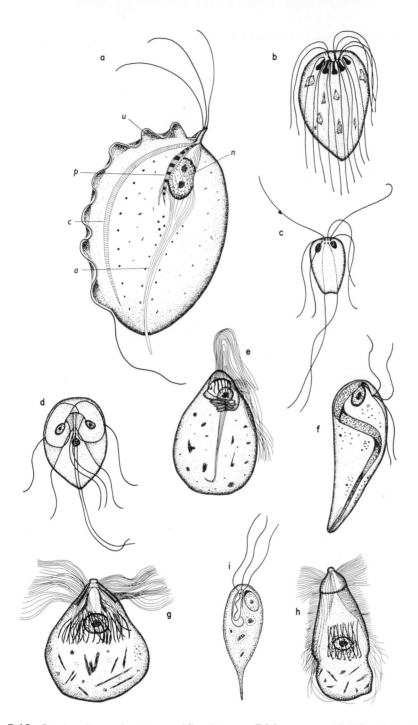

Fig. 5.12 Parabasalian and metamonad flagelates; **a**, *Trichomonas muris* (15 μm) from large intestine of mice, showing nucleus (*n*), axostyle (*a*), costa (*c*), parabasal fibre (*p*) and undulating membrane (*u̇*): **b**, *Coronympha octonaria* (40 μm) from *Calotermes*: **c**, *Hexamita intestinalis* (10 μm) from large intestine of frogs: **d**, *Giardia muris* (10 μm) from small intestine of mice: **e**, *Joenia annectens* (100 μm) from *Calotermes*: **f**, *Saccinobaculus ambloaxostylus* (100 μm) from *Cryptocercus*: **g**, *Barbulanympha ufalula* (300 μm) from *Cryptocercus*: **h**, *Trichonympha campanula* (200 μm) from *Zootermopsis*: **i**, *Chilomastix aulastomi* (25 μm) from the intestine of the leech *Haemopis*.

Noctiluca, which is diploid with gametic meiosis. In typical forms the zygote produced by fusion of two small biflagellate gametes swims with two longitudinal flagella and grows before developing a thick cyst wall, from which 1,2,4 or more flagellate cells emerge after meiosis and a longer or shorter rest period. *Crypthecodinium* shows a one-division meiosis, though two-division meiosis has been seen in other dinoflagellates (Beam and Himes, 1984). Asexual cysts are also formed and may provide a resting stage or may be temporary (Chapman *et al.*, 1982).

There are various other cytoplasmic features of interest in dinoflagellates. Paired pusules found in most freshwater and some marine species consist of systems of vesicles and tubules that surround the flagellar canals (Figs. 5.4, 5.7), and are believed to be osmoregulatory in function (Dodge, 1972); contraction of an associated fibre has been reported (Cachon *et al.*, 1983). Trichocysts and mucocysts similar to those in ciliates are often present (Fig. 5.11), and elaborate nematocysts similar to those in cnidaria are found in *Polykrikos* and a few other genera (Fig. 5.10e) (Greuet, 1972). Although only some dinoflagellates possess eyespots, these are sometimes very complex, as in *Erythropsodinium* (Greuet, 1977), where there is an ocellus with a lens and retinoid as well as pigment granules; simpler eyespots in other forms range between groups of globules near the sulcus or within a chloroplast to arrays of globules and membranes (Fig. 5.4). A number of photosynthetic species that produce blooms or 'red tides' (Iwasaki, 1979) in coastal waters produce toxins that may be water- or lipid-soluble and have haemolytic, neurotoxic (e.g. saxitoxin, brevetoxin) or gastrointestinal effects on fish and other vertebrates; most human cases of poisoning by dinoflagellate toxins result from eating shellfish that had concentrated the toxin (Steidinger and Baden, 1984). Finally, many dinoflagellates are bioluminescent, with a luciferin-luciferase system, e.g. *Gonyaulax* (Dunlap *et al.*, 1981).

A series of totally heterotrophic flagellates, most of them endobionts in the guts of animals, also have closed mitoses with an external spindle and in some cases condensed interphase chromosomes; these are the Parabasalia. They have a naked cell surface and are osmotrophic or phagotrophic, without a specific cytostome, and store glycogen. They lack mitochondria, even when free-living, but at least some of them possess hydrogenosomes (p. 18). Usually they have between four and thousands of flagella, though at least one amoeboid form seems to belong in this group. The basal bodies of the flagella are typically associated with a special array of fibres, best seen in trichomonads (Fig. 5.12a, 5.13); in this case three of the four flagella are free and extend forward and the fourth, recurrent, flagellum is associated with a reinforced fold of the body surface forming an undulating membrane (Fig. 5.14). A striated parabasal fibre is closely associated with a large parabasal golgi system; a dense pre-axostylar fibre connects to the axostyle, which is a curved sheet of microtubules that encloses the nucleus and extends to (or protrudes from) the posterior end of the cell; the same pre-axostylar fibre links to the pelta, which is a sheet of microtubules that curves around within a sort of collar that encloses the flagellar bases before turning back to meet the axostyle; a branched striated fibre called the atractophore extends towards the nucleus; finally there is the contractile costa, composed of longitudinal lamellae (p. 29), which runs beneath the undulating membrane. This assemblage of flagella, fibres and associated nucleus is called a

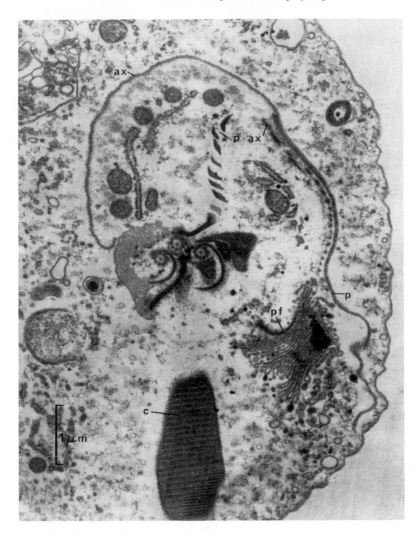

Fig. 5.13 Electron micrograph of a transverse section at the level of the basal bodies of *Trichomonas termopsidis* showing three basal bodies which have a close relationship with fibrous structures; these make connections at other levels with the costs (*c*), the parabasal fibre (*pf*) (with its associated dictyosomes) and through the pre-axostyle fibre (*p ax*) with the pelta (*p*) and the axostyle (*ax*). Micrograph by Hollande and Valentin (1968).

karyomastigont. Before division flagellar replication occurs, the two groups of flagella move to either side of the nucleus and an external spindle with a large bundle of continuous microtubules and diverging chromosomal microtubules develops with the atractophores acting as MTOCs (Fig. 2.20a). Sexual reproduction involving syngamy occurs in some forms, but is very variable (p. 76).

In a proposed evolutionary scheme for parabasalians Brugerolle and Taylor (1979) suggest that the stem form *Monocercomonas* (like *Trichomonas* but with

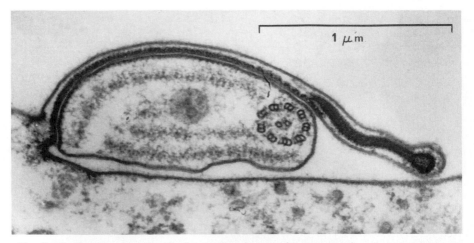

Fig. 5.14 Electron micrograph of a section through the recurrent flagellum and undulating membrane of ***Trichomonas termopsidis***. The fold of the body surface that forms the undulating membrane is closely attached to the expanded flagellar membrane for a large part of its width and is reinforced at the margin by a dense lamella. Micrograph by Hollande and Valentin (1968).

a free recurrent flagellum and no costa) may have led to trichomonads and also to *Histomonas* (a parasite of turkeys) with a single flagellum and to *Dientamoeba* (without flagella). Devescovinids, found in termites, are similar to trichomonads, but with a broad, ribbon-like, free, recurrent flagellum; some of these forms are curious in that the karyomastogont and terminal parts of the cell can rotate within the main body (Tamm and Tamm, 1974). Replication of karyomastigont systems occurs in Calonymphidae (Fig. 5.12b), and *Metacoronympha* contains up to several hundred karyomastigonts (each with nucleus, flagella and fibre systems) spirally arranged around its anterior end. In the hypermastigote flagellates the large numbers of flagella are arranged differently; in *Joenia* (Fig. 5.12e) a karyomastigont of 3–4 privileged basal bodies, associated with axostyle, parabasal system and atractophore occurs at one side of a large field of flagella, and acts as the organising centre of the cell. Two sets of such privileged basal bodies and root fibres occur one at either side of the nucleus in the multiflagellate *Barbulanympha* (Fig. 5.12g), and in *Trichonympha* (Fig. 5.12h) it appears that all that is left of the privileged group is a single apical basal body that plays a central role in division, although other anterior flagellar bases are connected to parabasal fibres (Hollande and Carruette–Valentin, 1971). Most of the 300 or so species of parabasalians, including all hypermastigotes, are found in the guts of wood-eating insects, especially termites (p. 280), and often bear symbiotic spirochaetes; all trichomonads and monocercomonads are probably parasitic, and some are of medical and veterinary importance, e.g. *Trichomonas vaginalis* and *Tritrichomonas foetus* (Kreier, 1977).

Another group of multiflagellate endobionts are the metamonad flagellates, but in these forms internal spindles occur within closed mitotic nuclei, or occasionally with polar fenestrae, and associated with flagellar bases outside the nucleus near the spindle poles. These naked flagellates lack both mitochondria and golgi bodies, which may indicate an early separation from other groups. They have one or several karyomastigonts, each of which has one or two pairs of naked flagella, with at least one flagellum of each karyomastigont being turned back as a recurrent flagellum. The two basal bodies of each flagellar pair are set orthogonally and the kinetosomes of each karyomastigont give rise to characteristic sets of fibres in the three component groups.

In retortamonads like *Chilomastix* (Fig. 5.12i) there are three free flagella and the recurrent flagellum beats in a ventral groove that contains a long cytostome where bacteria are ingested; the rest of the pellicle is underlain by microtubules, reinforced at either side of the ventral groove by kinetosome-associated microtubule bundles and on one side by a striated fibre, whilst bundles of microtubules extend laterally in front of and behind the anteriorly placed nucleus (Brugerolle, 1973); other members of this group have 2–6 flagella. *Trepomonas,*

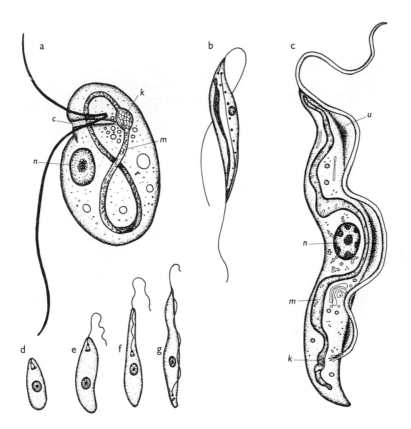

Fig. 5.15 Some representative kinetoplastids; **a**, *Bodo saltans* (10 μm) with nucleus (*n*), kinetoplast (*k*), mitochondrion (*m*) and cytostome (*c*): **b**, *Cryptobia helicis* (20 μm) from the seminal receptacle of the snail *Helix*: **c**, *Trypanosoma brucei* (20 μm) showing features of the bloodstream form, including nucleus (*n*), kinetoplast (*k*), mitochondrion (*m*) and 'undulating membrane' (*u*): **d**, amastigote (leishmanial) form of trypanosome: **e**, promastigote (leptomonad) form of trypanosome: **f**, epimastigote (crithidial) form of trypanosome: **g**, trypomastigote (trypanosomal) form of typanosome.

which is free-living, **Hexamita** (Fig. 5.12c), found free-living and endobioti-
cally, and **Giardia** (Fig. 5.12d), a parasite in vertebrates including man, are
examples of diplomonads; their double body has two karyomastigonts and
sometimes two cytostomes. A related group, the enteromonads, normally have
a single set of organelles, like half a **Hexamita**, although they are transiently
double before division, suggesting how diplomonads could have arisen from an
ancestral enteromonad (Brugerolle and Taylor, 1979). The recurrent flagella of
these forms are associated with the cytostomes, where these are present; microtubules, including those associated with the basal bodies, support the pellicle
and the lips of the cytostome, as in retortamonads, and a fibre runs to the
nucleus. **Giardia** is an osmotroph which clings to the gut wall, its complex
flagellar roots and other fibres forming an effective suction disc associated with
flagellar axonemes that remain intracytoplasmic for long distances (Holberton,
1973). Finally, the oxymonads, e.g. **Saccinobaculus** (Fig. 5.12f), have one to
many karyomastigonts, each with two pairs of basal bodies that are associated
with a dense pre-axostyle structure which connects to the anterior end of an

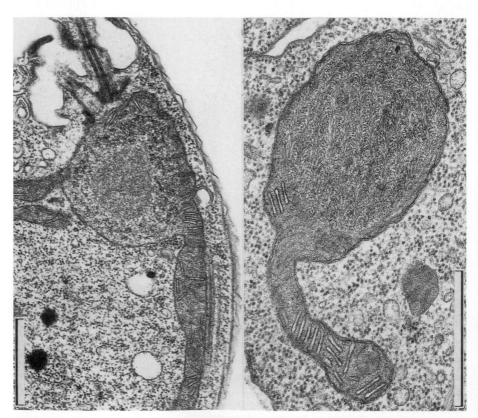

Fig. 5.16 Electron micrographs through the kinetoplast-mitochondrion, of **Bodo** sp. The
dilated kinetoplast region of the mitochondrion lies near the flagellar bases and contains large
amounts of fibrilla DNA: the mitochondrial cristae are discoidal. Micrographs by Brugerolle and
Mignot (1979). Scales 1 μm.

axostyle composed of many parallel sheets of microtubules (p. 28); undulations propagated down the axostyle(s) cause these flagellates to wriggle around in the gut contents of the insects they inhabit. Oxymonads are the only metamonads known to show sex, but this process is variable, some being haploid, some diploid (p. 76).

The kinetoplastid flagellates include the monoflagellate trypanosomes (Fig. 5.15c–g) and biflagellate bodonids (Fig. 5.15a,b), and are named from the presence of the kinetoplast, a large concentration (or sometimes dispersed masses) of fibrillar mitochondrial DNA that typically occurs near the flagellar bases (Fig. 5.16). The flagella arise from one or both of a pair of parallel basal bodies within a depression or flagellar pocket. Each flagellum contains a striated paraxial rod (Fig. 5.17), thin hairs may be present on the front or on both flagella, and either flagellum may form an undulating membrane by attachment along the side of the body; waves are often propagated from tip to base along the anterior flagellum (p. 107), while the posterior flagellum if present, trails behind or attaches the cell. The flagella have long transition zones and their basal bodies are linked to each other by striated fibres, and sometimes to the kinetoplast by dense material. In bodonids three groups of microtubules leave the basal bodies (Fig. 5.33a) and ascend the sides of the flagellar pocket; a small dorsal group runs with a dense lamina at the dorsal edge of the pocket and gives rise to dorsal pellicular microtubules, while the ventral group joins the ventral pellicular array, and a wider sheet of reinforced microtubules forms a preoral crest around the flagellar pocket before turning back to form the microtubular canal around the cytopharynx (Brugerolle *et al.*, 1979). In trypanosomes microtubular flagellar roots run along the flagellar pocket and join the array of microtubules that underlies the body surface membrane, which is naked apart from a glycocalyx (Fig. 5.17), and the flagellum may form an undulating membrane. The single mitochondrion forms a loop or network and contains discoidal cristae (Fig. 5.16), and a prominent Golgi body occurs between the nucleus and the flagellar bases; a contractile vacuole may be present, and occasionally trichocysts, but never plastids. Near the flagellar pocket there is typically a cytostome leading to a deep cytopharynx supported by microtubules. Mitosis is closed with an internal spindle and persistent nucleolus, but neither meiosis nor sexual reproduction have been confirmed, although genetic recombination has been reported. Most of the 600 species are parasites, although many bodonids are free-living in freshwater and marine habitats.

Bodo (Fig. 5.15a) is bacterivorous and especially abundant in organically polluted waters, and *Cephalothamnium* is a colonial form epizoic on crustaceans, to which it is attached by its posterior flagellum via a common secreted stalk. In this case, and in *Cryptobia* (Fig. 5.15b), a parasite in invertebrates and lower vertebrates, the posterior flagellum is attached along the side of the body. The trypanosomids are parasitic forms, often with polymorphic life cycles in which non-flagellate and flagellate stages may be present (Vickerman, 1976). *Leptomonas, Crithidia* and some related genera are found as gut parasites in insects, *Phytomonas* is found in plants (particularly latex-bearing species, between which it is carried by plant-sucking insects), and *Leishmania* and *Trypanosoma* are parasites of vertebrates (reviewed by Molyneux and Ashford, 1983), including man, being normally transmitted by blood-sucking invertebrates, usually insects or leeches. Four body forms are recognised, according to

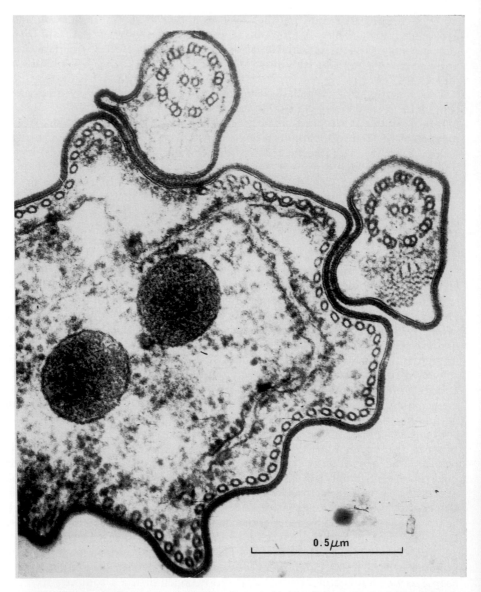

Fig. 5.17 Electron micrograph of a section through the bloodstream form of *Trypanosoma evansi* during the preparative stages for division. Two flagella are closely adherent to the body and a row of pellicular microtubules underlies the cell membrane, which carries a thick surface coat at this stage of the life cycle. Micrograph by K. Vickerman (1969).

the position of the basal body and kinetoplast and the course followed by the flagellum; these are amastigote, promastigote, epimastigote and trypomastigote (Figs. 5.15d–g). The main features of structure of the trypomastigote form of *Trypanosoma* are shown in Fig. 5.15c).

Species of *Leishmania* cause important human diseases, including the widespread kala-azar. The small amasigote parasites are found in vacuoles in the macrophages and lymphoid cells of infected vertebrates, where they grow and multiply, somehow escaping the digestive enzymes of lysosomes that fuse with the vacuoles (Chang, 1983). The parasites are transmitted by female sand flies (*Phlebotamus*, Diptera), which ingest parasitised cells when they suck the blood of infected hosts. Within the fly the parasites leave the white blood cells, transform to the promastigote type, and multiply by binary fission; flagellates migrate forward in the gut of the fly as far as the proboscis and are injected into a new host with saliva.

The life cycles of two species of *Trypanosoma* pathogenic in man will be described to illustrate different features of these organisms and of their relationships with their hosts (*see also* p. 278). *T. cruzi* is found in South and Central America in man and a diversity of mammals, between which it is transmitted by blood-sucking bugs, usually *Rhodnius*. Within the insect the ingested trypomastigote forms transform into epimastigotes which multiply repeatedly and eventually develop into small trypomastigote forms (metacyclic forms) in the hind gut of the bug. The vertebrate is infected by contaminative means, through faeces of the insect being rubbed into skin lesions or unprotected mucous membranes etc. The trypanosome enters lymphoid, macrophage, nerve and muscle cells of the vertebrate, lying in a vacuole in cells of the former type, but free in the cytoplasm in the tissue cells (Vickerman, 1974); intracellular stages multiply as amastigotes. These may subsequently transform through promastigote and epimastigote stages to form trypomastigotes which circulate in the blood before re-entering the host cells and commencing a new amastigote multiplicative phase. The human disease caused by this parasite, Chagas' disease, is often fatal in children, and chronic in adults.

The African human trypanosomiasis, known as sleeping sickness, is caused by *T. brucei*, of which two recognised sub-species, *T. brucei gambiense* and *T. brucei rhodesiense* are found respectively in western and eastern parts of tropical Africa. The vectors for these trypanosomes are species of the blood-sucking tsetse fly *Glossina*. Within the fly the trypomastigote forms multiply in the midgut region and undergo a migration forwards to the salivary gland via the proboscis. In the salivary gland epimastigote forms multiply and metamorphose to form infective metacyclic trypomastigotes, which may be injected into a new host at the next feed. Injected trypomastigotes circulate in the blood of the vertebrate and divide by binary fission; they do not have intracellular stages, but may later invade the cerebrospinal fluid and may fatally damage the brain. Other species of this type are important pathogens of domestic animals, particularly in Africa (Kreier, 1977; Molyneux and Ashford, 1983).

The euglenoid flagellates form a distinctive phylum, the main characteristics of which are illustrated by *Euglena* (Fig. 5.18a). Near the apical end is an invagination, the flagellar pocket, into which a contractile vacuole opens, and in the walls of which are two parallel kinetosomes each bearing a flagellum (a few species have four flagella); in some species only one flagellum extends out of the flagellar pocket, and when there are two long flagella one of these is usually directed backwards along the body (Fig. 5.18b,f). Each flagellum carries thin hairs, 2–3 μm long, usually in a one-sided array associated with a striated paraxial rod (Hyams, 1982), and often additional layered extraflagellar structures, all of which are well shown by the anterior flagellum of *Peranema*

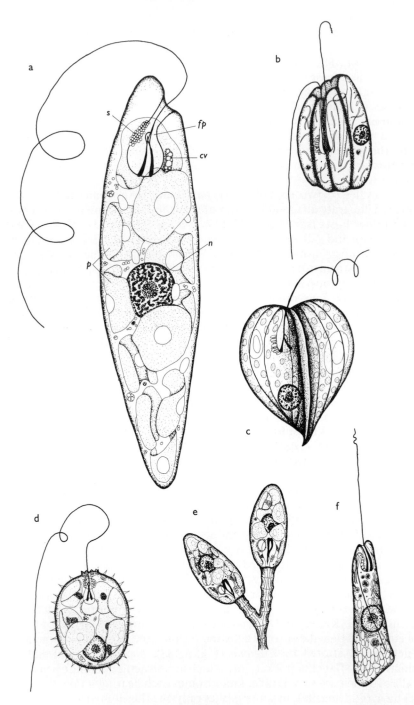

Fig. 5.18 Some representative euglenoid flagellates; **a**, *Euglena gracilis* (50 μm), with nucleus (*n*), stigma (*s*), plastids (*p*) with central pyrenoid, two flagella arising in a flagellar pocket (*fp*), the short one terminating at about the level of the photoreceptive flagellar swelling on the longer flagellum, and the contractile vacuole (*cv*); **b**, *Entosiphon sulcatum* (20 μm): **c**, *Phacus pleuronectes* (80 μm): **d**. *Trachelomonas hispida* (30 μm): **e** *Colacium vesiculosum* (20 μm), often epizoic on crustaceans: **f**, *Peranema trichophorum* (60 μm). Mainly based on information from Leedale (1967).

(Fig. 5.19). Flagellar rootlet microtubules support the walls of the flagellar pocket and contribute to the pellicle in the same patterns as in *Bodo* (see above). The anterior flagellum often has a swollen region containing paracrystalline material incorporating flavins (Diehn and Kint, 1970) situated opposite to a cluster of carotenoid-containing eyespot globules in the cytoplasm; the flagellar swelling is assumed to be a photoreceptor that is shaded by the eyespot to indicate the direction of illumination, but Cruetz and Diehn (1976) suggest that the polarisation of light may also be important. The mitochondria of euglenoids contain discoidal cristae with narrow stalks (p. 19), and the photosynthetic species have many green chloroplasts (p. 101). The stored carbohydrate is paramylon (a $\beta - 1{:}3$ glucan), which is deposited in crystalline granules of specific shapes. The pellicle is flexible in many species, where it consists of sub-plasmalemmal, interlocking longitudinal proteinaceous strips associated with microtubules (Fig. 5.20); in some species, e.g. *Menoidium*, the dense material of the pellicle is continuous and the pellicle rigid. Mucocysts are commonly found at the junctions of pellicular strips, and trichocysts near the cytostome in some species. The nuclear membrane and the nucleolus (endosome) persist throughout mitosis, in which chromosomes associate with an apolar intranuclear spindle; these chromosomes remain condensed throughout interphase. Sexual processes have not been confirmed in euglenoids.

About 1000 species of green and colourless species of euglenoids have been described from freshwater and occasionally seawater. They tend to be more abundant in smaller bodies of water with a high organic content, but some representatives of the group will almost always be found in freshwater habitats. Most forms swim or glide using flagella, but some move by peristaltic movements of the body associated with relative motion of pellicular strips (Suzaki and Williamson, 1986), possibly driven by microtubule links (Gallo and Schrevel, 1982). *Trachelomonas* (Fig. 5.18d) is a mobile loricate species, *Klebsiella* is loricate and *Colacium* (Fig. 5.18e) is a stalked species often epizoic on crustaceans; the secreted material in envelopes and stalk is mucilage (from mucocysts ?). The nutrition of members of this group is diverse (p. 69); of 37 genera described by Leedale (1967), 11 are photoautotrophs, 14 osmotrophs and 12 phagotrophs, many green species being facultative heterotrophs practising osmotrophy. The phagotrophs have a well-developed cytostome equipped with ingestive structures made largely of microtubules grouped into rods (e.g. *Peranema*, Nisbet, 1974) (Figs. 3.9, 5.18f), which may form a channel (e.g. *Entosiphon*, Mignot, 1963) (Figs. 5.18b, 5.21); even *Euglena* appears to have a cryptic cytostome partially lined by microtubules from one of the three flagellar roots (Surek and Melkonian, 1986). The colourless phagotroph *Isonema* has pellicle microtubules, cytostomial rods and a flagellar paraxial rod, but appears aberrant because it lacks pellicular strips and flagellar hairs, and has plate-like rather than discoidal cristae. A few parasitic euglenoids are known from invertebrates and amphibian tadpoles. It seems likely that the ancestral euglenoid evolved its cellular features as a heterotroph, and that symbiotic chloroplasts were acquired later (Leedale, 1978). Features of the nuclei, mitochondria and flagella suggest a close relationship with kinetoplastids, and some authors have placed them in the same taxon, the Euglenozoa (Cavalier-Smith, 1981). The group has been described by Leedale (1967, 1985) and Walne (1980) and there are other books devoted to the genus *Euglena* (Buetow, 1967, 1982; Gojdics, 1953).

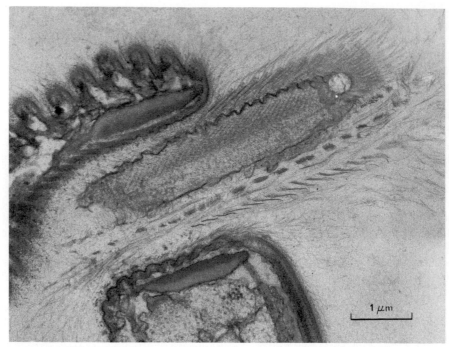

Fig. 5.19 Electron micrograph of a section of the flagellum of *Peranema* at the entrance to the flagellar pocket; the flagellum is complex, being surrounded by a variety of hairy projections and containing alongside the axoneme an intraflagellar rod that can be seen to have a striated pattern like that seen in similar rods in dinoflagellates and kinetoplastids. Micrograph by B. Nisbet.

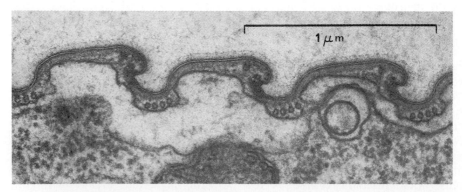

Fig. 5.20 Electron micrograph of a section of the pellicle of *Peranema* showing the ridges formed by dense material immediately beneath the cell membrane, and incorporating groups of microtubules. Micrograph by B. Nisbet.

The genus *Stephanopogon* was formerly classified with primitive ciliates, but ultrastructural studies place them with flagellates (Lipscomb and Corliss, 1982). *S. apogon* has 12 rows of short flagella and an anterior cytostome through which it ingests small organisms (Fig. 5.23b). Several similar nuclei with central nucleoli may be present, and the organism grows with repeated mitoses before undergoing palintomy in a cyst; mitosis is closed. There are mucocysts but no alveoli in the pellicle, the main component of which is a layer of microtubules without direct connection to kinetosomes. The latter lack any of the type of fibrils of the ciliate kinetid (p. 189), but have short roots running toward the next kinetosome, and microtubules radiating up toward the surface all around the basal body. The flagella lack external or internal specialisations and the mitochondria contain discoidal cristae but no kinetoplast. Corliss (1984) has placed this genus in a separate phylum, the Pseudociliata; the form of their mitochondria could suggest a relationship with kinetoplastids, and, if the report that there is an internal spindle and persistent endosome is confirmed, such a relationship would seem more justified.

The cryptophytes are generally unicellular with two almost equal flagella that emerge from an antero-lateral flagellar pocket or groove (Dodge, 1979; Gantt, 1980; Santore and Leedale, 1985). One flagellum usually carries two rows of mastigonemes and the other a single row. The flagellar bases are parallel and in

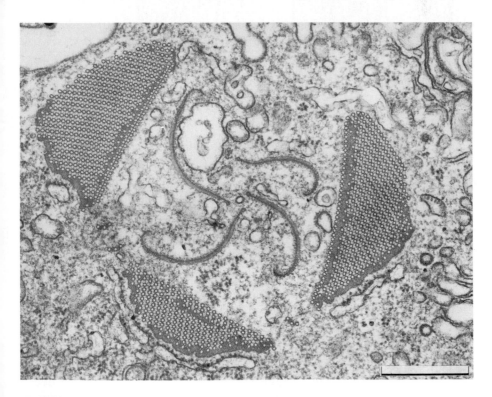

Fig. 5.21 Electron micrograph of a section through the cytopharyngeal rods of *Entosiphon*. Three rods, formed from microtubules, are arranged to make a channel containing some fibrous strands. Micrograph by D.J. Patterson. Scale 0.5 μm.

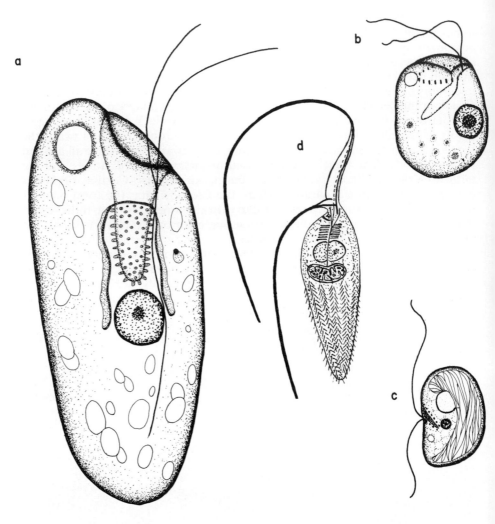

Fig. 5.22 Some representative cryptophytes and a proteromonad; **a**, *Chilomonas para-mecium* (30 μm): **b**, *Cyathomonas truncate* (20 μm), **c**, *Hemiselmis rufescens* (6 μm): **d**. *Proteromonas lacertae-viridis* (15 μm) from the lizard rectum.

Chilomonas one large microtubular root runs to the rear of the cell, passing through a groove in the nucleus, a second runs around the anterior of the cell and a striated root runs transversely across the body (Fig. 5.33); Roberts *et al.*, (1981) point out that this rootlet arrangement is reminiscent of the axostyle, pelta and costa, respectively, of trichomonad flagellates, and consider evolutionary relationships of cryptophytes with such forms (see p. 116). The cell membrane carries a fuzzy glycocalyx on its outer surface and at its inner side or at both inner and outer sides, in the thin pellicle (periplast), are thin (10 nm or so) rectangular or hexagonal proteinaceous plates (Kugrens and Lee, 1987). The (usually single) chloroplast has a unique structure and pigmentation and lies in a periplastidial compartment (p. 101). The nucleus, whose outer membrane is joined to the outermost plastid membrane, shows an unusual open (or partly so) mitosis without centrioles, in which the single plate of chromatin seen pierced by microtubules at metaphase separates into two parts at anaphase. The cell subsequently cleaves longitudinally, starting posteriorly. Sexual fusion has now been reported and details are awaited. In the anterior part of the cell is a contractile vacuole which opens into the flagellar pocket, a large Golgi body and a single, extended or branched, mitochondrion whose cristae appear like flattened tubules. Small ejectisomes occur at the junctions of pellicular plates and large ones line the flagellar pocket; these develop in Golgi vesicles and are formed of two sections of tightly-coiled ribbons which unroll upon extrusion.

One hundred or so autotrophic, brown, red, green, yellow and even blue, species in genera like *Cryptomonas* and *Hemiselmis* (Fig. 5.22c) occur in marine and freshwater habitats. *Chilomonas* (Fig. 5.22a) is probably saprotrophic (p. 57), but observed food vacuoles suggest that some species are phagotrophic. The red-tide ciliate *Mesodinium* contains a symbiotic cryptophyte and forms substantial blooms in estuaries and coastal areas (Lindholm,

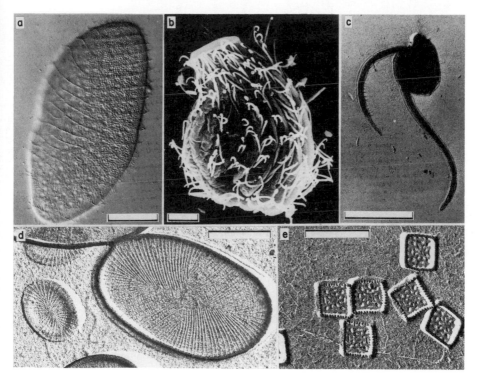

Fig. 5.23   **a**, Light micrograph of *Opalina* showing the metachronal waves formed by flagella (micrograph by D.J. Patterson). **b**, Scanning electron micrograph of *Stephanopogon apogon* from the ventral side (micrograph by G. Brugerolle). **c**, Electron micrograph of a crypotophyte, probably *Chroomonas* sp., showing hair patterns (micrograph by J.D. Dodge and R.M. Crawford). **d**, Electron micrograph of scales from the haptophyte *Chrysochromulina* (micrograph by R.M. Crawford). **e**, Electron micrograph of scales from the prasinophyte *Pyramimonas* (micrograph by R.M. Crawford). Scales: **a**, 100 μm; **b** and **c**, 5 μm; **d** and **e** 0.5 μm.

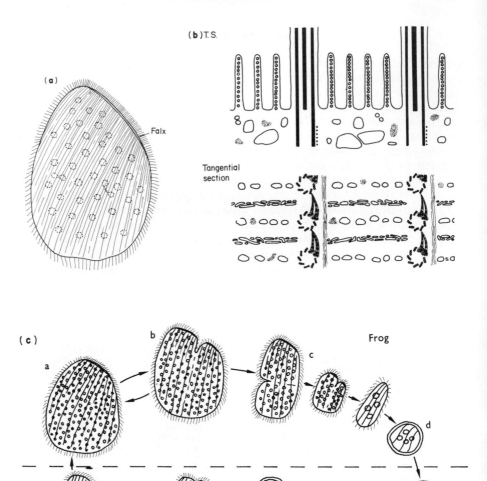

(b) T.S.

Falx

Tangential section

(a)

(c)

b

a

Frog

c

d

k

l

e

j

f

i

Tadpole

m

h

g

Fig. 5.24 *Opalina*. **a**, A general view showing the pattern of flagellar rows, numerous nuclei and the falx region. **b**, The structure of the pellicle seen above in transverse section and below in a tangential section grazing the surface at the level of the flagellar basal bodies; longitudinal ridges containing microtubules run between flagellar rows, and arrays of vesicles underlie them (based on electron micrographs by C. Noirot-Timothee, 1959, and H. Wessenberg, 1966). **c**, The life cycle of Opalina; growth in the frog rectum results in binary fission usually by longitudinal furrowing (*a,b*): multiple fission (*c*) leads to formation of a cyst (*d*) which is voided by the frog: the cyst hatches (*e*) when eaten by a tadpole and further division (*f,g*) results in the formation of uninucleate multiflagellate gametes which fuse (*h*) to form a zygote which encysts (*i*): when this cyst is eaten by another tadpole it may grow into a normal trophic form (*j,k,a*) or the small trophic form *k* may undergo division and asexual cyst formation (*l*), or the form that hatches from the zygocyst may develop directly into a gamont (*m*) (information from Wessenberg, 1966).

1985). A few species have a coccoid, thick-walled, trophic stage and biflagellate zoospores typical of the group (Bourrelly, 1970).

Opalinids (Fig. 5.23a, 5.24a) are common in the digestive tracts of fish and amphibians; about 400 species have been described. They are flattened with oblique longitudinal rows of flagella over the surface and with the intervening pellicle formed into longitudinal ridges supported by microtubules (Fig. 5.24b). In *Opalina ranarum* (Patterson, 1985) the naked flagella have a transitional helix and their basal bodies are connected within the rows by a fine fibrillar band at one side of the row and by dense curved connectives extending from one kinetosome to the next; there are no microtubular connections to the pellicle and no kinetodesmata of the type found in ciliates. In *Opalina* large numbers of vesicles and vacuoles that occur between and beneath the ridges are presumed to be pinocytotic and digestive; near them lie Golgi bodies and mitochondria with tubular cristae. In *Opalina* there may be several hundred similar nuclei which show closed mitosis with an internal spindle and no obvious MTOC. In some other genera there may be only two nuclei. Division of the body is normally longitudinal, but occasionally transverse. Wessenberg (1961) found that in spring *Opalina* divides rapidly (perhaps under control of host hormones) to produce small multinucleate cysts which are passed out with the faeces. When cysts are eaten by a tadpole they hatch in the gut and divide with meiosis to produce uninucleate multiflagellate macro- and microgametes which fuse in pairs. The zygote secretes a cyst wall and is shed into the water. Upon being eaten by another tadpole it hatches and either repeats the sexual stage or becomes a diploid multinucleate trophic form. The life cycle is outlined in Fig. 5.24c.

Two other flagellates from the gut of amphibians and reptiles show some affinities with opalinids; these are the proteromonads *Proteromonas* (Fig. 5.22e) and *Karotomorpha* (Brugerolle and Joyon, 1975). In both genera, and especially in *Karotomorpha*, the surface of the body has ridges supported by microtubules, and pinocytotic vesicles form between the ridges. The kinetosomes are set orthogonally in pairs, one pair in *Proteromonas* and two pairs in *Karotomorpha*. The flagella are smooth with a transitional helix, and in the anterior flagellum of *Proteromonas* a slender paraxial strand. The rhizostyle of *Proteromonas* is composed of two bundles of microtubules and a dense fibre; it descends from the kinetosomes through a large Golgi body, across the surface of the nucleus, and terminates on a large mitochondrion that has tubular cristae. In *Karotomorpha* the microtubular rhizostyle runs back past the Golgi to join pellicular microtubules. Mastigoneme-like hairs, formed in Golgi vesicles, clothe the posterior part of *Proteromonas*. These flagellates were once placed with kinetoplastids, but clearly they share few features with them; in addition to similarities with opalinids they share with heterokont algae the transitional helix, tubular cristae, a flagellar root which passes against the Golgi on its way to the nucleus, and tubular mastigonemes (Patterson, 1985).

The chrysophytes (golden-brown algae) provide a starting point for an account of the heterokont groups (p. 108), and *Ochromonas* (Fig. 1.2, p. 6) shows their basic organisation. The surface appears naked, with perhaps a thin glycoprotein coat and a few secreted fibres. Within the cell are one or two characteristic chloroplasts (see p. 101), whose pigmentation is dominated by fucoxanthin, they often partially surround the nucleus, with their outermost

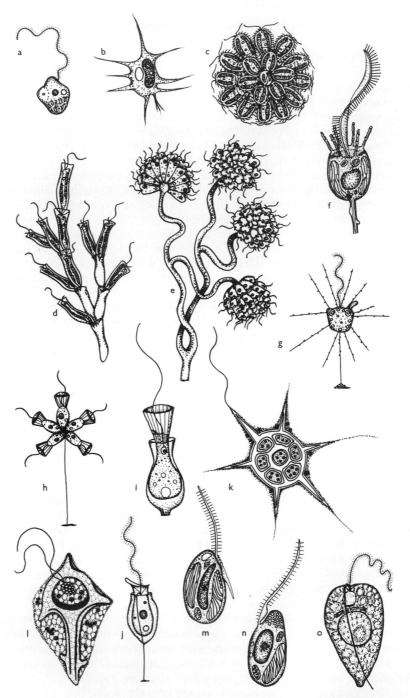

Fig. 5.25 Some representative chrysophytes and members of small flagellate groups; **a**, *Oikomonas termo* (15 μm): **b**, *Chrysamoeba radians* (10 μm): **c**, *Synura uvella* (cells 30 μm): **d**, *Dinobryon setularia* (cells 30 μm): **e**, *Anthophysa vegetans* (cells 5 μ): **f**, *Pedinella hexacostata* (10 μ): **g**, *Actinomonas mirabilis* (10 μm) (helioflagellate): **h**, *Codosiga botrytis* (15 μm) (Choanoflagellata): **i**, *Salpingoeca fusiformis* (15 μm) (Choanoflagellate): **j**, *Bicoeca exilis* (12 μm) (Bicoecida): **k**, *Dictyocha speculum (40* μm) (Silicoflagellida): **l**, *Hermesinum adriaticum* (15 μm) (Ebriida): **m**, *Bumilleria sicula* zoospore (15 μm) (Xanthophyceae): **n**, *Vischeria punctata* zoospore (10 μm) (Eustigmatophyceae): **o**, *Goniostomum semen* (50 μm) (Raphidiophyceae).

membrane continuous with the outer nuclear membrane. The two flagella emerge antero-laterally and almost at right angles; the longer flagellum bears two rows of tripartite mastigonemes, enabling the flagellum to pull the cell along in a characteristic manner (p. 35), and the smooth shorter flagellum typically has a dense swelling where it runs over a single layer of stigma globules at the anterior end of a chloroplast. There is a helical structure in the transition zone of the flagella and flagellar roots take the form of microtubules that run around the cell immediately under the membrane and a striated rhizoplast which extends past the Golgi body towards the nucleus and branches around it. A contractile vacuole occurs near the flagellar bases and mitochondria with tubular cristae occur throughout the cell, often associated with storage bodies of fat or the liquid polysaccharide chrysolaminarin (leucosin). Mitosis is open, with spindle fibres associated with rhizoplast fibres, and, although sex is unknown in *Ochromonas*, the cells are haploid with zygotic meiosis following fusion of isogametes in *Dinobryon*.

The group, totalling 600 or so living species, is largely comprised of flagellates, but includes coccoid and filamentous forms with cellulosic cell walls and naked amoeboid (Fig. 5.25b) or palmelloid forms, all with chrysophyte chloroplasts, some with and others without a typical flagellate zoospore stage. Although solitary biflagellates are common, colonial forms are very frequent, and in some cases the short flagellum may be reduced to a short stump or lost. One striking naked form with a single emergent flagellum is *Dictyocha* (Fig. 5.25k), common in marine plankton, and with a distinctive external silica skeleton of tubular elements associated with a central nucleus and a complex lobed body containing many chloroplasts and many Golgi bodies; they are frequently classed separately as silicoflagellates (Van Valkenberg, 1980). Silica is common in other chrysophytes, both in body scales (e.g. in the colonial *Synura*, Fig. 5.25c) and in endogenously produced walls of cysts (statospores) provided with a single pore and plug. Chrysophyte scales develop in ER vesicles; some species have organic scales and small scales are often present on the longer or both flagella. In *Dinobryon* (Fig. 5.25d) cellulose loricas are secreted so as to form colonies. Species of *Ochromonas* capture bacteria and practise phagotrophy at the same time as autotrophy, and other species like *Oikomonas* (Fig. 5.25a) and the colonial *Anthophysa* (Fig. 5.25e) are phagotrophs that lack chloroplasts. One species of *Ochromonas* has been found to have extrusomes. Andersen (1987) has argued that the silica-scaled algae have a number of features that set them apart from other chrysophytes in the class Synurophyceae.

The colourless phagotrophic bicoecids (Fig. 5.25j) are heterokont forms that appear closely related to chrysophytes (Moestrup and Thomsen, 1976), and like them have mitochondria with tubular cristae. They are unicellular and anchor themselves within a cup-shaped lorica by the posterior smooth flagellum, whilst collecting food with the anterior hairy flagellum (like Fig. 5.6R), whose tripartite mastigonemes are formed in ER vesicles; around the flagellar base they have a prominent lip reinforced by a broad microtubular flagellar root, smaller microtubular roots run to the pellicle, but there is apparently no rhizoplast. There are about 40 species of freshwater and marine bicoecids, some of which are colonial.

A number of phagotrophic, stalked helioflagellates, including *Actinomonas* (Fig. 5.25g), *Pteridomonas* and *Ciliophrys*, have a flagellum with tripartite mastigonemes that functions like the chrysophyte anterior flagellum; it propels a feeding current through a radiating array of contractile axopods supported by microtubule triads, each of which arises from an MTOC on the nuclear envelope. They have tubular cristae and may be related to chrysophytes through **Pedinella** (Fig. 5.25f), which contains chrysophyte-type chloroplasts but, like

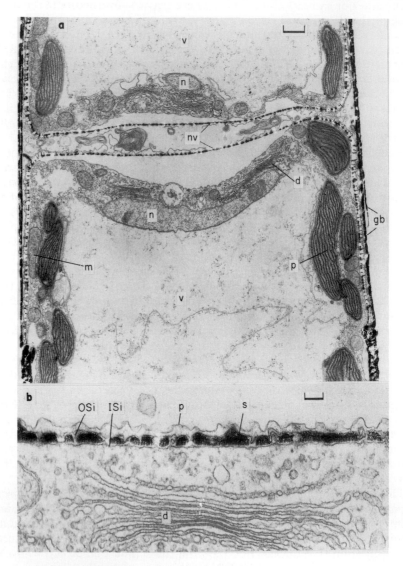

Fig. 5.26 Electron micrographs of sections through the diatom **Melosira varians** to show the formation of new shell valves. In **a** the deposition of silica in new valves (*nv*) between two newly divided cells is occurring in a zone enclosed by the overlapping girdle bands (*gb*) of the parental cell; the two nuclei (*n*), plastids (*p*), dictyosomes (*d*), mitochondria (*m*) and cell vacuoles (*v*) are also visible; micrograph by Crawford (1981). In **b** the outer (*OSi*) and inner (*ISi*) silicalemma membranes and plasmalemma (*P*) are shown in relation to the developing silica layer, and to the adjacent dictyosome (*d*); micrograph by R.M. Crawford. Scales 1 μm, 0.1 μm.

these helioflagellates, has a paraxial rod in the single flagellum, a second (barren) basal body, a similar contractile stalk, extrusomes with an amorphous core and a rhizoplast. Patterson and Fenchel (1985) discuss the possibility that these helioflagellates may form a link between the chrysophytes and the actino-phryiid heliozoans (p. 176), although the latter differ in important ways from typical chrysophytes. The dimorphid helioflagellates (p. 177) have different features and probably have different origins.

The diatoms, Bacillariophyta, form a large phylum of unicellular or colonial yellow-brown cells characterised by the presence of an outer covering (frustule) of silica (Werner, 1977). About 10 000 species are known from marine and freshwater, planktonic and benthic, epilithic, epipsammic and epiphytic habitats, and are responsible for some 20% of the earth's primary production; fossil diatoms are also abundant. They show many heterokont features and are considered to be close to chrysophytes. Each cell contains a nucleus, many mito-chondria with tubular cristae, fat and chrysolaminarin droplets, and one or more chloroplasts identical in structure and pigmentation with those of chryso-phytes (p. 101, Fig. 2.3e), but these structures surround a large central vacuole lined by a tonoplast (Fig. 5.26a). Golgi bodies are frequent, and separate from the silicalemma vesicles. Before division, larger or smaller parts of the silica shell are formed in internal silicalemma vesicles by deposition of silicic acid (Fig. 5.26); at division the two halves of the parental shell separate and comple-mentary new halves for each shell are assembled within the old shell with, if necessary, organic material to hold the parts together. The two membranes outside the silica layer reduce to an organic deposit and the membrane inside the shell becomes the plasmalemma. Since the new shell forms inside the old, the cells of many species progressively reduce in size. Many diatoms can secrete a gelatinous material which is used in adhesion and sometimes to form a perma-nent stalk. Mitosis is open, without centrioles, and the nuclei of vegetative cells are diploid because meiosis is gametic.

There are two major groups; centric diatoms are basically radially organised and often shaped like a petri dish with its lid (Fig. 5.27a), while pennate diatoms are bilaterally symmetrical or asymmetrical with longitudinal grooves (raphes) at either face (Fig. 5.27b). Colonial forms occur in both groups. The groups differ in sexual reproduction, which in pennate forms involves conjugation, whilst in centric diatoms the male gamete has a single anterior $(9 + 0)$ flagellum with mastigonemes and fertilises an oogonium; in both cases the zygote forms an auxospore within which the body is reorganised and its size restored. The pennate forms show gliding locomotion associated with a band of actin fila-ments beneath the raphe and mucilage secretion from it (Edgar and Pickett–Heaps, 1983), but centric forms also show movement (Medlin *et al.*, 1986). Most of the planktonic species are centric forms, whilst the pennate forms are typically benthic. Although diatoms are classical photoautotrophs, many species are facultatively heterotrophic, especially pennate forms.

The brown algae, Phaeophyta, are filamentous or parenchymatous multi-cellular plants with thick-walled cells that contain mitochondria and chloro-plasts with structure and pigmentation identical with chrysophytes, except that their pyrenoids are stalked; mitosis is open, with centrioles at the spindle poles (Dodge, 1973). The principal interest here is that reproduction involves biflagel-late gametes (both male and female in some groups, male only in oogamous

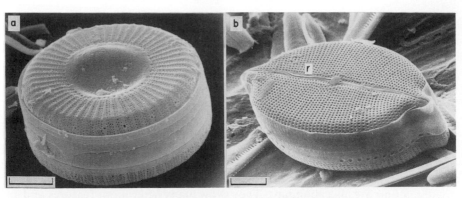

Fig. 5.27   Scanning electron micrographs of diatoms. **a**, The centric diatom *Cyclotella* and **b**, the pennate diatom *Mastogloia*, showing the raphe *r*. Micrographs by F.E. Round. Scales 5 μm.

types) and biflagellate zoospores, usually with haplodiploid alternation of generations. Zoospores and flagellate gametes are similar, typically having an anterior long flagellum with mastigonemes and a long, naked, trailing flagellum that is dilated where it runs over an eyespot located within a chloroplast (Fig. 5.28a). Almost all of the 1600 species are marine, and some grow very large (Irvine and Price, 1978).

The xanthophytes (yellow-green algae, Tribophyceae or Heterochlorida) (Hibberd, 1980a) share many features with chrysophytes, but the yellow-green colour is probably related to the absence of fucoxanthin. About two thirds of the 600 described species are unicellular and coccoid, and most of the remainder are filamentous or siphonous (coenocytic), leaving just a few amoeboid, palmelloid or flagellate forms. Some species have heterokont flagellate zoospores, while other species produce non-motile autospores. The naked zoospores (Fig. 5.25m) typically bear two flagella and contain two chloroplasts, one with an anterior eyespot; these and the nucleus, mitochondria, contractile vacuole and Golgi bodies are all very similar to those of *Ochromonas*, even down to small details; the cells contain fat, but probably other carbohydrates replace chrysolaminarin. Nuclear division in *Vaucheria* shows a completely closed mitosis with an internal spindle and pairs of centrioles outside the nucleus at each pole; confirmation from other species would be valuable. Many species reproduce sexually by biflagellate isogametes or anisogametes, but with oogamy in *Vaucheria*; meiosis is zygotic and the main stage haploid. The coccoid and filamentous forms are surrounded by a cell wall, formed of two halves in many species, which seems to be pectic in composition, though the wall of the coenocytic *Vaucheria* contains cellulose. Most xanthophytes are freshwater photoautotrophs, though coccoid forms are common in soils; *Vaucheria* has conspicuous species in marine, brackish and freshwater habitats. However, the fact that several species, both coccoid and filamentous, have been grown in the dark on organic substrates shows that they are facultative heterotrophs (Belcher and Miller, 1960).

Another small group, the eustigmatophytes, are comprised of yellow-green coccoid cells, but these are distinguished from xanthophytes by the lack of

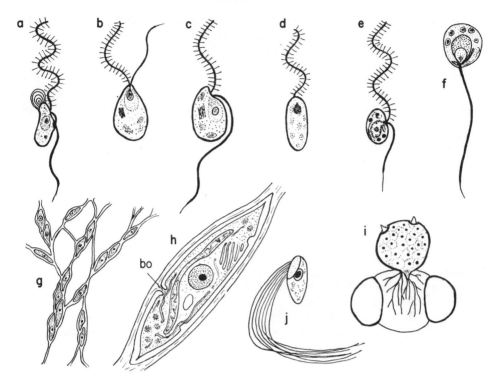

Fig. 5.28   Some flagellate zoospores, *Labyrinthula* and chytrids; **a** zoospore of *Fucus* (4 μm) (Phaeophyta): **b**, primary and **c**, secondary zoospore of an oomycete like *Saprolegnia* (spore cells 10 μm); **d**, zoospore of hyphochytrid (10 μm): **e**, zoospore of *Labyrinthula* (4 μm): **f**, zoospore of a chytrid like *Blastocladiella* (5 μm): **g**, cells of *Labyrinthula* within the channels of a slime net: **h**, a single cell of *Labyrinthula* showing a bothrosome (*bo*), surrounding membranes and cell organelles (cell 10 μm long): **i**, the chytrid *Rhizophydium* growing on a pine pollen grain (globular sporangium about 10 μm across: **j**, *Callimastix* (10 μm) a multiflagellated cell from the cow rumen, believed to be a chytrid zoospore.

chlorophyll *c* and the unique structure of their zoospores (Hibberd, 1980b). In fact, zoospores are rarely produced, since autospores are the normal means of reproduction, and sexual reproduction has not been seen. A typical zoospore (Fig. 5.25n) has a single anterior flagellum with mastigonemes, a transitional helix and a basal swelling against a large eyespot, which lies at the anterior end of the cell outside the chloroplast. A second basal body lies perpendicular to the first, and in two species bears a simple smooth flagellum. A rhizoplast and microtubular roots are present. The chloroplast is like that of xanthophytes, but lacks a girdle lamella and its outermost membrane is not continuous with the outer nuclear membrane; mitochondria have tubular cristae. The spherical, ovoid or stellate, vegetative cells are found in marine, freshwater and edaphic habitats; they have a single chloroplast with a large stalked pyrenoid and single Golgi bodies. The composition of the food storage bodies and of the cell wall are unknown. Although a 'peptone' medium enhanced the growth of *Monodus* in the light, it did not support growth in the dark.

Two groups amongst the flagellated fungi appear to belong here, the oomycetes and the hyphochytridiomycetes. The former have zoospores with heterokont flagella (Fig. 5.28b) and the zoospores and gametes of the latter have a single flagellum with mastigonemes and a barren basal body (Fig. 5.28d). In both groups the mitochondria have tubular cristae and mitotic nuclei that are closed or nearly so, with pairs of centrioles near the poles. The syncytial vegetative body extends to a greater or lesser extent into rhizoids or hyphae with walls of cellulose and/or chitin, but without septa, except at the tips where cross walls may cut off a sporangium from which flagellate zoospores emerge. Sexual forms are known to be diploid; oomycete sexual reproduction involves nuclear migration following apposition of two hyphae in a form of conjugation, but flagellate isogametes are produced in hyphochytrids; in each case the resting oospore germinates to release zoospores. The oomycetes are saprobic and parasitic forms, including water moulds like *Saprolegnia* and downy mildews like *Peronospora*, and are found in the sea, freshwater, soil and on plants. The hyphochytrids are simpler forms like *Hyphochytrium*, and are parasitic or saprobic in freshwater and marine habitats and in soils. Details of their macroscopic structure, life cycles and ecology will be found in mycology texts (e.g. Ainsworth *et al.*, 1973). The structure of their flagella and mitochondria are quite different from those of chytrids (p. 144), and support a relationship to heterokont algae, possibly xanthophytes, which also have closed mitosis with centrioles, and may be coenocytic with cellulose walls (p. 136).

The Raphidophyta (chloromonads) are naked, biflagellate, unicellular forms containing many yellow-green chloroplasts (e.g. Fig. 5.25o) (Heywood, 1980; Heywood and Leedale, 1985). The mitochondria, flagella and chloroplasts are all like those of chrysophytes, except that there is no eyespot, no pyrenoid, fucoxanthin is absent, and the posterior flagellum is long. Golgi bodies, a contractile vacuole and trichocysts are usually present and oil is the main food store. The nucleus is large and polar openings occur in mitosis, but sexual reproduction has not been confirmed. The flagellar basal bodies lie perpendicular to one another and give rise to a fibrous root that passes to the nucleus alongside a sheet of microtubules, and another root runs towards the anterior of the cell. Most of the 25 species are found in freshwater, but members of the marine genus *Chattonella* have some features that differ from the above; colourless forms have been described, but may not belong here.

Labyrinthula belongs to a group (slime nets) that has been placed in a separate phylum, the Labyrinthomorpha, because of its unique life-style; it was formerly classed among amoebae, but does not produce pseudopodia. It is mentioned here because it releases biflagellate zoospores of the heterokont type, which even posses an eyespot beneath the posterior flagellum (Fig. 5.28e) (Perkins and Amon, 1969). *Labyrinthula* and several related genera form complex colonies of spindle-shaped cells which glide along within a network of membranous slimeways (Fig. 5.28g) at speeds of up to 20 μm s^{-1}. Close examination shows that the slimeway is enclosed by a single membrane and that the cavity of the channel is connected to the cytoplasmic interior of the cell through a number of pores (the bothrosomes or sagenogenetosomes), the cells being surrounded by two membranes, a cell membrane and an inner channel membrane (Fig. 5.28h). Although the pores are partially blocked, cytoplasmic materials, but not organelles, pass into the slimeways, which contain actin filaments and probably myosin, the contractile activity of which is assumed to be responsible for the shuttling movements of the cells. Golgi bodies and mitochondria with tubular cristae are present, but not plastids; the nuclei show open orthomitosis during binary fission. Multiple fission may produce zoospores, and meiotic divisions produce flagellate isogametes; they also form cysts. The colonies are found on marine grasses and algae where they feed saprotrophically (possibly parasitically) secreting extracellular enzymes. A second group, the thraustochytrids, are parasitic on marine plants and have been placed in the same phylum; they form slime nets, but are non-motile (Perkins, 1973; Moss, 1985; Pokorny, 1985).

The green algae or chlorophytes have been classified in many ways; many authors have distinguished between the prasinophytes, whose cells and flagella are typically covered in scales, the conjugating algae, which never produce flagella and in which fertilisation results from conjugation, the charophytes which are complex, multicellular, oogamous forms in which the egg cell is enclosed in a complex envelope, and the remaining chlorophytes, which are flagellate, coccoid, filamentous or thalloid types not fitting these other descriptions. Corliss (1984), for example, separated these as four phyla, while others prefer to place them within a single phylum Chlorophyta (e.g. Sleigh *et al.*, 1984). Recent studies on nuclear and cell division and on the structure of motile flagellate cells suggests that members of the diverse class Prasinophyceae were the ancestral green algae and gave rise to three other classes, the Chlorophyceae, Ulvaphyceae and Charophyceae (Stewart and Mattox, 1978); ultrastructural studies have even separated (former) members of some genera, e.g. *Ulothrix* into separate classes (Sluiman *et al.*, 1980). Although this book is concerned principally with the microscopic and especially unicellular forms, it is important to place these in context within the whole group of perhaps 900 species of chlorophytes, as far as this is possible in an unsettled classification. The best-known genus of green alga is probably *Chlamydomonas*, which will be described first to summarise typical features before variation within the phylum is discussed.

Chlamydomonas has a single, large, cup-shaped chloroplast that fills much of the spheroidal cell and contains a large pyrenoid and a simple eyespot formed from two layers of pigmented globules (Fig. 5.29a); the ultrastructure, pigmentation and starch storage of the chloroplast are described on p. 101. Within the hollow of the chloroplast is the nucleus, two contractile vacuoles, a prominent Golgi body and mitochondria with flattened cristae. Two equal, homodynamic flagella, which have neither scales nor thick hairs, but may carry a few slender hairs, emerge at the body apex; their basal bodies are nearly at right angles and are connected by two striated fibres. Near the outer of these (the synistosome) originate four symmetrically-placed groups of microtubules (4-2-4-2 pattern), two accompanied by fibrous bands, which run around the cell surface beneath the cell membrane and form particular relationships with other cell structures (Melkonian, 1984c). Two barren kinetosomes may lie one on each side of the active ones, and in quadriflagellate species these also carry flagella. The flagella (as in all green algae and land plants) have a stellate linkage between the doublets in the transition zone (p. 106). The cell is surrounded by a rigid cell wall of secreted glycoprotein, not cellulose-like material as formerly thought. *Chlamydomonas* is haploid with zygotic meiosis (p. 74). During nuclear division the nuclear envelope persists almost completely, opening only at small polar fenestrae through which microtubules enter; after the daughter nuclei have

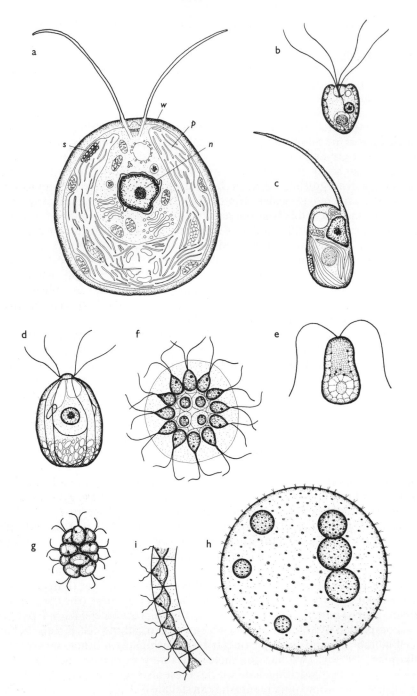

Fig. 5.29 Some representative flagellate chlorophytes; **a**, *Chlamydomonas reinhardtii* (10 μm), with cell wall (*w*), plastid (*p*), stigma (*s*) and nucleus (*n*) in the central cytoplasmic region: **b**, *Pyramimonas tetrarhynchus* (25 μm): **c**, *Pedinomonas minor* (6 μm): **d**, *Polytomella caeca* (15 μm): **e**, *Dunaliella salina (12* μm): **f**, *Gonium pectorale (cells 12* μm): **g**, *Pandorina morum* (cells 15 μm): **h**,, *Volvox globator* (colony 500 μm): **i**, arrangement of cells in the gelatinous wall of a *Volvox* colony. *Pedinomonas* and *Pyramimonas* are members of the Prasinophyceae, the others are members of the Chlorophyceae.

separated a band of microtubules (the phycoplast, Fig. 5.30g) develops in the plane of the cleavage furrow, and perpendicular to the spindle, so that as the cell cleaves the furrow separates these microtubules.

The Prasinophyceae (Norris, 1980) are mostly wall-less flagellates with many internal chlamydomonad features, but whose flagella and often cell bodies are covered with one or more layers of scales (which may fuse into a complete body coat); sometimes they are naked or walled cells clearly related by other features to scaly cells. They are rather diverse in root structures and features of mitosis and cell division, as befits their position between the first green alga – probably a simple uniflagellate form (Melkonian, 1984b) – and the more complex forms transitional to the other three chlorophyte classes. The naked, uniflagellate, scaleless *Pedinomonas minor* (Fig. 5.29c) has a closed mitosis and an unspecialised form of division, while the scaly quadriflagellate *Pyramimonas* (Fig. 5.29b) has an open mitosis. *Platymonas (Tetraselmis)* has two large contractile striated roots (rhizoplasts)(p. 28) that anchor the flagellar bases to both the nucleus and the lateral cell membrane (Robenek and Melkonian, 1979). Some flagellate cells have mucocysts or trichocysts. The Class includes multinucleate planktonic forms (*Halosphaera*) and multicellular, benthic, palmelloid forms, but both produce typical flagellate zoospores.

Cell division which involves furrowing or cell plate formation at the site of a phycoplast typifies the Chlorophyceae, including *Chlamydomonas*. They tend also to have an almost closed mitosis and almost symmetrical flagellate cells with cruciate rootlet arrangement and striated connectives between angled basal bodies; scales are not present on the bodies of flagellate stages. Flagellate variants on the *Chlamydomonas* pattern include *Dunaliella* (Fig. 5.29e) with no cell wall, *Polytomella* (Fig. 5.29d) with neither cell wall nor chloroplast (such forms are colourless osmotrophs, but never phagotrophs), and a series of colonial forms of various shapes and sizes, culminating in the hollow coenobia of *Volvox* (Fig. 5.29f–i) (Mignot, 1985). Chlorococcoid forms may be unicellular or colonial, with thick cell walls, and include freshwater forms like *Pediastrum* (Fig. 5.30c), *Scenedesmus*, *Chlorella* (also found as a symbiont, p. 283), and the fish parasite *Prototheca*; these and various filamentous forms (e.g. most *Ulothrix* species) reveal chlamydomonad affinities in the structure of flagellated gametes and often of flagellate zoospores, though some produce non-motile asexual spores. The zoospores and male gametes of the filamentous alga *Oedogonium* are multiflagellate, but retain chlamydomonad features and cell division is of the phycoplast type in spite of the unusual mode of filamentous extension (Pickett–Heaps, 1982).

The thalloid parenchymatous alga *Ulva* was found not to possess a phycoplast at cell division, although its cells furrow precociously as in Chlorophyceae, and mitosis is almost closed. Flagellate cells (zoospores and gametes) sometimes possess body scales, and have flagellar bases at 180° (opposite), linked by non-striated connectives and with cruciate roots. This combination of characters typifies the Ulvaphyceae, which appear to be rather primitive, in spite of their multicellular structure. The Class includes *Ulothrix zonata* (Sluiman et al., 1980), and the complex siphonous (non-cellular) algae like *Bryopsis* and *Codium*; these diploid forms and the coenocytic filamentous *Cladophora* (which shows an alternation of haploid and diploid phases) have closed mitosis and cruciate roots have been found in their flagellate cells.

The Charophyceae form haploid macroscopic plants composed of giant cells,

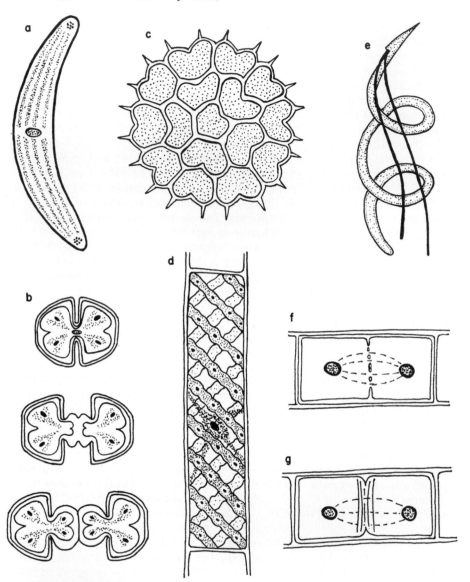

Fig. 5.30   Features of chlorophytes; **a**, the desmid **Closterium** (200 μm): **b**, stages in division of the desmid *Cosmarium* (cell 50 μm long): **c**, the colonial chlorococcoid *Pediastrum* (colony 60 μm across): **d**, a single cell of the filamentous conjugating alga *Spirogyra*, with spiral chloroplast (cell 100 μm long): **e**, a biflagellate sperm of *Chara* (length 30 μm): **f**, a diagram illustrating division with the formation of a phragmoplast prior to the formation of a cell plate: **g**, a diagram illustrating division with formation of a phycoplast structure of microtubules perpendicular to the spindle.

reproducing by flagellated male cells and enclosed oogonia. The asymmetric flagellated cells (Fig. 5.30e) have a single flat band of microtubules on which are borne two almost parallel basal bodies and a fibrous multi-layered structure (MLS). In this group the nuclear membrane disperses completely at mitosis, and following nuclear division the spindle microtubules persist and are interspersed with more microtubules and vesicles to form a phragmoplast (Fig. 5.30f) which takes part in the formation of a cell plate. These features of cell and nuclear division are shared with green land plants, and the flagellar features can be recognised even in the multiflagellate sperm of ferns and cycads. These ultra-structural studies confirm that charophytes like *Chara* and *Nitella* are near the line of evolution to land plants, but they will not be given further attention here. Several species of filamentous green algae from several former taxa have been found to have open mitoses and often phragmoplasts at division, and to possess asymmetrical flagellate cells with an MLS. In addition, the freshwater, haploid, conjugating algae like the familiar filamentous *Spirogyra* (Fig. 5.30d) and the unicellular and filamentous desmids (e.g. *Closterium*, Fig. 5.30a) also show an open mitosis and persistent spindle of the phragmoplast type, and are thought to stand near this line of evolution, but the absence of flagellate zoospores and the participation of amoeboid gametes in conjugation sets these forms apart, so they should probably remain as a separate class, the Conjugatophyceae. Desmids (Brook, 1981) are common in freshwater and typically have a bila-terally symmetrical body with a chloroplast in each half and a nucleus between them; separation of the two parts at division results in rapid growth of the missing half of the body and of its (largely) cellulose wall (Fig. 5.30b).

Haptophytes (prymnesiophytes) are best known as planktonic biflagellate cells, but there are also coccoid, amoeboid, colonial and filamentous forms, usually with flagellate stages (Hibberd, 1980c; Hibberd and Leedale, 1985). Formerly included with the chrysophytes, the group is characterised by the presence of two more or less equal, homodynamic, smooth flagella, one on either side of a haptonema (p. 33) of variable length (Fig. 5.31a,d), but reduced to a stump in *Isochrysis* (Fig. 5.31e). The body is covered by one or more layers of organic scales, made in vesicles of a large Golgi body, and sometimes there is also a layer of coccoliths (calcified scales) outside them (in coccolithophorids) (Fig. 5.31b,c). A system of vesicular spaces underlies the plasmalemma in many species, and some have mucocysts. Mitochondria, storage bodies and chloro-plasts are like those of chrysophytes except for the absence of a girdle lamella and usually of an eyespot; mitosis is open. The flagella emerge at an angle and have a short stellate region; no rhizoplast has been found, and various patterns of microtubular root have been reported, often in extensive sheets that extend into the cell. A variety of life cycles occurs (Hibberd, 1980). Some species of *Chrysochromulina* (Fig. 5.31a) show an alternation of flagellate and palmellate amoeboid phases. In *Pleurochrysis* a benthic haploid filamentous form may produce naked flagellate zoospores that settle to form a new benthic filament, or may release naked flagellate isogametes that fuse in pairs and secrete a thin zygocyst wall; within the cyst the zygote develops coccoliths and hatches as a diploid pelagic flagellate stage; this reproduces in the plankton and eventually encysts once more to undergo meiosis in the production of four naked zoospores that settle to form a new haploid benthic filament

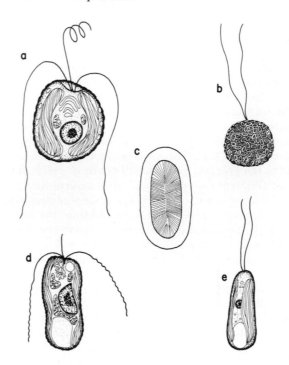

Fig. 5.31 Some representative haptophytes; **a**, *Chrysochromulina kappa* (6 μm): **b**, *Cricosphaera carterae* (15 μm): **c**, coccolith scale of *Cricosphaera carterae* (length 2 μm): **d**, *Prymnesium parvum* (6 μm): **e**, *Isochrysis galbana* (5 μm).

(Gayral and Fresnel, 1983). Most of the 450 species are marine and all are auto-trophic, although phagotrophy has been reported. Toxins released by blooms of *Prymnesium parvum* (Fig. 5.31d) have been responsible for fish kills (Shilo, 1967). Fossilised coccoliths form extensive chalk deposits.

The chytrids and blastocladias are small parasitic or saprobic organisms whose asexual zoospores and motile gametes are propelled by a single, naked, posterior flagellum (Fig. 5.28f), and have usually been referred to as flagellate fungi (Sparrow, 1973). These organisms are small, comprising only a coenocytic reproductive body or a reproductive body with tapering rhizoids (Fig. 5.28i), their secreted body wall containing chitin and glucans. They possess mito-chondria with flattened cristae and Golgi bodies, but no plastids, and the mitotic spindle is internal with paired centrioles placed orthogonally outside a closed nucleus in blastocladiids or one with polar fenestrae in chytrids. The uniflagellate zoospores that are released from sporangia have two kinetosomes (one barren) whose relative orientation and fibrillar attachments are variable (Barr, 1981). In chytrids a cluster of membranous organelles, the rumposome, is prominent in the zoospore, and a bundle of microtubules passes to it from the active kinetosome, while in blastocadias root fibres of different types link the active kinetosome with the nucleus. Sexual reproduction may involve isoga-

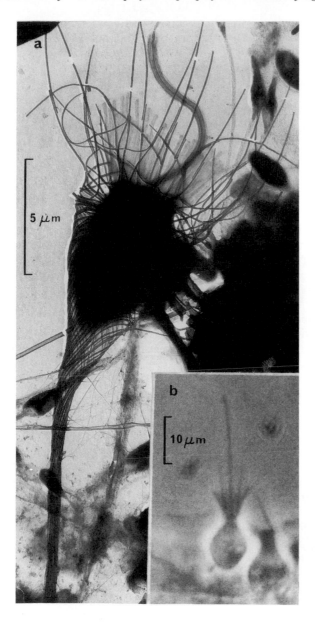

Fig. 5.32   a, Electron micrograph of the choanoflagellate *Acanthoeca spectabilis* from marine plankton; the dense cell body at the centre bears the flagellum and the collar filaments at its apex and is enclosed by a lorica formed from secreted silica bars; micrograph by B.S.C. Leadbeater (1972). **b**, Photomicrograph of *Salpingoeca*, sp. from fresh water.

mous or anisogamous flagellate gametes like the zoospores, or may be oogamous with a flagellate male gamete; the main body is haploid in chytrids, while blastocladias probably have alternate haploid and diploid phases. The zygote may germinate directly or form a resting spore. There are about 900 species saprobic on dead vegetation or parasitic on plants (e.g. *Synchytrium* on potatoes), on planktonic algae, (e.g. *Rhizophydium*) and occasionally on animals (e.g. *Coelomyces* on mosquitoes). It has recently been recognised that the multiflagellated *Callimastix* (Fig. 5.28j), found in the rumen of cattle, is the zoospore of a chytrid that grows on vegetation in anaerobic conditions. Chytrids may be the closest protist relatives of higher fungi.

The collar flagellates (Choanoflagellata) (Leadbeater, 1985) are recognised by the possession of a collar of fine pseudopods around the base of the single flagellum (Fig. 5.25h,i); the flagellum pushes water away from the cell (Fig. 5.6) and draws water through the filtering collar for feeding, for these are phagotrophs. In *Codosiga* the flagellum has two vanes of slender filaments, a transitional zone with neither helical nor stellate pattern, and a basal body associated with a perpendicular barren basal body and surrounded by bands of radiating microtubules that extend out to and run under the surface membrane, but there is no striated root (Hibberd, 1975). Within the cell a Golgi body lies in front of the central nucleus and mitochondria with plate-like cristae surround these; food vacuoles and (in freshwater species) contractile vacuoles are found near the posterior end of the cell. They may secrete a complex lorica of siliceous bars (Fig. 5.32) in some planktonic forms and a membranous sheath and often a stalk are secreted by others. There is no information on nuclear division or sex. In the life cycle of *Proterospongia* a sedentary solitary stage alternates with a motile stage in which many cells are arranged in a single layered colony (Leadbeater, 1983). About 150 species of marine and freshwater forms are known; colonial collar flagellates are believed to have been the ancestors of sponges.

The origins of the red algae, Rhodophyta, are obscure. They appear primitive, but rRNA sequencing studies suggest that they are not far removed from green algae, so are mentioned here for comparison with other autotrophic protists. The group includes some unicellular coccoid forms, though most are filamentous, parenchymatous or pseudoparenchymatous (Dixon, 1973). They are distinctive in the structure and pigmentation of the chloroplasts (p. 101) and the absence of either centrioles or flagella. They have polysaccharide (xylan, mannan) cell walls and their cells contain Golgi bodies, grains of floridian starch and mitochondria with flattened, plate-like cristae. Polar openings of the nucleus occur in mitosis, and the life cycles are often complex, involving haploid and diploid stages, asexual spores and unusual sexual cells. *Porphyridium* is a unicellular example (Dodge, 1973). Most of the 5000 species are marine.

Relationships among flagellated protists and non-flagellated algae

Most major characteristics that can be used to infer relationships between the groups considered in this chapter are listed in Table 5.1. Features of the nucleus and its division, and of the structure of the flagellum and its roots, as well as data derived from nucleotide sequences in rRNA or amino acid sequences in conserved proteins (p. 2), may be regarded as of primary importance; features of the mitochondria and chloroplasts, organelles acquired secondarily by symbiosis, should be treated more cautiously when considering relationships, for two organisms showing similar organelles may have inherited them from a common ancestor or could have acquired those symbionts independently. Consideration of these features leads to a recognition of clusters of groups, and some isolated groups, and allows the construction of a tentative phylogenetic tree.

In recent years attention has been focussed upon the possible value of detailed comparisons of the morphology of flagellar roots in studies of flagellate phylogeny. Such comparisons have proved useful in chlorophytes (Melkonian, 1984), and the impression given by the studies completed to date suggests that root structures may be well conserved and may provide support for a phylogenetic scheme relating most or all of the flagellate groups. However, accurate, detailed information is essential, and there are so far few adequate studies in some groups, so preliminary conclusions must be treated very cautiously, but should stimulate further research. Already it is possible to infer relationships between some of the isolated groups and larger clusters.

Our lack of the necessary detailed information makes proper comparisons of root patterns rather patchy, but several descriptions of 'absolute configurations' of flagellar roots have been published recently, and these, together with reconstructions based upon less detailed data, allow the construction of diagrams of comparative root patterns like those given in Fig. 5.33. The original figures often show three-dimensional reconstructions, but these have diverse orientations and, although they may be more complete, they do not allow such simple comparisons. Flagellar basal bodies normally occur in pairs (or fours); the members of a pair have various relative orientations, but this seems to be a 'plastic' feature, much more variable than the disposition of flagellar roots around the individual basal bodies. Thus, among heterokont flagellates there are forms with parallel basal bodies (Synurophyceae), perpendicular basal bodies (Chrysophyceae, Xanthophyceae) and with antiparallel basal bodies (oomycete zoospores), yet similar root patterns have been recognised in all of them (Andersen, 1987). Diagrammatic comparison is facilitated if the arrangement of the origins of flagellar roots is portrayed around the basal bodies as if these were parallel (Sleigh, 1988); to make this possible one basal body of a pair may need to be rotated with its attached roots (but not twisted) so that it comes to lie parallel to the other (Fig. 5.33β). Any transverse section of a basal body may be viewed from above or below, and published micrographs vary in this orientation; the diagrams here are believed to be drawn as seen from the flagellar tip, so that the imbrication of triplets in the basal body should always appear as in Fig. 5.33δ. It is not always possible to be certain of the orientation of basal bodies seen in published micrographs, so that some root patterns shown in Fig. 5.33 require confirmation before too much weight is placed on any conclusions. It should be recognised that Fig. 5.33 shows only examples that may represent the various groups, and that different members within several groups are known not to be identical in all features.

Some information on root patterns has been given earlier in this chapter, but discussion here will be based on the examples shown in Fig. 5.33 and derived from references listed in the legend; greater emphasis will be placed upon

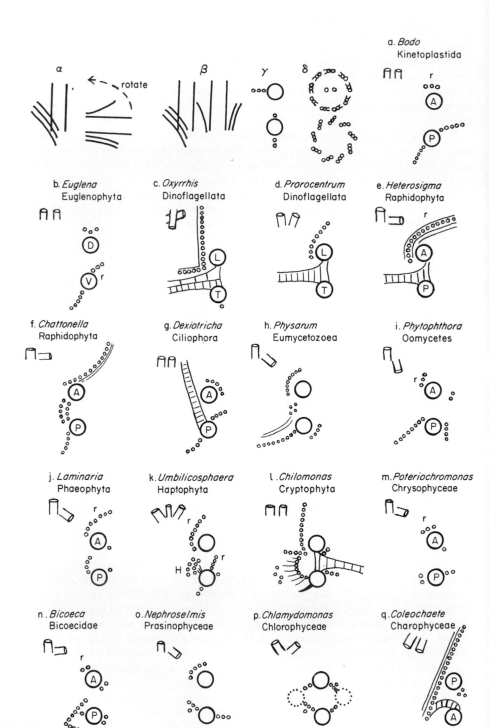

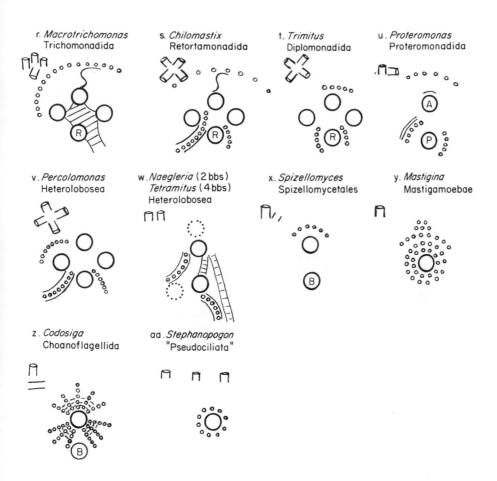

Fig. 5.33 A comparison of the patterns formed by flagellar root fibres in examples from different groups. The root maps are prepared as shown in α–γ and described in the text; all maps are drawn as viewed from the tip of the flagellum, i.e. with orientations of the axonemal doublets and basal body triplets as shown in δ. The maps are derived from micrographs and diagrams in the following sources; **a**, *Bodo*, Brugerolle *et al.* (1979): **b**, *Euglena*, Surek and Melkonian (1986): **c**, *Oxyrrhis*, Roberts (1985): **d**, *Prorocentrum*, Roberts (pers. comm.): **e**, *Heterosigma*, Vesk and Moestrup (1987): **f**, *Chattonella*, Mignot (1976): **g**, *Dexiotricha*, Peck (1977): **h**, *Physarum*, Wright *et al.* (1979): **i**, *Phytophthora*, Hardham (1987): **j**, *Laminaria*, O'Kelly and Floyd (1984): **k**, *Umbilicosphaera*, Inouye and Pienaar (1984): **l**, *Chilomonas*, Roberts *et al.* (1981): **m**, *Poteriochromonas*, Schnepf *et al*, (1977): **n**, *Bicoeca*, Moestrup and Thomsen (1976): **o**, *Nephroselmis*, Moestrup and Ettl (1979): **p**, *Chlamydomonas*, Melkonian (1977): **q**, *Coleochaete*, Sluiman (1983): **r**, *Macrotrichomonas*, Hollande and Valentin (1969): **s**, *Chilomastix*, Brugerolle (1973): **t**, *Trimitus*, Brugerolle (1986): **u**, *Proteromonas*, Brugerolle and Joyon (1975): **v**, *Percolomonas*, Fenchel and Patterson (1986): **w**, *Naegleria*, Dingle and Fulton (1966) with two basal bodies and *Tetramitus*, Balamuth *et al.* (1983) with four basal bodies: **x**, *Spizellomyces*, Barr (1980): **y**, *Mastigina*, Brugerolle (1982): **z**, *Codosiga*, Hibberd (1975): **aa**, *Stephanopogon*, Patterson (pers comm.). Key to lettering on basal bodies; *A*, anterior: *B*, barren: *D*, dorsal: *F*, flagellated: *L*, longitudinal: *P*, posterior: *R*, recurrent: *T*, transverse and *V*, ventral; *H*, leads to a haptonema, and *r*, indicates that the root marked is ribbed (*see* text). Modified from Sleigh (1988). The relative orientations of the two more basal bodies are shown at the top of each map.

microtubular roots than fibrous roots. Such fibrous roots, usually with prominent cross striations, are present in many groups, most commonly extending from beneath the basal bodies towards the nucleus, and often in company with microtubules. Because of their position these roots are difficult to portray on root maps; they are also erratic in their occurrence or detection, for some roots that have been revealed by fluorescent antibodies to root proteins have been difficult to locate by electron microscopy. The striated roots shown in Fig. 5.33 are those that are lateral in position and are prominent and typically present in the examples shown; these striated roots do not include fibres that only link the basal bodies with the nucleus, for separate fibres of this type exist in these and other cases.

The surface membrane of most flagellates is underlain by cytoskeletal microtubules in bands or sheets. Some of these cortical microtubules originate from near the basal bodies in most flagellates, and often a ribbon of microtubules which runs beneath the plasmalemma gives rise to an array of lateral microtubules (i.e. is 'ribbed') to provide the main cytoskeletal system of the cell cortex. The ribbon of microtubules that arises at the dorsal or (organism's) right side of the dorsal or anterior basal body is usually the group that is ribbed in this way (Fig. 5.33), although sometimes a left, middle/ventral group may also be ribbed, and often other unribbed ribbons may join the cortical array. Where the dorsal ribbon appears not to be ribbed, it may be a rather broader band of microtubules, as in *Percolomonas*, *Chilomonas*, *Oxyrrhis* and perhaps *Coleochaete*, although only in the first of these cases does it appear to occupy a cortical position, and in that case it may, in fact, be ribbed; the other large ribbons may carry additional layers of material, forming more complex roots like the multi-layered structure (MLS) of some green algae and land plants. Where there is no dorsal microtubule ribbon, the dorsal basal body is linked by a fibrous strand to the cortical array of microtubules, as in *Chilomastix* and probably in *Proteromonas*, and to the equivalent endoplasmic axostyle/pelta system in trichomonads. In some flagellates with a permanent cytostome, the right and left ribbons on the posterior (ventral/recurrent) basal body usually underlie the lateral lips of a ventral cytostome posterior to the flagellar bases (e.g. in *Chilomastix*, *Trimitus*, *Tetramitus* and *Percolomonas*), but in *Bodo* and *Euglena* the ventral right ribbon underlies the wall of the tubular cytostome, while the ventral left ribbon joins the cortical array, being ribbed in *Euglena* and lining a groove for the posterior flagellum in *Bodo*. In some flagellates without permanent cytostomes (e.g. *Poteriochromonas*, *Phytophthora*, *Umbilicosphaera*), the ventral ribbons of microtubules often form a loop behind the posterior flagellar base as if a cytostome were present. Microtubule ribbons still occupy sub-plasmalemmal positions in forms with cell walls (e.g. *Chlamydomonas*). Groups of microtubules found at the right side between the two basal bodies, and often only loosely connected to the posterior basal body by a filamentous link, generally run up and/or down perpendicular to the surface, some often extending down to or even beyond the nucleus, and the same microtubules or others of the group may extend up to the cell surface, and in the case of haptophytes they protrude at the cell surface as the core of the haptonema. In many flagellates the downwardly-running microtubules, most of which link to larger cell organelles, especially the nucleus, are rather randomly arranged or poorly described and have often been omitted from the maps shown in

Fig. 5.33; the microtubules attached to basal bodies in chytrids appear to be of this type, and all microtubules attached to the basal body in **Mastigina** run to or around the nucleus except for a single lateral bundle. The microtubules attached to basal bodies of **Stephanopogon** and **Codosiga** extend towards the cell surface, but do not appear in distinct bundles.

The diversity of examples shown in Fig. 5.33 suggests a number of clusters of protists. The heterokonts form one group, including mycetozoa, and show some similarities with haptophytes and **Nephroselmis** (which could lead to other chlorophytes). Cryptophytes share features with haptophytes and raphido-phytes share features with dinoflagellates. **Bodo** and **Euglena** are very similar and share some features with ciliates, heterolobosea and even metamonads. Metamonads, heterolobosea and proteromonads show similarities, with tricho-monads rather less similar. Chytrids, **Mastigina**, **Codosiga** and **Stephanopogon** at present seem rather isolated.

These root patterns, together with flagellar, nuclear, mitochondrial and plastid characters and data from rRNA sequencing, were used to draft the speculative phylogenetic scheme shown in Fig. 5.34. Patterns of mitochondrial cristae appear to be consistent with the rRNA sequence data, so the protistan taxa were grouped according to the presence or absence of mitochondria and the type of cristae. In view of the key role given to types of mitochondrial cristae, it should be pointed out that it is not always possible to classify mitochondrial cristae into standard categories with confidence, and the support given by rRNA studies is very reassuring. The origins of groups from ancestral protists are also related to types of root patterns by suggesting that groups with similar root patterns originated in close proximity to one another, even if they came from different mitochondrial 'fields'.

A large cluster is based upon the heterokonts that possess not only the typical heterodynamic flagella, usually with tripartite mastigonemes and a helical fibre in the transition zone, but also a set of 4–6 microtubular roots, mitochondria with tubular cristae and usually plastids with all three of chlorophylls a, c_1 and c_2. Near this cluster one might tentatively place the foraminiferids (see p. 168) and possibly the sporozoa (p. 222). Ciliates and dinoflagellates, whose rRNA sequences suggest a near relationship, share the possession of tubular cristae with heterokonts and have a comparable pellicle organisation and closed mitotic spindle, but ciliates are diploid with internal spindles and dinoflagellates are haploid with external spindles. The proteromonads and opalinids also have tubular cristae and a helical fibre in the transition zone, and may lie between heterokonts and a cluster containing heterolobosea, kinetoplastids and eugle-noids. Euglenoids and kinetoplastids share features of flagella and roots as well as internal spindles and discoidal cristae; heterolobosea also have discoidal cristae, internal spindles and somewhat similar flagellar roots, but lack stacked Golgi, which may suggest that they are earlier offshoots from the same ancestry. **Stephanopogon**, which is difficult to relate to any other group on the basis of its flagellar roots, also has discoidal cristae and so occupies an uncertain position near euglenoids. Metamonads lack mitochondria and Golgi bodies and usually have internal spindles; they appear to be a very ancient group with a characte-ristic root array, traces of which are found in parabasalians (also without mitochondria, but with very well-developed Golgi and extranuclear spin-dles), heterolobosea and perhaps proteromonads. Haptophytes often have

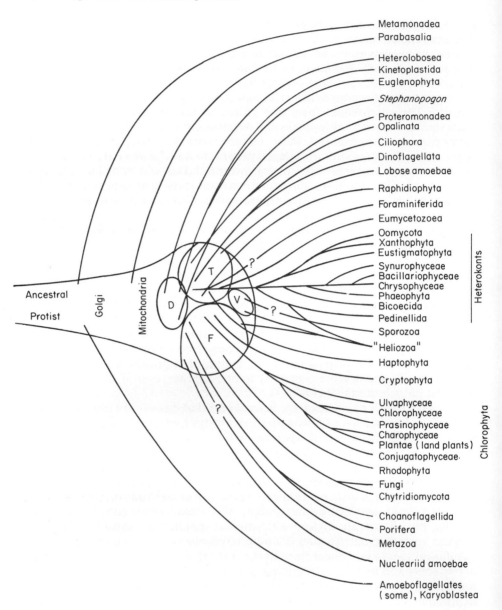

Fig. 5.34   A speculative evolutionary scheme of flagellate groups and some related protists. It is assumed that some groups evolved from ancestral eukaryotes before mitochondrial symbionts were acquired (see also Fig. 1.1), and that different forms of mitochondrial cristae appeared in different sectors of the ancestral radiation; forms originating from one sector may have similar cellular (e.g. flagellar root) features to forms originating nearby but in a different sector (see text). *D*, cristae discoidal: *F*, cristae flattened: *T*, cristae tubular and *V*, cristae vesicular.

homodynamic flagella, with a stellate pattern in the transition zone in at least one example, and in these ways resemble chlorophytes, but they have tubular cristae, a flagellar root pattern and plastids that place them closer to heterokonts. The cryptophyte flagellar root map is comparable with that of haptophytes, but cryptophytes appear an isolated group with unique plastids in eukaryotic symbionts and flat cristae. Chlorophytes form a rather compact cluster, possessing homodynamic flagella, a stellate structure in the transition zone and flat cristae, but their root pattern suggests an origin not far from the heterokont cluster, in spite of the presence of different plastids and mitochondria. Analysis of rRNA sequences and presence of flat cristae suggests that red algae and fungi are fairly closely related to green algae and land plants. If (true) fungi evolved from chytrids, then some evidence of flagellar pattern is available, but is unhelpful except in suggesting a possible origin fairly close to collar flagellates among forms with flat cristae. Red algae share some features with conjugating green algae, but appear rather isolated; the very close similarity of their plastids to cyanobacteria suggests a rather recent symbiosis, and perhaps a relatively recent origin for the group, which is consistent with most rRNA data but not with earlier ideas which suggested that they were among the oldest eukaryotes. Little evidence connects metazoa with any of these other groups with flat cristae, except for rRNA sequence data which places them closest to green algae and land plants. *Mastigina* lacks both mitochondria and Golgi bodies and, like *Pelomyxa* and some other amoeboid forms, may have diverged early like metamonads. Other amoeboid groups are shown in Fig. 5.34 to accord with the type of cristae and supposed relationships with flagellate groups.

In addition to groups included here there are numerous single genera (including *Apusomonas* (Vickerman *et al.*, 1974), *Colponema* (Mignot and Brugerolle, 1975a), *Cercomonas* (Mignot and Brugerolle, 1975b), *Cyathobodo* and *Pseudodendromonas* (Hibberd, 1976), *Reckertia* and the cyanelle-containing genera *Cyanophora*, *Glaucocystis* and *Gloeochaete* (see Moestrup, 1982) and *Phalansterium* and *Spongomonas* (Hibberd, 1983), as well as other protists with flagellate stages, that must in time be fitted into the picture.

Divergent views have arisen among taxonomists working on protists about the taxonomic rank to be assigned to the groups considered here. Many protistologists have been led by the recognition of fundamental differences in the nuclear and cytoplasmic organisation of many of the groups to regard such groups as separate phyla – perhaps as many as 30 among flagellated forms. However, the discussion above picks out a few clusters of groups (as well as some isolated groups) which it may be preferable to regard as phyla, while constituent groups within these clusters may be regarded as classes. For example a phylum Heterokonta could comprise about 10 classes, and a phylum Chlorophyta might contain about five classes, while the Dinoflagellata and Cryptophyta would presumably constitute independent phyla. Other people might prefer to regard each of these clusters as kingdoms in the confederation of protists, but this may seem unnecessarily inflationary? This is not regarded as a suitable place to formally propose a reclassification of flagellate groups, but a summary list of the major groups is given in Table 5.2 with examples used in this book.

Table 5.2   A list of the groups mentioned in this chapter, with examples.

Phylum Dinophyta: about 2000 living species in eight orders:
  Order Prorocentrales, e.g. *Prorocentrum, Exuviella*
  Order Dinophysiales, e.g. *Ornithocercus, Dinophysis*
  Order Gymnodiniales, e.g. *Amphidinium, Gymnodinium, Oxyrrhis, Polykrikos*
  Order Noctilucales, e.g. *Noctiluca, Cymbodinium, Craspedotella*
  Order Peridiniales, e.g. *Ceratium, Peridinium, Heterocapsa, Gonyaulax, Crypthecodinium*
  Order Blastodiniales, e.g. *Blastodinium, Haplozoon*
  Order Syndiniales, e.g. *Syndinium*
  Order Phytodiniales, e.g. *Dinothrix*

The Orders Ebriida (e.g. *Hermesinum, Ebria*) and Ellobiopsida (e.g. *Ellobiopsis*) belong here.

Phylum Parabasalia: about 550 species in two orders:
  Order Trichomonadida, e.g. *Monocercomonas, Trichomonas, Histomonas, Dientamoeba,*
                  Devescovina, Metacoronympha
  Order Hypermastigida, e.g. *Lophomonas, Joenia, Trichonympha, Barbulanympha,*
                  Holomastigotoides

Phylum Metamonada: about 200 species in two classes and three orders:
  Class Anaxostylea
    Order Retortamonadida, e.g. *Chilomastix, Retortamonas*
    Order Diplomonadida, e.g. *Enteromonas, Trepomonas, Hexamita, Giardia*
  Class Axostylea
    Order Oxymonadida, e.g. *Oxymonas, Saccinobaculus, Pyrsonympha*

Phylum Kinetoplasta: about 600 species in two orders:
  Order Bodonida, e.g. *Bodo, Cephalothamnium, Cryptobia, Rhynchomonas*
  Order Trypanosomatida, e.g. *Trypanosoma, Crithidia, Phytomonas*

Phylum Euglenophyta: about 1000 species in six orders:
  Order Eutreptiales, e.g. *Eutreptia, Distigma*
  Order Euglenales, e.g. *Euglena, Astasia, Trachelomonas, Colacium, Phacus*
  Order Rhabdomonadales, e.g. *Rhabdomonas, Menoidium*
  Order Heteronematales, e.g. *Peranema, Entosiphon*
  Order Sphenomonadales, e.g. *Petalomonas, Notosolenus*
  Order Euglenomorphales, e.g. *Euglenomorpha*

The Genus *Stephanopogon* – a single isolated genus at present – belongs near here

Phylum Cryptophyta: about 200 species in two orders; flagellates are in the
  Order Cryptomonadales, e.g. *Cryptomonas, Chilomonas, Hemiselmis*

Phylum Opalinata (Slopalinata): about 400 species in a single order:
  Order Opalinida, e.g. *Opalina, Cepedia, Zelleriella*

The Order Proteromonadida, an isolated order at present, with about six species, e.g.
Proteromonas, Karotomorpha, belongs near here

Phylum Heterokonta: about 10 classes, most regarded as phyla by Corliss (1984):
  Class Chrysophyceae: about 600 living species in eight orders, including:
    Order Ochromonadales, e.g. *Ochromonas, Dinobryon, Anthophysa*
    Order Chrysamoebidales, e.g. *Chrysamoeba*
    Order Pedinellales, e.g. *Pedinella*
    Order Dictyochales, e.g. *Dictyocha*
  Class Synurophyceae: a single order
    Order Synurales, e.g. *Mallomonas, Synura*
  Class Bacillariophyceae: about 10 000 living species in two orders
    Order Biddulphiales (Centrales), e.g. *Coscinodiscus, Rhizosolenia*
    Order Bacillariales (Pennales), e.g. *Nitzschia, Navicula, Bacillaria*

Table 5.2 cont.

Class Phaeophycea: about 1600 species of brown algae, not described here
Class Xanthophyceae (Tribophyceae): about 650 species in five orders
   Order Tribonematales, filamentous, e.g. *Tribonema*
   Order Vaucheriales, siphonous, e.g. *Vaucheria*
   Order Mischococcales, coccoid, e.g. *Ophiocytium, Botrydiopsis*
   Order Chlamydomyxales, amoeboid and palmelloid
   Order Heterochloridales, flagellate, e.g. *Olisthodiscus, Heterochloris*
Class Eustigmatophyceae: about 10 species in one order
   Order Eustigmatales, e.g. *Monodus, Ellipsoidion, Vischeria*
Class Oomycotea, about 800 species in four orders, e.g. *Saprolegnia, Leptomitus,*
  Lagenidium, Peronospora
Class Hyphochytridiomycotea, about 25 species, e.g. *Hyphochytrium*
Class Raphidophyceae: about 25 species in one order:
   Order Vacuolariales, e.g. *Vacuolaria, Chattonella, Gonyostomum*
Class Labyrinthulea: probably only one genus, *Labyrinthula*
Class Thraustochytridea: several genera, e.g. *Thraustochytrium*

The Order Bicoecida (Bicosoecida), with 30–40 species, e.g. *Bicoeca*, probably belongs in
this phylum, along with a number of genera of helioflagellates, e.g. *Actinomonas,*
Pteridomonas, Ciliophrys.

Phylum Chlorophyta: about 8000 living species in six classes:
  Class Prasinophyceae: orders include:
   Order Pedinomonadales, e.g. *Pedinomonas*
   Order Halosphaerales, e.g. *Halosphaera*
   Order Pyramimonadales, e.g. *Pyramimonas, Tetraselmis*
  Class Chlorophyceae: orders include:
   Order Volvocales, e.g. *Chlamydomonas, Volvox, Polytoma, Carteria*
   Order Chlorococcales, e.g. *Scenedesmus, Pediastrum, Chlorella*
  Class Oedogoniophyceae: Order Oedogoniales, e.g. *Oedogonium*
  Class Ulvaphyceae: orders include:
   Order Cladophorales, e.g. *Cladophora*
   Order Codiales, e.g. *Codium*
  Class Conjugatophyceae: large class, about 5000 desmid species, orders include:
   Order Desmidiales, e.g. *Closterium, Cosmarium, Micrasterias*
   Order Zygnematales, e.g. *Spirogyra*
  Class Charophyceae: Order Charales, e.g. *Nitella, Chara*

Phylum Haptophyta (Prymnesiophyta): about 450 living species in one class:
  Class Prymnesiophyceae: with four orders
   Order Isochrysidiales, e.g. *Isochrysis*
   Order Coccosphaerales, e.g. *Coccolithus, Hymenomonas*
   Order Prymnesiales, e.g. *Prymnesium, Chrysochromulina, Phaeocystis*
   Order Pavlovales, e.g. *Pavlova*

Phylum Chytridiomycota (Chytridiomycetes): about 900 species in four orders:
  Order Blastocladiales, e.g. *Blastocladiella*
  Order Monoblepharidiales, e.g. *Monoblepharella*
  Order Chytridiales, e.g. *Rhizophydium, Chytridium, Synchytrium*
  Order Spizellomycetales, e.g. *Spizellomyces, Neocallimastix, Olpidium*

Phylum Choanoflagellata: about 140 species in three families, e.g. *Codosiga, Salpingoeca,*
Acanthoeca, Proterospongia

Phylum Rhodophyta: about 5000 species of red algae in two classes:
  Class Bangiophyceae includes simpler forms e.g. unicellular *Porphyridium*
  Class Floridiophyceae includes more complex forms.

Note added in proof.

A new phylum of protists, the Hemimastigophora, represented at present by a single species of **Hemimastix,** has recently been described from soils in Australia and Chile by Foissner, Blatterer and Foissner (1988). It has two opposite kineties of about 12 simple flagella, single vesicular nucleus whose nucleolus persists through mitosis, tubular mitochondrial cristae, complex extrusomes and a pellicle with epiplasm underlain by microtubules, but no permanent cytostome. For the present, at least, it is probably best considered like **Stephanopogon** as a representative of a separate group.

Reference: Foissner, W., Blatterer, H. and Foissner, I. (1988). The Hemimastigophora (**Hemimastix amphikineta** nov. gen., nov. spec.), a new protistan phylum from Gondwanian soils. *European Journal of Protistology,* **23,** 346–83.

6
Amoeboid protists

The amoeboid protists are characterised by the possession of pseudopodia. They have traditionally been distinguished from one another by the form of the pseudopods, the presence of skeletons, shells and tests that determine body shapes and by features of the nucleus and cell organelles visible in the light microscope. It was suggested in Chapter 1 that the original protists were amoeboid, and that some protists developed flagella, most acquired mitochondrial symbionts and some plastid symbionts. Some groups whose main life form is amoeboid have flagellated gametes or distributive stages, and must therefore have been derived from forms that had developed flagella; other amoeboid forms may have evolved from pre-flagellate or non-flagellate stock, or have lost flagella. When we know enough about them, the flagellated forms will be integrated into the evolutionary radiation discussed in Chapter 5, but in the absence of detailed information about their flagellar characters they are best considered on the basis of their more visible amoeboid characters. Non-flagellate forms must form part of a radiation parallel to that of flagellate forms.

The impact of ultrastructural studies on understanding of the relationships among flagellates was clearly evident in Chapter 5. Such studies on amoebae are proving equally important. However, fewer structural characters that can be unequivocally recognised have yet been found in non-flagellate forms, and biochemical characters will probably have to be used to support phylogenetic conclusions in some cases. It is also true that fewer amoeboid protists have been adequately studied than flagellates, and we are at an earlier stage in understanding how most amoeboid groups fit into the protistan evolutionary scheme. Much of what can be said here about relationships will therefore be even more tentative than suggestions made in Chapter 5, although much interesting evidence has been published. Present knowledge suggests that the amoeboid protists, like the flagellates, comprise a number of evolutionary lines that may be only distantly related to one another, and may have arisen from different sectors of the protistan evolutionary tree. It is convenient to consider them together, but essential to be aware that their origins are almost certainly polyphyletic.

The most obvious feature about most living amoeboid protists is their pseudopodia, either because of their motion, or their extent, or both. Four basic types of pseudopodial structure are generally recognised, as illustrated in Fig. 6.1. The most familiar are the lobopodia, a variable category including cylindrical and flattened types and typified by *Amoeba*, filopodia, which are hyaline, filiform and often branched, reticulopodia, which form anastomosing networks, and axopodia which are supported by a structured array of axial

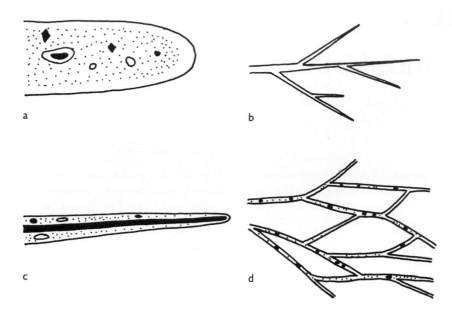

Fig. 6.1   Forms of pseudopodia found in amoeboid protozoa; **a**, a lobopodium: **b**, filopodia: **c**, an axopodium: **d**, reticulopodia.

microtubules. The first three are varieties of rhizopodia, and the amoebae that possess them are called rhizopods, while those with axopodia are actinopods. The lobopodial forms include the syncytial slime moulds in which the growing margins fan out as broad lobopods but the older parts of the plasmodium form tubular networks. The groups of amoebae will be considered according to the character of their pseudopods, and features of the structure, movement and function of different pseudopod types will be dealt with in the relevant section. The classification used will follow that of Levine *et al.* (1980), except where amendment seems necessary because of new information, such as that provided for 'naked' amoebae by Page (1987).

Features of amoeboid groups

Amoebae with lobopodia

The tubular lobopodia of *Amoeba*, with flowing axial streams of granular endoplasm within ectoplasmic tubes, have been the subject of much research, and the mechanism of their movement has been the subject of long controversy. Observations by Mast (1926) led him to conclude that the streaming endoplasm was propelled by contraction of the whole ectoplasmic tube that surrounded it. Attempts to test this view experimentally led Goldacre and Lorch (1950), and more recently Wehland *et al.* (1979), among others, to conclude that only the rear ectoplasm was actively contractile, supporting a tail contraction theory of amoeboid movement (Fig. 6.2a). Other evidence from flow patterns and the

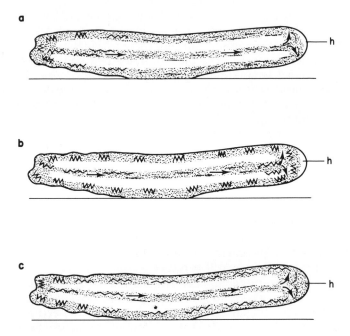

Fig. 6.2 Three explanations for the observed motion of the lobopodia of ***Amoeba***. Observations show that endoplasm streams forward along the axis of the lobopod and transforms to ectoplasm at the advancing tip, while gelated ectoplasm is solated at the tail as it is transformed again to endoplasm. The tail contraction model a suggests that actomyosin in the tail ectoplasm contracts and pushes endoplasm forward through a tube of relaxed ectoplasm by pressure exerted only at the tail. The fountain zone theory **b**, suggests that contraction of actomyosin as endoplasm moves to the surface to form ectoplasm at the anterior end draws forward a core of gelated endoplasm through a tube of ectoplasm. The cortical contraction model **c**, suggests that the ectoplasm anteriorly is contracting isometrically, maintaining tension on the endoplasm, while isotonic contraction of the posterior ectoplasm is putting increasing pressure on the tail endoplasm to squeeze it forwards. *h* = Hyaline cap. (See Grebecki, 1982.)

contractility of isolated amoebal cytoplasm led R.D. Allen (1961, 1973) to propose a fountain zone theory, whereby contraction at the advancing tips of pseudopods draws the endoplasm forwards (Fig. 6.2b). Completing the circle of argument, A. Grebecki (1982, 1986a) concludes from a new examination of all the data that Mast was essentially correct, and that the whole cell cortex provides the motive force for movement by exerting contractile pressure on the endoplasm, while control of locomotion is exerted at the pseudopodial tips (Fig. 6.2c). This is a satisfying conclusion because all parts of the ectoplasm can be shown to contract during locomotion (Grebecki, 1982), and the pseudopodial tips are the behaviourally responsive sites to, for example, light stimuli (Grebecki and Klopocka, 1981).

The cell cortex of *Amoeba* contains actin microfilaments, which run parallel to the plasma membrane and to the longitudinal axis in the anterior regions, and are best developed in the lateral walls, but form an unoriented network in the rear part (Stockem *et al.*, 1982); filaments are not seen in the endoplasm, where actin is assumed to be depolymerised or in small oligomers. Thick filaments of myosin are only found in the uroid region at the posterior end, although myosin is present throughout the length of the cell; there is evidence that myosin need not be highly polymerised to take part in amoeboid contraction. Grebecki concluded that isometric contraction of the ectoplasm occurs in the antero-lateral parts and isotonic contraction posteriorly. The uroidal contracting cytoplasm solates, its actin filaments are shortened (or depolymerised) and pass forward in the fluid endoplasm. Anteriorly the endoplasm comes to the surface at the pseudopod tip and gelates as actin filaments reanneal (or repolymerise) to form a longitudinal network interacting with myosin and with actin-binding proteins, some of which are linked to the plasma membrane, to form a contractile cytoskeleton, whose contraction requires MgATP and is regulated by Ca^{2+} concentration (Taylor and Condeelis, 1979). The increase in internal hydrostatic pressure produced by this contraction results in cytoplasmic flow towards any area where the endoplasm can extend, i.e. the thin-walled tips of pseudopods or points of weakness (or relaxation) such as can be created experimentally in pseudopod walls by light, chemical, or mechanical means. A combination of contraction and tip relaxation will produce a fountain-like movement of the cytoplasm unless the anterior ectoplasm is anchored to the substratum, thereby allowing the ectoplasmic shortening and endoplasmic flow to produce forward locomotion (Grebecki, 1986a). Sites of attachment of the ectoplasmic cytoskeleton to the plasma membrane that are bound to anchorage sites on the substratum, or to extracellular particles, move backwards along the body surface as the ectoplasmic tube contracts; at the same time, particles bound to membrane areas not attached to the cytoskeleton move forwards with the extending pseudopod, providing a supply of fluid membrane material for the extending pseudopodium (Grebecki, 1986b).

Some other large lobose amoebae, like *Chaos carolinense*, behave in much the same way as *Amoeba*, but others, and the smaller forms in particular, show varied patterns of locomotion, which can be explained by the same general mechanism of generalised ectoplasmic contraction. Some smaller amoebae, called limax amoebae, are slug-shaped and monopodial and move smoothly like a small version of *Amoeba* (e.g. *Hartmannella*), while other cylindrical types move in a jerky eruptive fashion, presumably because of spasmodic relaxation of the pseudopodial tip (e.g. *Vahlkampfia*, Fig. 6.3f; *Naegleria*). Others are more flattened and produce broad pseudopodia; sometimes their pseudopodial surface carries small sub-pseudopodia that may be digitiform (e.g. *Mayorella*, Fig. 6.3c) or filiform (e.g. *Acanthamoeba*), and outside the membrane of some forms is a multi-layered, flexible, cell coat which is often wrinkled as the amoeba moves (e.g. *Thecamoeba*). *Acanthamoeba* has been used extensively in studies of contractile proteins – actins and two types of myosins (Pollard *et al.*, 1982); it is also interesting because it contains a microtubular cytoskeleton that has not been found in most other lobose amoebae (Preston, 1985). The peculiar large amoeba *Pelomyxa palustris* (Fig. 6.3e) is monopodial, but slow-moving, and often shows a bidirectional fountain flow of cytoplasm – perhaps it rarely forms attachments to the substratum, or perhaps the prominent uroid dominates contraction.

Other basically lobose types move in more specialised ways. The testate amoebae extend rather slender lobopodia from the mouth of the test and use

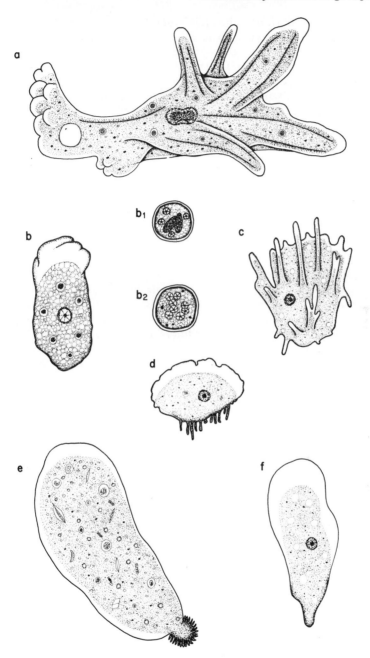

Fig. 6.3 Representative amoebae with lobopodia; **a**, *Amoeba proteus* (500 μm): **b**, Entamoeba histolytica (10–20 μm), with a cyst of the same species at **b₁** (5–20 μm) containing four nuclei and typical 'chromatoid bodies' compared with a cyst of *E. coli* containing eight nuclei at **b₂** (10–30 μm): **c**, *Mayorella bulla* (100 μm): **d**, *Flabellula citata* (20 μm): **e**, *Pelomyxa palustris* (2 mm): **f**, *Vahlkampfia limax* (30 μm).

these to haul the body along. Thus *Difflugia* (Fig. 6.4d), which has a heavy test agglutinated with sand grains, extends a long lobopod ahead of the cell and, after attachment of this lobopod to the substratum, a strong longitudinal bire-fringence due to the presence of thin and thick microfilaments develops between the attachment site and the base of the pseudopod; the lobopod contracts strongly to pull the cell towards the attachment site; another lobopod may then repeat the process (Eckert and McGee–Russell, 1973). Syncytial slime moulds, mentioned above, produce very extensive networks of tubules that branch and rejoin (Fig. 6.6); these tubular elements have gelated ectoplasmic zones around fluid endoplasm which constantly shuttles to-and-fro, often at speeds in excess of 1 mm s^{-1}. The shuttle flow is produced by cyclical contraction and relaxation of the ectoplasm in different parts of the network; actin filaments are associated with invaginations of the surface membrane into the ectoplasmic zone, and these interact with myosin to produce contraction (Wohlfarth–Bottermann, 1974).

Among these lobose amoebae *Pelomyxa palustris* (Fig. 6.3e) seems set apart from other sarcodines by features which have suggested to some that it is the most primitive eukaryote. It is placed in a separate phylum, the Karyoblastea, by Corliss (1984), or in the order Pelobiontida within the class Lobosea by Levine *et al.* (1980). Ultrastructural evidence is scanty (because of sand grains in the cytoplasm – Daniels, 1973); clearly it is multinucleate, but reports of a closed nuclear division lacking spindle microtubules need confirmation; meiosis is unknown. There are no Golgi bodies and no mitochondria, although the cyto-plasm contains at least two types of symbiotic bacteria (p. 282), as well as glycogen granules. Curiously, non-motile cilium-like projections containing microtubules have been seen in electron micrographs. Several authors have suggested a possible relationship of *Pelomyxa* with mastigamoebae – uniflagel-late forms like *Mastigina*, which has a distinctive flagellar structure and lacks both mitochondria and Golgi bodies (Brugerolle, 1982); microtubules arising from dense material beneath the basal body form a conical array that surrounds the nucleus of *Mastigina* (just as in the 'chytrid' *Blastocladiella*, Barr, 1981), and there is a lateral microtubular root, and probably an internal mitotic spindle. *P. palustris* is microaerophilic and algivorous.

Typical amoebae in the class Lobosea contain mitochondria with branching tubular cristae, Golgi bodies with stacked cisternae, and nuclei whose mitosis is open without centrioles, but with centrospheres in some small and testate forms. Naked gymnamoebae in this group include the carnivores *Amoeba*, *Chaos*, *Thecamoeba* and *Mayorella* (some with surface scales), and smaller, generally bacterivorous, forms like *Hartmannella* and *Acanthamoeba*. The pseudopods of testate arcellinids emerge from a single opening of a cell covering that may be rigid and tectinous (*Arcella*, Fig. 6.4a–c) or agglutinated with sand grains (*Difflugia*, Fig. 6.4d) and sometimes diatom frustules (*Centropyxis*, Fig. 6.4e,f); forms with a flexible coat bearing scales (*Cochliopodium* Fig. 6.4g) have been classed here, but some consider they properly belong with naked forms. Testate amoebae are algivorous or omnivorous and are usual found in freshwaters and soil, including organically polluted places. Both the parasitic *Entamoeba* (Fig. 6.3b) and the marine trichosids (that secrete a fibrous test, sometimes with calcareous granules, from whose many apertures lobopodia emerge), show a closed intranuclear orthomitosis. *Entamoeba* lacks mito-

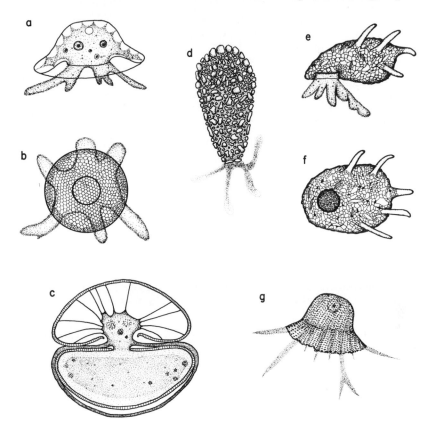

Fig. 6.4 Representative testate amoebae; *a,b,* ***Arcella vulgaris*** (50 μm) showing features of the shell and pseudopodia seen from the side and above, and in **c** the formation of a new shell within a mass of protruded cytoplasm precedes cell division and the separation of two shelled progeny (see Netzel, 1971): **d,** ***Difflugia oblonga*** (250 μm): **e,f,** ***Centropyxis aculeata*** (100 μm) from the side and from below: **g,** ***Cochliopodium bilimbosum*** (40 μm).

chondria and has a simple endomembrane system; it could be a primitive form or it may have suffered reduction because of its parasitic life-style. Two species of **Entamoeba** are common in the human gut, and may be distinguished by the number of nuclei in their cysts (Fig. 6.3b$_1$, b$_2$); *E. histolytica* causes amoebic dysentery and occasionally may ulcerate the large intestine and pass in the portal vein to the liver where they may produce abscesses whose rupture can be fatal. **Acanthamoeba** species also have medical significance as causative agents of chronic meningitis.

The slime moulds, class Eumycetozoea, with about 700 species, produce spores in 'fruiting bodies' at the head of a stalk; when these spores hatch they release small amoebae with filiform sub-pseudopodia and whose mitochondria have tubular cristae. There are two well-known subclasses. In the normal growth cycle of the cellular slime moulds (Dictyosteliia) the bacterivorous amoebae grow and divide repeatedly until some environmental feature causes some amoebae to secrete 'acrasin' (cyclic AMP in **Dictyostelium discoideum**,

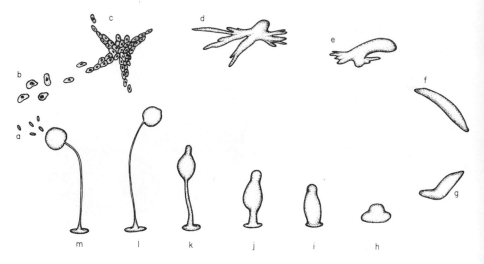

Fig. 6.5 The life cycle of the cellular slime mould *Dictyostelium discoideum* described in the text; the amoebae **b**, are 5–15 μm long and the migratory grex stage **g** 1–2 mm long.

Mato and Konijn, 1979); this stimulates aggregation of the amoeboid cells which stream together to form pseudoplasmodia; after a period of migration as a slug-like grex, cells of the mass (remaining as independent cells) differentiate under the influence of a morphogen (Morris *et al.*, 1987) into erect stalk cells that vacuolate and secrete cellulose and cells that climb to the top of the stalk and there encyst as spores (Fig. 6.5). The encystment of individual amoebae as microcysts occurs under some unfavourable conditions, and under other conditions cellulose-walled macrocysts are formed around giant cells; sexual fusion and meiosis are believed to occur during the formation and subsequent division of the macrocyst (O'Day and Lewis, 1981). There is no flagellate stage, and polar openings occur in mitotic nuclei of these forms. *Dictyostelium* spp. have become important research organisms in studies on differentiation, communication and movement; their cells contain microtubules, like *Acanthamoeba*, and they may be more closely related to such amoebae than to other slime moulds.

Members of the second subclass, the syncytial slime moulds (Myxogastria) have more complex life cycles (Fig. 6.6); haploid amoebae emerge from spores and may multiply as bacterivorous amoebae, but these may fuse in pairs, or may first (if flooded with water) transform to flagellates that fuse in pairs; in either case the amoeboid zygote grows, as it feeds on bacteria and fungi, becomes multinucleate and forms an extensive plasmodium, showing shuttle streaming of endoplasm and fan-shaped lobopodial fronts. The plasmodial network (described above) eventually produces stalked (or occasionally sessile) sporangia with numerous spores, or cytoplasmic masses may form sclerotia (thick-walled resting structures). The flagellate cells usually have two unequal apical flagella, lacking mastigonemes. Nuclear division varies, with open mitosis in some, polar openings in some and closed intranuclear pleuromitosis in others; meiosis occurs in spore formation. *Physarum* has been studied extensively,

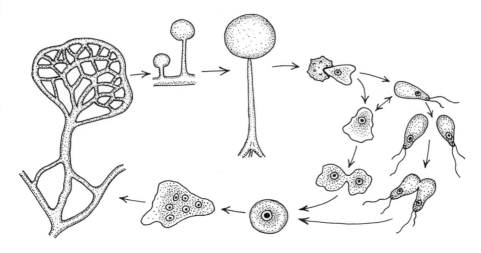

Fig. 6.6   The life cycle of a syncytial slime mould as described in the text; the amoeboid cells are 10 µm or so long, and the plasmodium, only part of which is shown, may extend over several cm.

principally in research on motility. Members of a third subclass, the Protostelia, form small syncytial plasmodia which do not show shuttle streaming; minute fruiting bodies are formed with one or a few spores, and some forms have flagellate stages. The fruiting bodies of these forms and the growth habit of syncytial slime moulds led to their adoption as fungi, but most of their cellular and biochemical characters are quite different from those of members of the kingdom Fungi.

Members of a small group, the class Acarpomyxea, may form small plasmodia (e.g. *Leptomyxa*, Fig. 6.7f) or more usually highly branched unicellular amoebae. Neither fruiting bodies nor flagellate stages occur, but the pseudopods contain microfilaments and the mitochondrial cristae are tubular. They occur in soils and may be important, but are little known at present. Page (1987) classes them as lobose amoebae.

Another class, the Plasmodiophorea, also contains forms that produce small plasmodia, but these species are intracellular parasites in higher plants (Fig. 6.7d). Spores hatch to release zoospores with two anterior heterodynamic flagella lacking mastigonemes; these zoospores are infective and transform to amoebae that grow within host cells to produce plasmodia that eventually divide to produce a mass of uninucleate spores. These forms have mitochondria with tubular cristae and mitosis with a persistent nucleolus and polar openings near adjacent centrioles; meiosis may occur in sporulation. *Plasmodiophora* causes club-root disease of cabbages and is one of a number of economically important species of this small class.

Superficially similar to small members of the class Lobosea are the schizopyrenid amoebae like *Vahlkampfia*. However, recent research has led Page and Blanton (1985) to establish a separate class, the Heterolobosea, for the schizopyrenids and acrasids. In these forms the monopodial amoebae move in a strongly eruptive manner, and the cells contain mitochondria that are enclosed

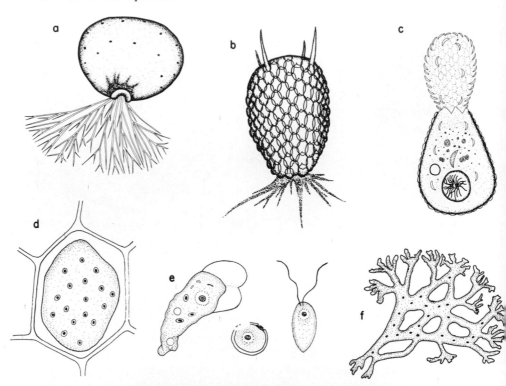

Fig. 6.7 Amoebae from some smaller groups; **a**, *Gromia oviformis* (2 mm) with filopodia: **b**, *Euglypha alveolata* (150 μm) with siliceous scales, spines and filopodia, and **c** the same during formation of a new shell by the fitting together of preformed scales at the surface of a mass of protruded cytoplasm prior to binary fission: **d**, *Plasmodiophora brassicae*, a multinucleate plasmodium shortly before sporulation within a cabbage root cell (host cell 100 μm): **e**, *Naegleria gruberi* amoeboid cell (30 μm) and flagellate and cyst forms at the same scale: **f**, a multinucleate plasmodium of *Leptomyxa reticulata* (200 μm).

by a rough ER cisterna and have discoidal cristae, have nuclei with a closed intranuclear orthomitosis, but do not have any stacked Golgi systems. The acrasids, e.g. *Acrasis*, produce aggregates of amoebae that form stalked fruiting bodies containing spores. Most schizopyrenids, e.g. *Naegleria* (Fig. 6.7e) and *Tetramitus*, as well as one or two acrasids, produce flagellate stages in diluted media (p. 262); the paired apical flagella lack mastigonemes and have both a striated rhizoplast and two or three bands of microtubules (Patterson *et al.*, 1981; Balamuth *et al.*, 1983). *Naegleria fowleri* is a causative agent of acute amoebic meningitis. The marine quadriflagellate *Percolomonas* (Fenchel and Patterson, 1986) appears to be related to the Heterolobosea, and has a flagellar arrangement and rootlet pattern (Fig. 5.33) reminiscent of retortamonads; the nuclear division is also similar in both groups (and in euglenoids).

Amoebae with filose pseudopodia

The pseudopodia in these forms are hyaline, slender and often branching (Fig. 6.7a). The clear appearance of the filopods is due to the exclusion of cyto-

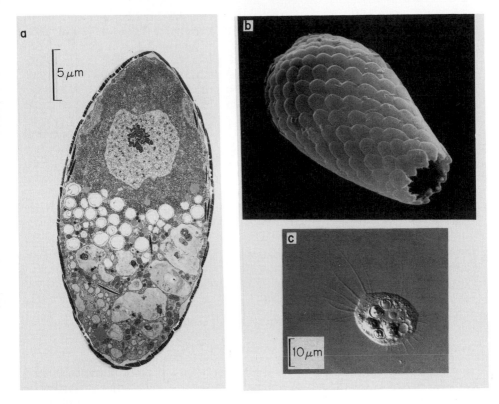

Fig. 6.8 Micrographs of amoebae with filose pseudopodia; **a,** Electron micrograph of an oblique longitudinal section of *Euglypha rotunda* showing the scaly shell enclosing a nucleus in dense cytoplasm above and vacuoles, newly secreted scales and a few mitochondria are seen below: **b,** Scanning electron micrograph of the shell of *E. rotunda,* formed of imbricated siliceous scales: **c,** light micrograph of *Nuclearia moebiusi* (20 μm). (Micrographs **a** and **b** from Hedley and Ogden, 1973 and **c** from Patterson, 1983).

plasmic granules from these parts of the cell. The pseudopodia do not contain microtubules, so it must be assumed that the mechanism of pseudopodial extension, support and contraction involves the dense arrays of microfilaments that they contain. These amoebae constitute the class Filosea, containing two orders, the testate Gromiida and the naked Aconchulinida. Studies on two testate forms reveal considerable differences, probably indicating a need for reclassification. *Gromia* (Fig. 6.7a) has a thick, multi-layered test with protein and polysaccharide components and a single opening; it is marine and multinucleate, with prominent microbodies, mitochondria with tubular cristae and numerous Golgi bodies, apparently involved in wall secretion; small gametes with a single posterior flagellum and a second orthogonal basal body are produced, normally after gamontogamy, but details of flagellar rootlets or nuclear division are not available (Hedley and Wakefield, 1969). **Euglypha** (Fig. 6.7b,c; 6.8) carries imbricated silica scales held together by polysaccharide cement; it occurs in freshwater and soil, is uninucleate, has mitochondria whose cristae are vesicular or dilated tubules, microtubules in the superficial cytoplasm but not in the filopods, a single large Golgi body and vesicles containing siliceous plates that will later be positioned around the daughter cell at division (Fig. 6.7c; 6.8b) (Hedley and Ogden, 1973); the only recorded form of division is binary fission, and, although microtubules surround the nucleus between polar MTOCs, the mitosis is a closed one with an internal spindle (Ogden, 1979).

Naked forms with filose pseudopodia include a couple of families of medium-sized herbivorous amoebae, the vampyrellids, that are able to digest a hole in the cell wall of healthy algal cells (e.g. **Spirogyra**) and suck out their contents (Lloyd, 1926), and the nucleariids, that may engulf small algae or eat the contents of damaged **Spirogyra** cells (Cann, 1986). Species of **Nuclearia** examined by electron microscopy have a mucous coat and show mitochondria with unusual flattened and unbranched cristae, numerous Golgi bodies with stacked cisternae around the several nuclei, closed nuclear division with internal spindles, a distinctive contractile vacuole complex, no extranuclear microtubules and no extrusomes (Patterson, 1983). Species of *Pompholyxophrys* are spherical with fine radiating pseudopods and surface beads of silica, suggesting an affinity with heliozoans, but neither microtubules nor extrusomes are present in the pseudopods; cytoplasmic and nuclear features indicate a relationship of this and some similar genera to nucleariids (Patterson, 1985). Vampyrellid cytoplasm and mitochondria (vesicular cristae) are different from those of nucleariids; the present class Filosea appears to be a heterogeneous group needing reclassification.

Amoebae with granuloreticulose pseudopodia

In this case the fine pseudopodia not only branch, but anastomose to form a complex network, the threads of which contain many fine granules that may be seen to stream in either direction along the threads. The whole reticular network is constantly changing in extent and in detailed configuration by extension or retraction of the individual threads or by changes in their lateral attachments to one another (Fig. 6.9). Electron micrographs of reticulopodial strands show that they are supported by microtubules without any specific arrangement or obvious linkages (Fig. 6.10). Studies of motility by video-enhanced microscopy has shown that microtubules or groups of microtubules may extend or shorten (probably by sliding) to change the array of microtubules or even extend or retract filopodia, and microtubules may also move sideways to change the pattern of linkage in the reticulopodial network (Travis *et al.*, 1983). The same authors noticed that bidirectional streaming occurs exclusively along the microtubular fibrils, and involves both particulate organelles and hyaline cytoplasm, moving at speeds of up to 10 μm s^{-1} or more. Changes in the extent of the network are likely to involve microtubule assembly and disassembly as well as active sliding of the microtubules. Motility is inhibited and microtubules disassembled by colchicine, but movement is also disorganised by cytochalasin, presumably by disruption of actin filaments concerned with adhesion of the pseudopods to the substratum (Travis and Bowser, 1986). Movement of the whole cell is achieved by continued migration of the reticulum in one direction, carrying the cell body with it.

Most members of the class Granuloreticulosea belong to the large order Foraminiferida, with over 35 000 species, 80% of them fossils from marine and brackish water deposits dating since the palaeozoic, but Levine *et al.*, (1980)

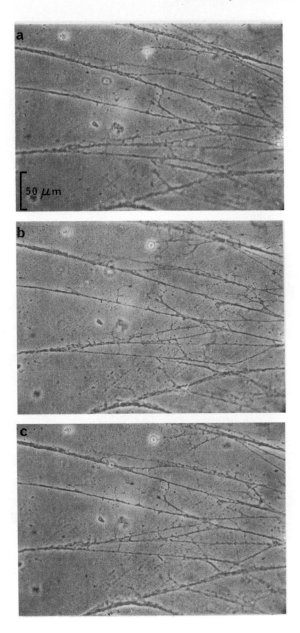

Fig. 6.9 Photomicrographs of part of the pseudopodial reticulum of a foraminiferan taken at 10 second intervals.

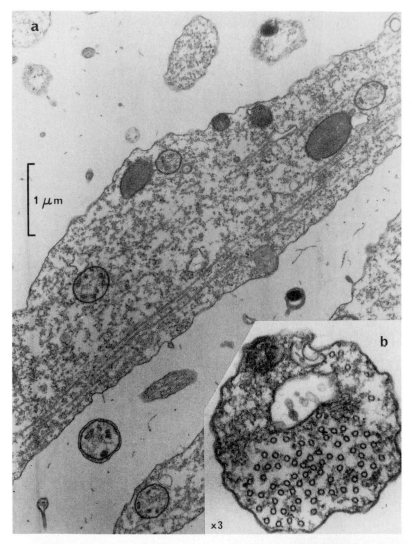

Fig. 6.10 Electron micrographs of sections through pseudopodia of the foraminiferid *Allogromia*. In a the pseudopodia are of various sizes and are sectioned at a variety of angles; they show longitudinal microtubules, occasional dense bodies (extrusomes ?) and mitochondria with few cristae. In **b** the transverse section of a single pseudopod shows sectioned microtubules clearly. Micrographs by R.H. Hedley.

place various, largely freshwater, forms in two small orders. The Athalamida, including *Biomyxa* and *Rhizoplasma*, often have a thin fibrous envelope (but not a firm test) from which typical reticulopods may emerge, and are often multinucleate. Monothalamids like *Microgromia* have a single-chambered organic test with one or several openings, and are said not to show an alternation of generations, but the status of the group needs confirmation. Foraminiferids have tests with one or many chambers and a life cycle with a haplodiploid alternation of generations, examples of which were described in Chapter 4 (p. 78); these cycles are diverse but relatively few have been studied. The number and ploidy of the nuclei varies at different times in the life cycle, division of the nuclei involving closed intranuclear pleuromitosis, with centrioles sometimes present outside the nucleus during gametogenesis (only). Some forms show nuclear dimorphism in agamonts (p. 78). Granular extrusomes, food vacuoles and mitochondria with tubular cristae occur in the reticulopods as well as in the cell body, but nuclei and Golgi bodies are restricted to

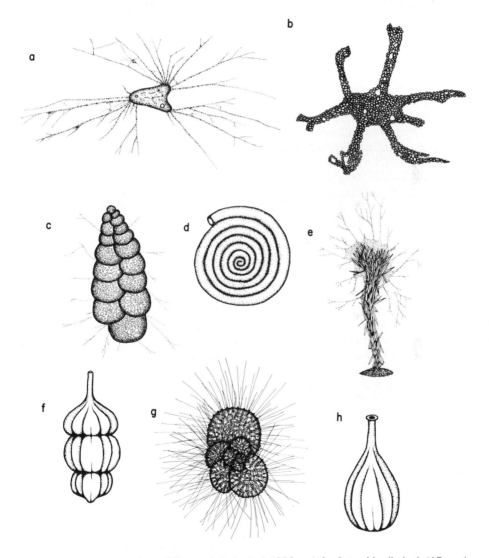

Fig. 6.11 Foraminiferids; **a**, *Allogromia laticollaris* (600 μm): **b**, *Astrorhiza limicola* (15 mm): **c**, *Textularia agglutinans* (1 mm): **d**, *Cornuspira involvens* (1 mm): **e**, *Haliphysema tumanowiczii* (2 mm): **f**, *Nodosaria* sp. (800 μm): **g**, *Globigerina bulloides* (600 μm) complete with spicules (cf. Fig. 6.12a): **h**, *Lagena striata* (500 μm).

the cell body. The gametes are occasionally amoeboid, or more usually hetero-dynamic biflagellates. Hedley *et al.*, (1968) found that the gametes of *Boderia* possessed a flagellar pattern typical of heterokont flagellates (p. 108), with a short naked flagellum and a longer flagellum with two opposite rows of tripartite hairs; foraminifera may therefore be related to heterokont flagellates such as xanthophytes, which also have tubular cristae in their mitochondria and internal mitotic spindles.

Single-chambered membranous tests of acid mucopolysaccharides are formed by such foraminiferids as *Allogromia* (Fig. 6.11a), which may be found in the holdfasts of laminarian algae. Cytoplasm emerges at one or more specialised apertures to form the pseudopodial reticulum and often also a layer of cytoplasm over the outside of the test. Tests into which sand grains are incorporated may have one or several chambers, and may have a regular or irregular shape; some grow large, but the distinction between irregular forms and xenophyophorea (p. 174) is poorly known. In some cases the shell seems to be a secreted layer upon which arenaceous fragments are superimposed, but in other forms these components seem more intimately mixed. Arenaceous tests may be hard and brittle or soft and flexible; in the rigid shells the mucopolysaccharide cement may contain calcium and iron salts. The incorporated particles in these textularinids may be an unselected collection of mineral grains covering a considerable size range, as in *Astrorhiza* (Fig. 6.11b) or *Cribrostomoides* (Fig. 6.12f), or may be a very selective collection of fragments, e.g. echinoderm plates in *Technitella thompsoni*, or sponge spicules in *Technitella legumen*. Sponge spicules form the main part of the test in *Haliphysema* (Fig. 6.11e), which is found commonly among tufts of coralline algae and Bryozoa on British shores.

Foraminiferids with calcareous shells mostly have many chambers whose walls are formed from an organic layer reinforced with calcite. The shells of these forms are composed either of minute crystals which give the shells a shiny appearance like porcelain, or of minute granules or radially arranged crystals, which form hyaline shells. In a few families the shells are made of aragonite or of secreted silica spicules. The porcellanous shells are normally without perforations and all pseudopods emerge from one aperture, e.g. miliolids, while hyaline shells are normally perforated and small pore pseudopods emerge through the perforations as well as larger pseudopods from the aperture. Examples of these foraminiferids are shown in Figs. 6.11 and 6.12; they seldom exceed 1–2 mm in diameter, but the abundant fossils produced and their small size make them valuable in geological statigraphy, especially for the identification of rock strata in exploratory drilling, e.g. for oil deposits. Dimorphism of multichambered shells occurs in some species; the first chamber (proloculus) is smaller in shells of one generation than in the other, and it has been established that in *Elphidium* (Fig. 6.12) the zygote forms the smaller proluculus found in the agamont generation, while in *Spirulina* the shells with the smaller proloculus belong to the gamonts.

The majority of foraminiferids live on the sea floor, particularly in shallower waters where they are common on weeds, rocks and softer substrata. These forms generally have flattened shells, and a pseudopodial net that spreads over the substratum in search of bacteria, diatoms, other algae and similar small prey (Lee, 1980); when symbionts are present they are generally diatoms (p. 283). By

Fig. 6.12 Micrographs of foraminiferids; **a,b**, two views of the calcareous shell of *Globigerina bulloides* (SEM): **c,d**, two views of the calcareous shell of *Elphidium crispum* (SEM): e, the sculptured proteinaceous shell of *Oolina melo* (SEM): **f**, sand fragments agglutinated in the shell of *Cribrostomoides jeffreysii* (SEM): **g,h**, high power scanning micrographs of *Amphistegina madagascariensis* showing (**g**) normal reticulopodial strands passing among debris on the shell surface, and (**h**) small pseudopodia that emerge from pore plates at the mouths of shell perforations: **i**, light micrograph of three specimens of *Julienella foetida*, a dominant member of a community on sublittoral sands of Ghana, with cm scale. (Micrographs **a–f** from Murray, 1971, **g,h** from Jorgen Hansen, 1972).

contrast, there are also a relatively small number (~ 30) of genera of important planktonic foraminiferids, such as ***Globigerina*** (Figs. 6.11g, 6.12). These forms are large, usually with symbiotic zooxanthellae, and often possess many long calcareous spines which aid the capture of large active prey, such as copepods, small medusae or ***Sagitta*** (Anderson *et al.*, 1979).

The class Xenophyophorea includes deep-sea species that occupy a branched organic test, often many cm long, within which are also found membrane-bound faecal pellets. The protoplasmic strands within the test comprise a multi-nucleate plasmodium which contains numerous crystals of barite ($BaSO_4$), 1–5 μm in length. Members of this class were formerly placed within the Foraminiferida, and their agglutinated tests may be confused with those of some textularinids. Although little is known about their living parts, they probably do not have granuloreticulose pseudopodia, and are now placed in a separate class (Tendal, 1972). The high concentrations of barite may be important in geochemical cycling of barium in the deep ocean (Gooday and Nott, 1982). Large textularinids, the Komokiacea, several mm long, are found in similar habitats of the deep ocean.

Amoebae with axopodia

The pseudopodia of actinopods are axopodia, formed from an outer layer of flowing cytoplasm (rheoplasm) surrounding an axial bundle of microtubules (stereoplasm); these microtubules are cross-linked to form a variety of patterns and grow from a variety of MTOCs. In some cases slender filopods may also be produced and veil-like sheets of cytoplasm may be formed to enclose captured prey of these predatory amoebae (p. 26). The prey is first trapped by the release of the contents of extrusomes borne on the axopods. Siliceous, organic or strontium sulphate skeletal elements may be present as internal or external spicules or perforate shells. Reproduction may involve binary fission or the formation of flagellate swarmer cells, and sexual processes occur in some groups. Their mitochondrial cristae are usually tubular, but in some groups there are vesicular or flattened, plate-like cristae. Members of all groups have well-developed Golgi systems. Actinopods tend to be spherical and planktonic, often with symbiotic algae, and probably represent a life-form arrived at by convergence from several different phylogenetic origins, their life-form being dependent on the support of radial pseudopodia by bundles of microtubules presumably derived from persistent asters of different types of nuclear spindle MTOCs. The elementary division into 'radiolaria', typically marine forms with substantial internal skeletons, and heliozoa, typically naked freshwater forms, is now seen to be quite inadequate; not only do we now recognise three classes of forms with skeletons, the Acantharea, Polycystinea and Phaeodarea, perhaps with different origins, but the present Class Heliozoea appears to contain up to six groups whose ancestry is different (Febvre-Chevalier, 1985; Smith and Patterson, 1986). Few heliozoa are, in fact, planktonic, for most of them remain attached to surfaces by stalks or axopodia.

The microtubules of the stereoplasm are labile, being subject to assembly and disassembly under cellular control, probably through changes in calcium ion concentration, to achieve physiological changes in axopod length. Re-extension of axopods after retraction takes place slowly, at speeds of a few μm min^{-1}, and

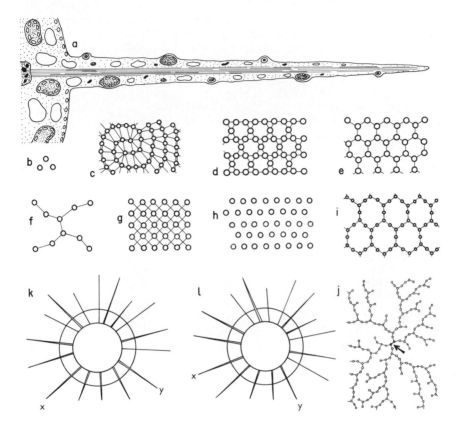

Fig. 6.13   The structure and role of axopodia and their stereoplasmic axial cores. **a**, A diagram of a longitudinal section of an axopodium of **Actinosphaerium** showing the extent of the axial core of stereoplasm from the surface of a nucleus at the left to the tip of the axopod. **b–j** Diagrams showing the arrangement of microtubules seen in cross sections of the stereoplasm in axopodia of **b**, the helioflagellate Ciliophrys: **c**, the actinophryid **Actinosphaerium**: **d**, a centrohelid, e.g. **Acanthocystis**: **e**, hexagonal arrays occur in 'centrohelids' like **Actinocoryne** and taxopodids like **Sticholonche**, though the angle between links may be unequal and the hexagons of alternate rows may be tilted in different directions: **f**, the *x*-shaped array of 10 microtubules found in cryptoaxoplastid radiolarians, **g**, the 'double square' lattice of microtubules, each with four links, found in **Dimorpha**: **h**, alternating rows of microtubules, apparently without links, characterise Phaeodarea: **i**, twelve-membered rings, in which microtubules with 3 links alternate with microtubules with 2 links, are found in nassellarid and periaxoplastid radiolarians: **j**, the pin-wheel array found in centroaxoplastid radiolarians is based on extensions from the ends of a central *x*-shaped array (arrowed) and constructed of microtubules with 2 and 3 links – only part of the array is show. The relative lengths and positions of axopodia of an **Actinosphaerium** moving over the substratum are shown at **k** and at **l** 10 minutes later; the heliozoan is moving to the right by a combination of rolling and gliding.

a similar speed of retraction is found in many forms, but retraction in Actinophrys can by rapid and in centrohelids it takes the form of a quick contraction, being complete in perhaps 20 ms. These retraction movements are used to draw prey, trapped by extrusomes, to the cell body for phagocytosis. The slow movements may also be used for locomotion of larger forms over

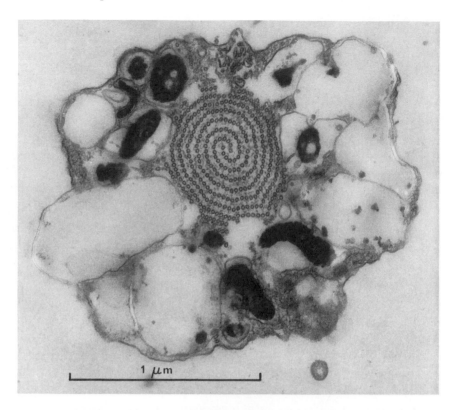

Fig. 6.14   Electron micrograph of a section through an axopodium of ***Actinosphaerium nucleofilum***. Note the two intercoiled spirals of microtubules and the vacuolated cytoplasm. Micrograph by A.C. Macdonald.

surfaces (Fig. 6.13k,l), while smaller forms may glide by transport of membrane binding sites (p. 14). The motion of extrusomes along axopodia may also depend on these membrane sites (Bardele, 1976), but these movements may be actin-based, as are the streaming movements of rheoplasm that carry particles up and down the axopodia (Edds, 1975).

The most familiar heliozoa are probably the freshwater actinophryids *Actinophrys*, which is uninucleate, (Fig. 6.15a) and the multinucleate *Actinosphaerium* (*Echinosphaerium*), but these may be the only genera in the order. The axis of the axopodia is formed of a unique double interlocking spiral array of microtubules joined by short links within spirals and long links between spirals (Figs. 6.13c, 6.14). The MTOCs for these microtubule bundles lie on the surface of the central nucleus in *Actinophrys* (Fig. 6.15a) or on the surface of a more peripheral nucleus in multinucleate species. Mitochondria with vesicular cristae occur in the main body and in the rheoplasm, where they are accompanied by extrusomes (mucocysts) with amorphous dense centres (Fig. 6.14). Nuclear division is almost closed, but polar openings occur from metaphase onwards and allow entry of microtubules from diffuse cytoplasmic centrospheres, while the nucleolus persists. The cells are naked, with a granular central region and a frothy surface layer of vesicles, but there is no skeletal material, although silica scales are secreted at the surface during encystment (Patterson, 1979). Reproduction is by binary fission and by division following sexual fusion within a cyst (p. 79). The helioflagellate *Ciliophrys* (p. 133) is sometimes classed with actinophryids; it shares features with both actino-phryids and pedinellid heterokonts, which appears to suggest that actinophryids evolved from heterokont flagellates.

The centrohelid heliozoans differ in several important respects from actino-phryids. They tend to be smaller cells with slender axopodia supported by small groups of microtubules each with four links of equal length and forming arrays of hexagons and triangles (Fig. 6.13d). The axopodia retract repeatedly; their microtubule bundles grow from a MTOC at the centre of the cell – a tripartite discoidal 'centroplast' (Fig. 6.16), and the nucleus is displaced to one side. Mito-chondria with markedly flattened cistae (Fig. 2.3, p. 19) appear to be confined to the cell body, but extrusomes in the form of distinctive structured kinetocysts (Fig. 6.15c) move rapidly up and down the axopodia and also occur at the body surface. The body surface is deeply invaginated to form a system of lacunae and carries siliceous scales and spines in *Acanthocystis* (Fig. 6.15b) or organic spines in *Heterophrys*, the shapes of these secreted elements varying with the species. Another group of superficially similar, but pobably not closely related, 'centrohelids' includes the marine genera *Gymnosphaera, Actinocoryne* (Fig. 6.15d) and *Hedraiophrys*, of which the first two are multinucleate and the last two have living contractile stalks. In these genera the axopods arise from a microfibrillar (non-tripartite) 'axoplast' at the centre of the cell; the micro-tubules of the axopods and stalks each have three similar links so that they form hexagonal arrays (Fig. 6.13e). Their mitochondria have tubular cristae, and both mucocysts and specific types of kinetocyst are present. They vary in the extent of contractility, presence of spicules etc., but the stalked species, at least, reproduce by budding or multiple fission to produce small free heliozoans (Febvre–Chevalier, 1980). Other stalked centrohelids may belong to either group, but need more study. The freshwater dimorphid helioflagellates (Fig. 6.15e), that have flagella without mastigonemes and contractile axopods with structured kinetocysts, may belong here, but their relationship to other flagellates is obscure. In *Dimorpha* the axopod microtubules form a square array (Fig. 6.13g) and grow from an axoplast connected to the kinetosomes of a pair of flagella and lying near the nucleus. In *Tetradimorpha* the axopod micro-tubules form an irregular pattern and originate from a layered centroplast connected to the kinetosomes of the four flagella and some distance from the nucleus. In both cases the mitochondria are packed with tubular cristae, and in *Dimorpha* mitosis is open with paired centrioles.

In two other small orders of heliozoans the axopodial microtubules arise from the nuclear envelope, but these forms produce biflagellate swarmers without mastigonemes and have tubular cristae; they share features with both dimor-phids and actinophryids, and their relationships are, for the time being, obscure. Members of the Order Desmothoracida typically secrete a spherical, extracellular, latticed, organic capsule at the head of a secreted stalk. *Clathru-lina* (Fig. 6.15f) (Bardele, 1972) is the best known genus, and *Hedriocystis* (Brugerolle, 1985) is somewhat similar. Axopods, and sometimes filopods, emerge from pores in the lattice and carry structured kinetocysts; microtubules

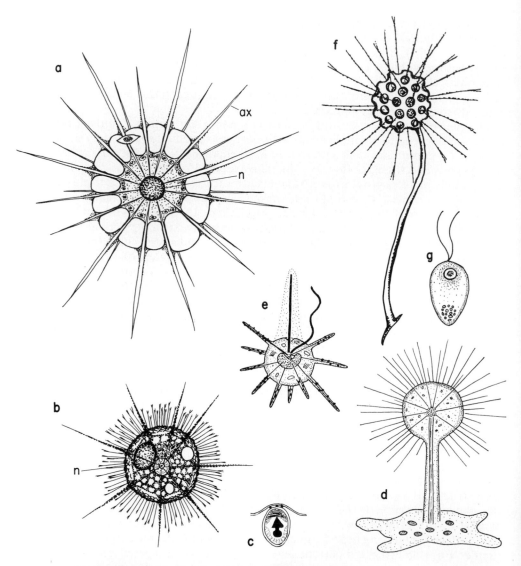

Fig. 6.15   Representative heliozoans; **a**, *Actinophrys sol* (40 μm) with nucleus *n*, axopodia *ax*, and vacuolated surface: **b**, *Acanthocystis chaetophora* (50 μm) with nucleus at one side of the centroplast and bifid silica spines as well as axopodia: **c**, a kinetocyst of *Acanthocystis* (2 μm): **d**, *Actinocoryne contractilis* (body 30 μm) with nuclei in basal cytoplasm and central axoplast which gives rise to stereoplasmic axes of both stalk and axopods: **e**, *Dimorpha mutans* (20 μm) with an axoplast between the flagellar bases and the nucleus, and one straight flagellum surrounded by mucoid material: **f**, *Clathrulina elegans* with a secreted capsule (50 μm) and **g** a flagellated 'swarmer' of *Clathrulina* (25 μm).

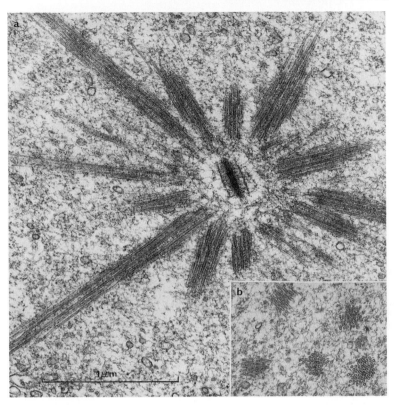

Fig. 6.16 Electron micrographs of sections of *Raphidiophrys ambigua* showing the bundles of microtubules which support the axopodia. In **a** the bundles of microtubules radiate around a centroplast, and in **b** sections across several bundles are shown. Micrographs from Bardele (1971).

in the axopods are not regularly linked and arise from MTOCs at one pole of the nuclear surface. Reproduction involves the release of small migratory cells (Fig. 6.15g) that develop two flagella at the pole opposite to the axopod MTOCs, migrate, and later transform to heliozoans which secrete a capsule around the body and a stalk around a bundle of microtubules. In other genera the stalk is shorter or absent, the capsule may be siliceous and the swarmers uniflagellate. The curious taxopodid heliozoan *Sticholonche* is a bilaterally symmetrical, pelagic, marine form with axopodia that move like oars. The large body carries 14 rosettes of silica spines and many fixed dorsal and motile lateral axopodia. Each motile axopod originates in a cup-shaped pit in the thick envelope of the elongate nucleus, and is anchored by non-actin contractile filaments that swing the axopods in their basal sockets (Fig. 6.17); the microtubules form hexagonal arrays as in *Actinocoryne* (Fig. 6.13e).

Radiolaria (Anderson, 1983) are planktonic oceanic forms, whose skeletons were familiar to nineteenth century microscopists, but whose biology was little

180   *Amoeboid protists*

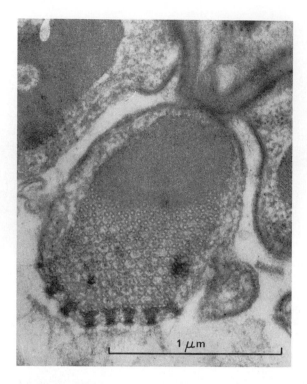

Fig. 6.17   Electron micrograph of a near-transverse section at the base of an axopodium of the taxopodid heliozoan *Sticholonche zanclea*; the hexagonally-arranged microtubules of the axopodial axis originate from a dense region which articulates with a specialised socket at the surface of the perinuclear capsule. Micrograph by Holland, Cachon, Cachon-Enjumet and Valentin (1967).

studied until recently. They are complex protists whose cells are specialised into differentiated regions, and are generally placed in two classes, which may not be closely related. In one class, the Polycystinea, the cells range in size from small solitary nassellarids (50–100 μm in diameter and supposedly bacterivorous) to large spumellarids (several mm in diameter and feeding on larger phyto- and zoo-plankton); colonial species may contain hundreds or even thousands of cells in a gelatinous mass several cm in diameter or in long threads; most species

Fig. 6.18   Radiolarians and an acantharian; **a**, *Aulacantha scolymantha* (1 mm) (Phaeodarea) with a skeleton of separate radiating tubular rods, a prominent central capsule with three openings (an upper, fluted, astropyle and two lower parapyles), a nucleus within the capsule, and an endoplasmic region with granular cytoplasm containing phaeodellae (dark pellets, probably of waste material) and oil droplets enclosed within an outer vacuolated calymma zone: **b**, *Acanthometra elastica* (300 μm) (Acantharea) with regularly arranged spines each formed of a single crystal of $SrSO_4$, a central capsule enclosing cytoplasm with symbiotic algae and an outer region of reticulopodia: **c**, skeleton of *Hexacontium asteracanthion* (120 μm) (Polycystinea, Centraxoplastida); **d**, skeleton of *Pipetta fusus* (Polycystinea, Periaxoplastida): **e**, skeleton of *Cycladophora pantheon* (Polycystinea, Nassellarida): **f**, small colony of *Collozoum inerme* (up to 10 mm or more long) (Polycystinea, Spumellarida), an exoaxoplastid with numerous central capsules in a common vacuolated calymma zone.

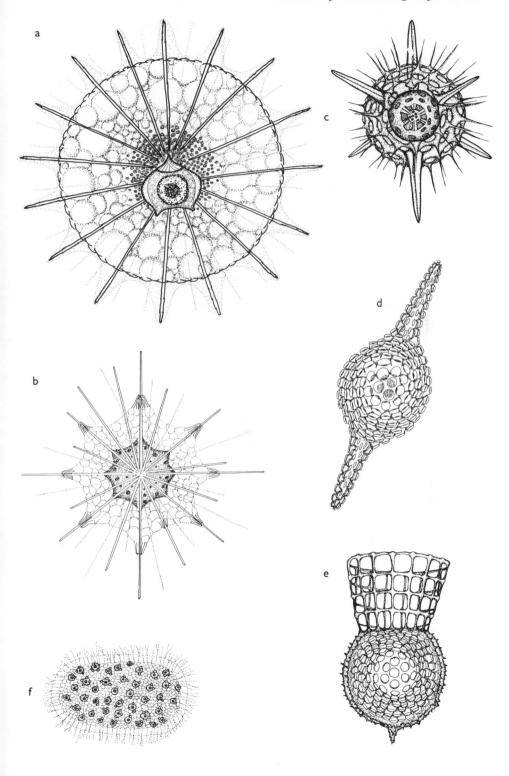

in this class have an internal siliceous skeleton of solid elements and many contain algal symbionts (usually dinoflagellates (p. 283), but occasionally prasinophytes in some spumellarids). In the other class, the Phaeodarea, cells are usually of intermediate size, and typically have a skeleton of hollow silica elements connected by organic material; they apparently lack symbionts, but possess a brown granular mass, the phaeodium, outside the main pore of the capsule, and most have oil droplets. Radiolarian skeletons predominate in the ooze of certain areas of the ocean foor and are found as fossils, dating from as early as the Cambrian.

Within the complex body the cytoplasm is separated into two regions by a membranous central capsule, which is perforated by three pores in Phaeodarea (Fig. 6.18a) and many pores in Polycystinea, either small scattered (1 μm) pores in the Spumellarida or a cluster of small pores at one pole in the Nassellarida. The cytoplasm within the capsule is granular and somewhat vacuolated; it contains one or more nuclei and numerous mitochondria, many Golgi and other vacuoles, as well as food storage materials, principally oil and fat droplets, which may aid buoyancy. Outside the central capsule is a thin layer of compact cytoplasm surrounded by an extensive, highly-vacuolated region, the calymma, with few mitochondria and no Golgi bodies (Anderson and Botfield, 1983). Within the calymma are food vacuoles containing planktonic animal and plant prey caught by the pseudopodia; also, in many species, symbiotic algae live in the extracapsular region and apparently enable these protists to survive long periods without catching prey, provided they are illuminated. The vacuolated layer is believed to provide a variable buoyancy which can be controlled to achieve a certain amount of vertical movement or depth regulation; the fluid in the vacuoles is presumably less dense than sea water, since a cell deprived of most of the calymma sinks rapidly, rising again on regeneration of the calymma. Pseudopodia, in the form of fine filopodia or axopodia (Fig. 6.1c,d), extend outwards from the calymma and from projecting spicules, the latter providing anchorage for the feeding net of pseudopods which may trap vigorous prey like copepods. In these cases strong rhizopodial pseudopods with large microfilament bundles first hold the prey and then break it open to allow invasion by vacuolate pseudopods that engulf prey fragments and carry them away by cytoplasmic streaming for digestion in phagosomes in the calymma.

The microtubular axes of axopods arise from MTOCs that are usually within the capsule and differ in form in different groups; the arrangement of the microtubules within the bundles also varies, as does the pattern of emergence through the capsule, where there is normally a cytoplasmic channel through a specialised collar structure called the fusule. In Phaeodarea a pair of pores without fusules (the parapyles) each surround a large bundle of microtubules which arise from a dense fibrillar axoplast just inside the pore, and are arranged in alternate rows without links (Fig. 6.13h); the large pore (astropyle) has a fluted capsule margin lined by unlinked microtubules, and appears to act as a cytopharynx (Cachon and Cachon, 1973). A cup-like axoplast in the endoplasm of nassellarids is the MTOC for large bundles of microtubules in 12-membered rings (Fig. 6.13i) that leave the capsule through fusules in a group of polar pores. The spumellarids have been subdivided by Cachon and Cachon (1985) into four groups which differ in the form of the MTOC and the arrangement of the microtubules of the stereoplasm, bundles of which often pass in channels through the nucleus. In the cryptoaxoplastids small axoplasts on the surface of the thick nuclear envelope are the origins of small groups of microtubules with a double-Y pattern (Fig. 6.13f). Centroaxoplastids have an axoplast at the centre of the cell, in a membrane-enclosed cavity within the nucleus, and radiating bundles of microtubules in a pin-wheel array (Fig. 6.13j), which is an extension of the cryptoaxoplastid pattern. A deep cup-shaped axoplast at one side of the nucleus in periaxoplastids gives rise to radiating axopods with microtubules in 12-membered rings (Fig. 6.13i), and the same type of microtubule array is found in the exoaxoplastids, whose MTOCs are scattered on the exterior surface of the capsule.

The skeleton is usually composed of a network of elements whose lattice forms a spherical (many spumellarids) or pyramidal (many nassellarids) framework to which radial spicules may be attached; occasionally there may be only isolated spicules. Components of the skeleton are formed within an appropriately-shaped cytoplasmic sheath, the cytokalymma (Anderson, 1981), the silica being deposited upon an organic matrix within a vacuolar cavity surrounded by a membrane resembling the silicalemma of diatoms (p. 135). These skeletons are among the most exquisite of natural structures and are often exceedingly complex; some of them consist of several concentric spheres, representing successive stages of growth of the organism – the central capsule may grow to enclose one or more of these spheres, and the central sphere may become surrounded by nuclear lobes. Some examples of these skeletons are shown in Figs. 6.18 and 6.19.

A similar organisation of the body is found in the marine, and almost exclusively planktonic, Acantharea, whose radiating spicules are composed of strontium sulphate. Typically, numerous small nuclei (derived from the single nucleus of the initial vegetative stage) are found in a central mass surrounded by symbiotic haptophyte or dinoflagellate algae, this whole region being enclosed within a delicate central capsule. An outer cortical region contains a network of reticulopodia enclosed within a limiting filamentous layer, attached by contractile myonemes to the spicules. Axopodia and more slender filopods extend from the surface and catch microplanktonic prey; the axopodial microtubules in this case arise from a central axoplast, and are arranged in hexagonal arrays (Fig. 6.13e) in large bundles which pass through fusules in the capsular pores. In *Acanthometra* (Fig. 6.18b) the 20 precisely arranged radial spicules penetrate the central capsule and meet at the centre of the organism.

Reproduction in radiolaria and Acantharea is imperfectly known; it is said to involve multiple division to produce biflagellate zoospores, or possibly gametes. These do not contain symbionts, and it is assumed that symbionts are reacquired in each generation. The ultrastructure of these flagellate cells is not known. In all these groups the mitochondrial cristae are tubular and the mitotic nucleus is closed with paired centrioles outside the nucleus near polar plaques on the nuclear membrane. Those radiolaria and Acantharea that contain symbionts are typically found in the illuminated surface layers, and more abundantly in tropical oceans, particularly the more eutrophic parts. Phaeodarea and some nassellarids are typical of deeper waters, where they may be dependent on 'marine snow' (p. 264) for food.

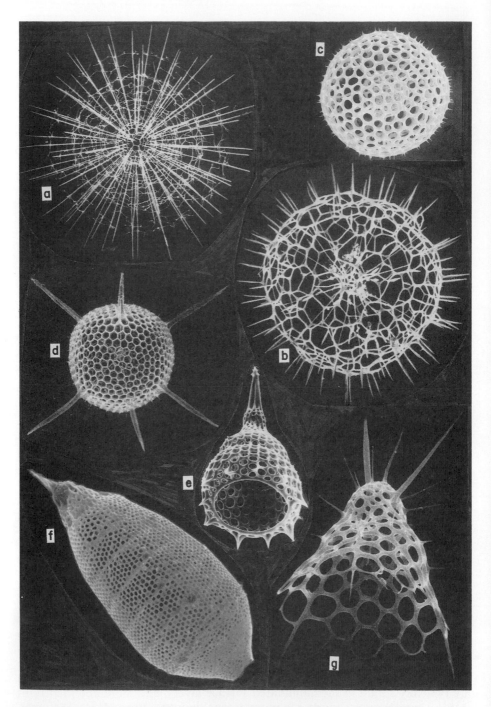

Fig. · 6.19 Scanning electron micrographs of silica skeletons of *Polycystinea*; **a**, the cryptoaxoplastid *Arachnosphaera oligacantha* (50 μm): **b**, the centroaxoplastid *Actinospaera capillaceum* (250 μm): **c**, the centroaxoplastid *Thecosphaera* sp: **d**, the periaxoplastid *Heliosoma irregulare* (200 μm): **e**, the nassellarid *Lamprocyclas maritalis*: **f**, the nassellarid *Eucyrtidium acuminatum* (200 μm); **g**, the nassellarid *Pterocanium trilobium* (60 μm). Micrographs by J. and M. Cachon.

An outline classification of amoeboid protists

The classification of amoeboid protists has not yet reached a stable state, so it is not possible to give an outline that would be accepted by all specialists in the group. The listing in Table 6.1 is used to present examples from the various groups and indicate their relationships, rather than to indicate the ranks of taxa. Corliss (1984) suggested elevating the status of many groups, and recognised some 12 phyla of amoeboid protists; further studies will show whether this is justified.

Table 6.1 A list of the groups mentioned in this chapter, with examples.

Phylum Rhizopoda: about 48 000 described species in eight classes:
  Class Karyoblastea (Pelobiontea), a single genus, ***Pelomyxa***
  Class Heterolobosea: two orders:
    Order Schizopyrenida, e.g. ***Naegleria, Vahlkampfia, Tetramitus***
    Order Acrasida, e.g. ***Acrasis***
  Class Lobosea:
    Sub-class Gymnamoebia: three or four orders, including:
      Order Euamoebida, e.g. ***Amoeba, Thecamoeba, Mayorella, Hartmannella, Chaos***
      Order Leptomyxida, e.g. ***Leptomyxa, Flabellula***
      Order Acanthopodida, e.g. ***Acanthamoeba***
    Sub-class Testacealobosia: three orders, e.g. ***Arcella, Difflugia, Centropyxis, Cochliopodium, Trichosphaerium***
  Class Eumycetozoea: about 700 species of slime moulds in three sub-classes:
    Sub-class Dictyostelia, cellular slime moulds, e.g. ***Dictyostelium***
    Sub-class Myxogastria, syncytical slime moulds, e.g. ***Physarum***
    Sub-class Protostelia, e.g. ***Cavostelium, Protostelium***
  Class Plasmodiophorea, about 40 species, e.g. ***Plasmodiophora, Polymyxa***
  Class Filosea: about four orders in two sub-classes:
    Sub-class Aconchulinia, e.g. ***Nuclearia, Vampyrella, Pompholyxophrys***
    Sub-class Testaceafilosia, e.g. ***Gromia, Euglypha***
  Class Granuloreticulosea: 37 500 described species, 80% fossils:
    Order Athalamida, e.g. ***Biomyxa, Rhizoplasma***
    Order Monothalamida, e.g. ***Microgromia***
    Order Foraminiferida, e.g. ***Allogromia, Astrorhiza, Cribrostomoides, Haliphysema, Textularia, Rotaliella, Elphidium, Globigerina***
  Class Xenophyophorea: about 35 species so far, e.g. ***Stannophyllum***

Phylum Actinopoda: about 3500 living species in four classes:
  Class Heliozoea: about 200 species:
    Order Actinophryida, e.g. ***Actinophrys, Actinosphaerium***
    Order Centrohelida, e.g. ***Acanthocystis, Raphidiophrys, Gymnosphaera, Actinocoryne*** (***Dimorpha*** and ***Tetradimorpha*** may belong here)
    Order Desmothoracida, e.g. ***Clathrulina, Hedriocystis***
    Order Taxopodida, e.g. ***Sticholonche***
  Class Polycystinea: about 10 000 described species, 70% fossils:
    Order Spumellarida, e.g. ***Thalassicolla, Collozoum, Hexacontium***
    Order Nassellarida, e.g. ***Pterocanium, Eucyrtidium, Lamprocyclas***
  Class Phaeodarea: about 800 species in six orders, e.g. ***Aulacantha, Challengeron, Planktonetta***
  Class Acantharea: about 200 species in five orders, e.g. ***Acanthometra***.

7

The ciliates

Ciliate protists belong to a single phylum, the Ciliophora, and are characterised by the possession of two types of nucleus, the sexual process of conjugation (p. 81), a basically equatorial division plane in binary fission (p. 91), and a form of pellicular organisation in which cilia and their intracellular attachments (the infraciliature) are dominant features. There has been extensive evolutionary exploitation of the ciliate pattern of organisation, and this is most clearly reflected in changes in the ciliature and infraciliature. It is thought that the characteristic nuclear dualism evolved within the group after the appearance of the main features of pellicular structure. The 7000 plus species include parasites and free-living members from marine, freshwater and edaphic habitats, and many life cycles include resting or reproductive cysts.

The classification of ciliates has undergone several major revisions in the last decade or so, and no clear consensus has yet been reached. Some points of view represented in the different schemes will be described at the end of this chapter after the diversity of the phylum has been discussed; the names used for the component groups are intended to allow them to be related to any of the main schemes currently in use, and these groups will be introduced in the sequence given in the classification scheme proposed by a Committee of the Society of Protozoologists (Levine *et al.*, 1980). The status accorded to the different groups by different authorities is shown on pp. 219–220.

The nuclei of ciliates

Most ciliates possess nuclei of two types, named macronuclei and micronuclei (Raikov, 1982). Viable strains of ciliates without micronuclei are known, but all ciliates possess at least one macronucleus and in some ciliates many macronuclei are present; the number of micronuclei is often greater than the number of macronuclei. Micronuclei are small and compact and are normally diploid with large amounts of histone protein but little or no RNA (no nucleoli). At binary fission micronuclei divide mitotically within persistent nuclear envelopes, and in the processes of conjugation and autogamy (p. 81) a micronucleus undergoes meiosis before giving rise to the gametic pronuclei. Following sexual processes it is normal for the macronucleus to degenerate and for a new macronucleus to be formed from division products of the micronuclear synkaryon (p. 82). Macronuclei are larger and are normally highly polyploid, estimates of DNA content ranging from 16n to 13 000n having been recorded from polyploid macronuclei of various ciliates, n being the DNA content of a gametic nucleus, i.e. a haploid G_1 micronucleus. During the formation of a macronucleus from a micronuclear synkaryon, rapid replication of chromosomes occurs and the macronucleus becomes polyploid, with in some cases a stage when polytene chromosomes are visible. In the fully formed macronucleus the chromatin is confined to many small dense bodies containing numerous 10–12 nm thick microfilaments representing chains of nucleosomes; as in other eukaryotes this chromatin contains five types of histones, but it also contains much non-histone protein. The numerous less-dense nucleoli contain much RNA and many ribosome components. It has been shown that protein synthesis in the cytoplasm depends upon RNA of macronuclear origin. Formation of DNA as well as RNA occurs in most macronuclei, and in some examples endomitotic replication of macronuclear chromosomes has been observed in which distinct threads appear, split longitudinally, and complete separation shortly before the macronucleus divides.

Features of the diversity of different patterns of macronuclear organisation, development, replication and division have been thoroughly reviewed by I. B. Raikov (1969, 1982). In *Tetrahymena* and *Paramecium* the macronucleus probably contains many autonomous and complete subnuclei which may replicate as a unit and segregate as a unit during macronuclear division. In hypotrichs and some other ciliates polytene chromosomes appear during development of the macronucleus from the micronucleus, and some chromosome loss may occur at this stage; the polytene chromosomes then fragment transversely into separate bands, with the loss of much DNA, leaving multiple copies of short lengths of the genome, which may be free genes, and which may be randomly segregated at division. Replication of DNA in these hypotrichs involves special replication bands (p. 46). Other ciliates show other patterns of macronuclear organisation. During division of the macronucleus, which takes place within an intact nuclear envelope, the nucleus elongates by the extension of a 'pushing spindle', or other array of microtubules, that probably corresponds to the continuous fibres of a normal mitotic spindle (Tucker, 1967; Tucker *et al.*, 1980). Following the sexual processes of conjugation or autogamy the parental macronucleus degenerates, often with fragmentation, and is replaced by a newly formed macronucleus; normally this is provided by division of the micronuclear synkaryon, but, in the absence of such a macronuclear anlage, one or more fragments of the degenerating parental macronucleus may regenerate and form a complete new macronucleus.

Various authors have alluded to somatic functions of micronuclei, usually because experimentally-produced amicronucleate ciliates often have limited viability. Ng (1986) has considered evidence for somatic roles of micronuclei, notably in the development of a normal oral apparatus, but, in the absence of clear evidence for RNA synthesis in the micronucleus, the means by which micronuclear genes might exert any function are obscure.

In three families of karyorelictid ciliates, and one or two genera of uncertain affinity, the macronuclei are diploid and are unlike the macronuclei of typical ciliates in both structure and behaviour. The micronuclei of these ciliates are typical and lack RNA. The macronuclei are rather small and are generally vesicular with a large central nucleolus and peripheral granular chromatin. A single round of DNA replication appears to occur early in macronuclear development, so that each macronucleus normally contains the same amount of DNA as is present in a G_2 micronucleus. However, the macronuclei never divide; they are

a Binary
fission

b Typical
higher
ciliate

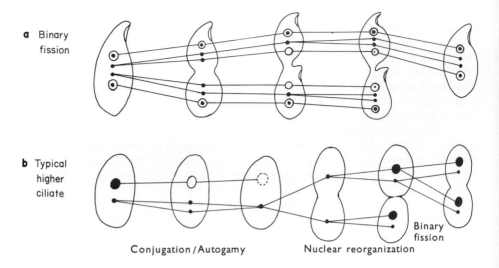

Conjugation / Autogamy   Nuclear reorganization

Binary
fission

Fig. 7.1   Comparison of the behaviour of nuclei during the life cycles of **a**, the karyorelictid *Loxodes striatus* and **b** a typical ciliate. In *L. striatus* the two macronuclei and two micronuclei are all diploid, but macronuclei never divide, so that new macronuclei are always formed from micronuclei and additional micronuclear divisions are necessary to maintain the full number of nuclei through successive cell cycles. In a typical ciliate the macronucleus disperses during conjugation; at the same time the micronucleus undergoes meiosis and fertilisation, and then a new macronucleus is formed from the micronuclear synkaryon; during binary fission both macronuclei and micronuclei divide. Information for **a** from Raikov (1982).

segregated into the daughter cells at cell division, and their number is maintained by transformation of micronuclei directly to macronuclei by development of nucleoli and condensation of chromatin. Between successive cell divisions the micronuclei undergo several mitoses to maintain the numbers of both macronuclei and micronuclei; both mitosis of micronuclei and transformation of micronuclei to macronuclei may occur at any part of the cell cycle. There are often several to many macronuclei, and frequently two macronuclei for each micronucleus, so that many micronuclear mitoses are necessary to maintain the complete nuclear complement. A species with two nuclei of each type is *Loxodes striatus* (Fig. 7.1a), while *Trachelocerca coluber* has four macronuclei and two micronuclei, and up to thirty-five macronuclei have been counted in *Remanella multinucleata*. Conjugation seems not to occur in karyorelictids.

For some time it was believed that the multiflagellate *Stephanopogon* (p. 127) was a primitive ciliate with only one type of nucleus, and therefore that the nuclear dualism found in typical ciliates must have evolved within the phylum. The unique character of ciliate nuclear dualism still argues for its development after ciliates diverged from other protists, but evidence is not available about whether the characteristic ciliate pellicle structure evolved before nuclear dualism. A tendency towards multinuclearity characterises evolution in several protistan groups, in association with increase in size and in synthetic activity. The several nuclei of an ancestral ciliate may then have differentiated into generative nuclei (micronuclei) and somatic nuclei (macronuclei), with in one group a pattern in which the generative nuclei retain the ability to divide to produce macronuclei and micronuclei, while the somatic nuclei lose the power to divide. This incidentally is the stage of nuclear dualism reached by some foraminifera, and may be valuable because it permits the nuclear material to become more active in the synthesis of RNA without the need to conserve the ability to replicate DNA. The formation of more conventional macronuclei involved the replication of genomes in somatic nuclei, leading to a polyploid state which required the retention of the ability to replicate DNA repeatedly within the macronucleus. The ciliate pattern of nuclear dualism with polyploidy provides a more advantageous mechanism than polyploidy without nuclear dualism, as seen in the phaeodarean *Aulacantha*, because of the presence of the separate genetic store of the micronuclei.

Organisation of the ciliate pellicle

Pellicle structure in *Tetrahymena*

In many ciliates the cilia are only present as feeding organelles or on part of the body surface, but this is believed to result from reduction and specialisation of the ciliature. The pellicle structure of *Tetrahymena* provides a good illustration of the arrangement of cortical organelles in a ciliate which bears cilia over almost all of its surface (Allen, 1967, 1969). The body ciliature of *Tetrahymena* shows features found in many of the more primitive ciliates and the basic components of the mouth ciliature of more complex ciliates are represented in the buccal cavity of *Tetrahymena*.

The body cilia occur in longitudinal rows called kineties, the whole body array of which constitutes the kinetome. Each ciliary base is associated with several intracellular fibrils, an adjacent membranous invagination just anterior to each basal body (kinetosome) called the parasomal sac, and a pair of membranous vesicles, all of which together form the basic unit of pellicular organisation, which is called the kinetid, in this case a monokinetid, because a single kineto-some is present (Fig. 7.2). A surface membrane covers the whole organism, including the cilia (p. 29); beneath this membrane in most places are two further membranes which can be seen to form the outer and inner walls of flattened vesicles or alveoli extending over the body surface, and arranged in pairs around the base of each cilium, one on the left side and one on the right. The outer vesicle membrane lies close to the cell membrane, and the space within the vesicle is variable, perhaps because of different fixation techniques. The membranes between adjacent vesicles are seldom complete, and the alveolar cavities of a whole kinety may be continuous. The innermost membrane is underlain by a dense layer of cytoplasm called the epiplasm.

The basal body of the cilium is closely surrounded by the membrane vesicles at its outer end near the transition from basal body to cilium, and at its inner end is associated with four ectoplasmic fibril systems. The largest fibril is that which contributes to the kinetodesmos, a fibre which runs along the (ciliate's) right side of the row of kinetosomes of a kinety. A kinetodesmal fibril arises at the right of the anterior side of the kinetosome, being connected to one or more of triplets 5, 6, 7 and 8 by amorphous material. From this anchorage the fibril passes forwards towards the right and outwards towards the innermost

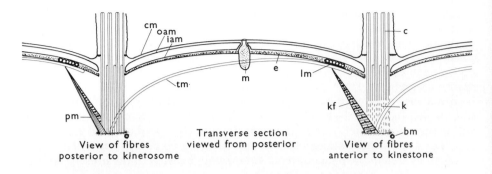

View of fibres
posterior to kinetosome

Transverse section
viewed from posterior

View of fibres
anterior to kinestone

Surface view of fibre systems,
pellicular alveoli and mucocysts shown dotted

Fig. 7.2   Pellicle structure in ***Tetrahymena***, seen in transverse section (above) and in surface view (below). *bm*, Basal microtubules: *c*, cilium: *cm*, cell membrane: *e*, epiplasm: *k*, kinetosome: *kf*, kinetodesmal fibre: *iam* and *oam*, inner and outer alveolar membranes: *lm*, longitudinal microtubules: *m*, mucocyst: *pm*, postciliary microtubules: *tm*, transverse microtubules. Cilia of adjacent rows are not necessarily opposite one another. A parasomal sac normally occurs immediately anterior to each basal body, but is not shown here.

membrane of the pellicle, where the it runs alongside the kinetodesmal fibril of the next anterior kinetid before terminating (Fig. 7.2). Each kinetodesmal fibril is striated with a periodicity of about 30 nm (repeat periods of 15 to 50 nm have been reported), and is made up of a close bundle of thin longitudinal filaments; it is widest at the kinetosome and tapers anteriorly.

A set of 7 to 12 longitudinal microtubular fibrils is found between the inner-most pellicle membrane and the epiplasm a little to the right of the kine-todesmos. These fibrils also do not run the length of the organism, but form an overlapping series of shorter fibrils. They do not make any direct connection with the kinetosomes, although they are associated with two other sets of micro-tubular fibrils which arise at basal bodies.

At the anterior of the kinetosome on the left a group of 5 to 6 transverse microtubules is seen in association with amorphous material around triplets 4 and 5; these microtubules run outwards towards the surface and transversely towards the left, so that they pass immediately below the epiplasm and terminate close to the longitudinal microtubules near the kinetodesmos of the adjacent kinety. Another group of 4 to 8 postciliary microtubules originates at the posterior side of the kinetosome on the right close to triplet 9 and runs towards the surface and backwards towards the right, where they also terminate close to the longitudinal microtubules. One or several basal microtubular fibrils have also been found running longitudinally beneath and to the left of the kine-tosomes of a kinety and making contact with each basal body; these occupy an equivalent position to nematodesmal fibrils of some other ciliates.

The association of the terminations of the transverse, postciliary and kine-todesmal fibrils with the epiplasm and longitudinal microtubules of the pellicle, and also with the dense material around the inner end of the kinetosome, suggests an anchorage function for these fibres. These fibril systems, together with the attachment of the distal end of the basal body to the pellicular membranes at the cell surface, can form a firm tetrahedral anchorage for the ciliary base (Sleigh and Silvester, 1983). The basal microtubules are not as straight as those of the other systems, and may have some other function. The regularly arranged ciliary territories are linked in rows by the kinetodesmal fibrils, by the basal microtubules and by their regular connections to the longi-tudinal bands of microtubules, so that all of these fibres may also have impor-tant morphogenetic functions.

Extrusomes called mucocysts lie in the ectoplasm beneath the spaces between pellicular alveoli along both primary meridians, connecting the rows of kine-tosomes, and the secondary meridians between adjacent kineties (Satir, Schooley and Satir, 1973) (p. 17). The parasomal sacs also lie on the primary meridians, and extend inwards for about half the length of the basal body; in *Paramecium* these invaginations are associated with other, endoplasmic, membranous vesicles (Patterson, 1978).

Pellicle structure in other ciliates

Patterns comparable with this are found in other ciliates, Fig. 7.3 (*see also* Grain, 1969; Pitelka, 1969 and Lynn, 1981). In many other holotrichs (ciliates with cilia over the whole surface) the kinetodesmal fibres are larger and longer; e.g. in *Paramecium* (Fig. 7.3b) they extend anteriorly past 4 or 5 kinetosomes before terminating (Pitelka, 1965; Allen, 1971), and they are even longer in some astome and apostome ciliates, so that the kinetodesmos in these forms is a much larger bundle of fibrils. Although the kinetodesmos may appear in the light microscope as a continuous fibre, it seems always to comprise a series of overlapping component fibrils. In many ciliates the kineties consist of

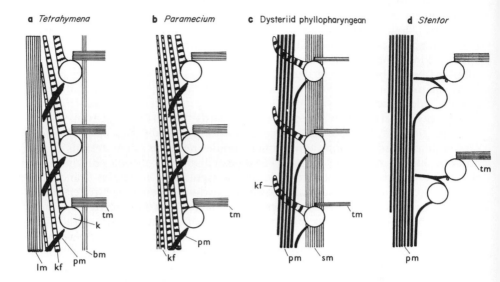

a *Tetrahymena*   **b** *Paramecium*   **c** Dysteriid phyllopharyngean   **d** *Stentor*

Fig. 7.3 Comparison of patterns of ciliary root fibres seen in surface views in four ciliates. *sm*, Subkinetal microtubules; other abbreviations as in Fig. 7.2. Many kinetids in *Paramecium* spp. are dikinetids, but a monokinetid example is shown here for comparison with *Tetrahymena*. The alignment of rows of transverse microtubules at their origin near the kinetosomes is shown schematically; although this row is aligned radially to the kinetosome in some cases (e.g. *Tetrahymena*, Fig. 7.2), in others, including *Paramecium* and *Stentor*, the row of microtubules may be aligned tangentially to the circumference of the kinetosome (and on both basal bodies of a dikinetid), whilst in the example shown in Fig. 5.33g there is a tangential row on the anterior kinetosome and a radial row on the posterior kinetosome of the dikinetid. The post-ciliary row is always aligned radially.

dikinetids, with paired basal bodies; thus, in large species of *Paramecium*, for example, two mature kinetosomes may occur in each kinetid. The kinetodesmal fibrils and postciliary microtubules are then associated with the posterior kinetosomes, which may or may not carry cilia, and the transverse microtubules with the anterior kinetosomes, which normally carry cilia. Lynn and Small (1981) argue that the dikinetid is the ancestral form of kinetid. Groups of nematodesmal microtubules approximately in the position of the basal micro-tubules of *Tetrahymena* are more prominent in some cases, for example in cyrtophorids (Fig. 7.3c; Lom and Corliss, 1971). The nematodesmata may be ribbon-like rows of 5 or more microtubules which may run forward or back along the kinety, overlapping with nematodesmata of several other kinetids, or they may be bundles of hexagonally- or square-packed microtubules which extend into the cytoplasm at a steeper or shallower angle, especially near the cytostome.

In many cases the postciliary or nematodesmal microtubules are the best developed longitudinal elements of the infraciliature. In heterotrichs like *Blepharisma* (Kennedy, 1965), *Stentor* (Bannister and Tatchell, 1968) and *Spirostomum* (Ettienne, 1970), as well as in karyorelictids like *Remanella* (Raikov, 1978), postciliary microtubules are the dominant infraciliary compo-nents associated with the dikinetids. In these forms a postciliary ribbon of 9 or more microtubules runs backwards from the posterior kinetosome of each dikinetid to join with several other ribbons from more anterior kinetids in the same kinety to form a stack of rows of microtubules, a postciliodesma (Gerassimova and Seravin, 1976; Seravin and Gerassimova, 1978) (the km fibre of some authors). This large bundle of microtubules lies along the right side of the kinety in the position of the striated kinetodesmos of other ciliates, but is clearly a different structure. In such contractile ciliates as *Stentor* and *Spirostomum* it is

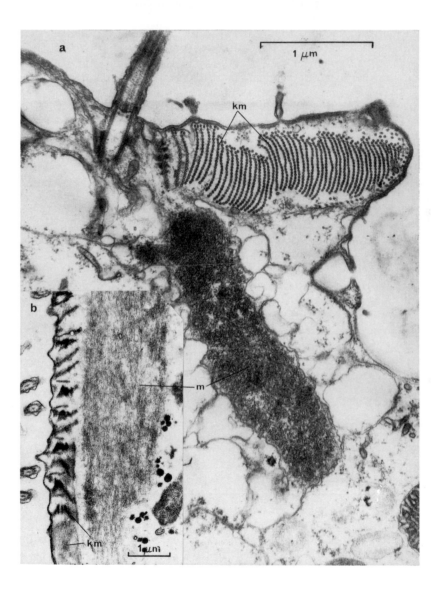

Fig. 7.4   Electron micrographs of sections of contracted specimens of ***Stentor coeruleus*** to show the structure of pellicular fibre systems. **a**, A transverse section of the body passing through a ciliary base, a stack of rows of postciliary microtubules (*km*) and a myoneme (*m*); in the longitudinal section **b** portions of the *km* and *m* fibres are seen. Micrograph by Bannister and Tatchell (1968).

believed that these postciliodesmata are concerned with extension following contraction (Huang and Pitelka, 1973; Huang and Mazia, 1975). Contraction in these two ciliates is brought about by shortening of myonemes (M fibres) (p. 29), which are large bundles of microfilaments situated deeper in the body than the postciliodesmata (Fig. 7.4) (Bannister and Tatchell, 1968; Huang and Mazia, 1975). These bundles may be an elaboration of the infraciliary lattice of microfibrillar tracts found in *Paramecium* at the boundary between ectoplasm and endoplasm.

The extent and direction followed by the kinetodesmata and all three types of microtubules associated with somatic kinetosomes are rather variable in different groups, as is the orientation of the transverse microtubule ribbon with respect to the perimeter of the kinetosome (Lynn, 1981). Similarly, the development of alveolar sacs is variable, and some authors would exclude them as typical components of the kinetid. The extent of the epiplasm is also variable; in many cases it is complex and contains many proteins that form arrays of resistant structures that may have morphogenetic roles.

Body ciliature may be reduced to specialised cilia, such as the dorsal bristle cilia and ventral cirri of hypotrich ciliates like *Euplotes* (Gliddon, 1966). Such cirri are examples of somatic polykinetids in which the kinetosomes are arranged in a hexagonal pattern interlinked at several levels; kinetosomes at the right side carry postciliary ribbons and kinetodesmal fibrils and those at the left carry transverse ribbons oriented tangentially to the kinetosome (Fig. 7.5a). These microtubules run in the ectoplasm and provide the firm anchorage needed when these large cirri are used for propulsion. In other ciliates body cilia may be missing, at least for much of the life of the ciliate, as in peritrichs like *Vorticella*. In suctorians the sessile adult lacks cilia, but gives rise to a motile ciliated larva.

Many ciliates possess extrusomes (p. 17) in the pellicle. Thus in *Paramecium* trichocysts are found in similar positions to the mucocysts of *Tetrahymena*, but only between kinetids along the ciliary meridians; mature undischarged trichocysts of *Paramecium* show a paracrystalline structure of a protein, Trichynin, which suddenly elongates on discharge, with extension of the paracrystalline lattice by a factor of about eight (Bannister, 1972; Hausmann *et al.*, 1972a,b).

Mouth ciliature

Ciliates typically have a single cytostome; exceptions include some primitive karyorelictid ciliates e.g. *Kentrophoros* (Fig. 9.12b), which is said to ingest food anywhere over a large oral area, some mouthless astome ciliates (p. 209) and the suctorians (p. 205), which usually have multiple ingestion sites at the tips of tentacles. The cytostome is located anteriorly or ventrally (rarely posteriorly) and is usually associated with modified somatic cilia or specialised oral ciliature (Corliss, 1979; Lynn, 1981; Small and Lynn, 1985). Commonly the cytostome is located in a depression whose walls bear cilia; this depression is called a vestibulum when the ciliature is essentially an extension of the somatic kineties, even if modified, or a buccal cavity when it contains specialised compound oral cilia.

While the fibrils associated with somatic kinetosomes generally remain in the ectoplasmic region of the ciliate, the kinetosomes of cilia around the mouth give rise to infraciliary fibres which form patterns characteristic of different groups.

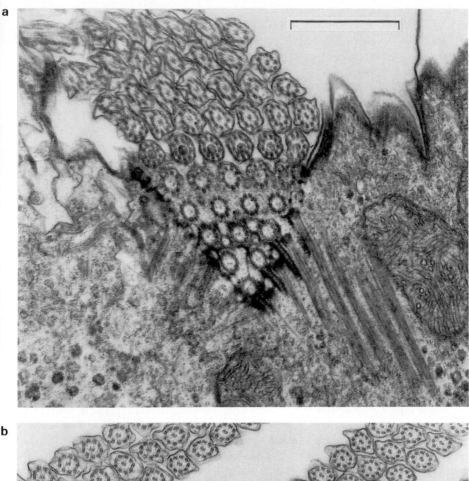

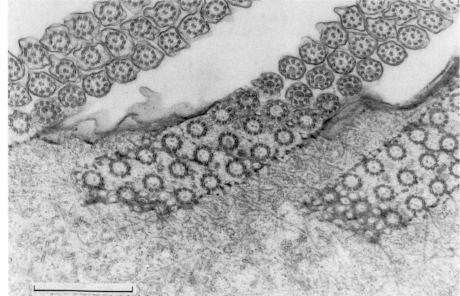

Fig. 7.5 Electron micrographs of compound ciliary organelles of *Euplotes eurystomus*. **a**, A section passing obliquely through the base of a cirrus showing the arrangement of the kineto-somes and the microtubular roots; **b**, a section through the basal region of the membranelles showing·interconnecting fibres. Micrographs by R. Gliddon. Scales 1 μm.

In many ciliates these cilia provide the origins of nematodesmal bundles of hexa-gonally- or square-packed microtubules which extend for a specific distance into the endoplasm. In some groups these bundles form prominent rods that support the walls of a tubular cytopharynx; this may be straight (rhabdos-type), and lined with longitudinally-running microtubules derived from transverse microtubules of circumoral ciliature, in some taxa, or curved (cyrtos-type), and lined with postciliary microtubules, in other taxa. In other forms too cyto-pharyngeal ribbons of microtubules, arising as postciliary and/or transverse groups, are found beneath the surface membranes around the cytostome, and may subsequently extend into the endoplasm of the cytopharynx region. A fila-mentous reticulum, which may represent a specialisation of the infraciliary lattice of somatic regions, often occurs in close association with kinetosomes and other organelles of the mouth region; in peritrichs it forms a close mesh-work with dense nodes and may support the walls of the buccal cavity. Both the infraciliature and the external motile cilia of the mouth region show special features concerned with the collection of food and subsequent digestive processes (p. 24). The complex pellicle structure is modified around the mouth and is interrupted at the cytostome, where the body surface is formed by a single unit membrane.

The ciliature involved in food collection is generally formed of dikinetid or polykinetid structures, although there are a few examples of oral structures made up of monokinetids (Lynn, 1981). Dikinetids vary in fibrillar attach-ments; kinetodesmal fibrils are almost always absent, postciliary ribbons of microtubules are always present, attached to the posterior member of the pair, transverse microtubules may be present on either, both or neither of the kine-tosomes, and nematodesmata arise from dikinetids in some groups. Oral poly-kinetids are most commonly formed of three (but sometimes more) rows of kinetosomes with square packing, often with postciliary ribbons on the posterior row and transverse ribbons on the anterior row of kinetosomes; the packing of kinetosomes in a few groups is hexagonal and in these cases only postciliary ribbons of microtubules seem to be present. Within these oral poly-kinetids, which are normally called membranelles because of their flattened, tongue-like shape, the kinetosomes are connected together by filaments at two or three levels, and adjacent polykinetids are linked by extensions of these fila-ments or by deeper-running groups of microtubules that intermingle (Fig. 7.5b).

Circumoral or perioral arrays of somatic dikinetids may form the feeding ciliature in some groups or may supplement more specialised oral ciliature in some others. In the latter, paroral ciliature at the (ciliate's) right of the mouth is normally formed of a special row of dikinetids, or occasionally one or more polykinetids, and at the left there are three or more adoral polykinetids. A typical pattern of oral ciliature comprising a paroral row of dikinetids and three adoral polykinetids occurs in *Tetrahymena* (Fig. 7.6); the close-set cilia of the paroral row, borne on the anterior kinetosome of each dikinetid, form an undulating membrane at the right of the buccal cavity, with three membranelles at the left. Microtubule ribbons interconnect these structures and contribute to a post-oral (cytopharyngeal) fibre. In polyhymenophorans there may be many more membranelles and often a large undulating membrane (see Fig. 3.4, p. 63).

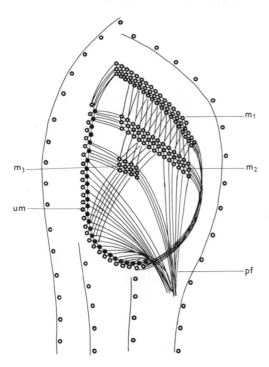

Fig. 7.6   The arrangement of kinetosomes and associated fibres in the oral region of ***Tetrahy-mena***. Kinetosomes of the three membranelles (m_1, m_2, m_3) and the undulating membrane (*um*) lie in the buccal cavity and are surrounded by somatic kineties; some root fibres interconnect the compound ciliary structures and others run down into the cytoplasm as a post-oral fibre (*pf*). The barren kinetosomes of the undulating membrane are shown as filled circles; rows of postciliary microtubules from these barren kinetosomes underlie oral ribs (*or*) that reinforce part of the posterior wall of the buccal cavity.

The formation of a new mouth, stomatogenesis, is a very important component of ciliate morphogenesis, usually associated with division (pp. 90, 91); it is also variable, but the type is usually characteristic of a particular group and is regarded as an important phylogenetic and taxonomic feature. In some genera (e.g. the hypotrich ***Diophrys***) an extensive reorganisation of the ciliature occurs before division, with stomatogenesis in both proter and opisthe, rather than formation of a new mouth only in the opisthe, which is more usual. In different types of stomatogenesis kinetosomes for new mouth ciliature may originate from the anterior parts of somatic kineties (telokinetal), from post-oral somatic kineties (parakinetal, as in ***Tetrahymena***, p. 91), from parental oral ciliature (buccokinetal) or from a site remote from somatic or oral ciliature (apokinetal).

Characteristics of different ciliate groups

These are described by Corliss (1979) and Small and Lynn (1985).

Gymnostome ciliates, literally naked-mouthed, may have little or no ciliature related specifically to the mouth, though the circumoral

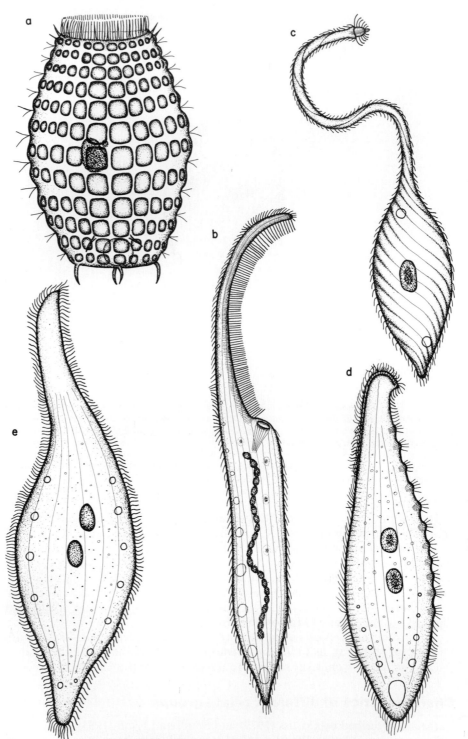

Fig. 7.7 Representative ''gymnostome ciliates''; **a**, the prostomatid *Coleps octospinus* (100 μm): **b**, the haptorid *Dileptus anser* (300 μm) with a prominent and permanent sub-terminal cytostome: **c**, the haptorid *Lacrymaria olor* (200 μm) with a flexible contractile neck and terminal cytostome: **d**, the pleurostomatid *Loxophyllum helus* (200 μm) with large batteries of marginal toxicysts: **e**, the pleurostomatid *Amphileptus claparedei* (130 μm).

ciliature may be more closely packed than normal somatic cilia and nematodes-mal bundles of microtubules associated with them may support a rhabdos-type cytopharynx, so that the mouth is seldom truly naked. Groups often described as gymnostomes include karyorelictids, prostomatids, haptorids and pleuro-stomatids.

Karyorelictids are usually elongate and flattened ciliates, sometimes ciliated only on the ventral surface, and typically interstitial in marine sands (Fig. 9.12). They have supposedly primitive diploid macronuclei (p. 187) and the dominant features of the infraciliature are the large postciliodesmata. The oral area is typically apical or in a ventral slit, the cytostome is superficially located and surrounded by longer cilia, occasionally in specialised arrays (e.g. *Geleia*), and nematodesmata of circumoral cilia support a simple cytopharynx. Some members of this group are very large, e.g. *Trachelocerca* (Fig. 9.12a) and may carry symbiotic bacteria, e.g. *Kentrophoros* (Fig. 9.12b). Members of the genus *Loxodes* are found in freshwater sediments and plankton, and show a remark-able geotactic behaviour; in *L. striatus* and *L. magnus* the Muller's bodies serve as 'statocyst organelles' associated with one of the kineties, and influence ciliary propulsion in a manner that is determined by oxygen tension, so that *Loxodes* can follow the oxycline by swimming up when oxygen is absent or very low and swimming down when oxygen levels are higher (Fenchel and Finlay, 1984, 1986).

Coleps (Fig. 7.7a) is a common member of the prostomatid group. These are spherical or cylindrical ciliates whose mouth is apical or nearly so and is usually equipped with toxicysts (p. 17) to assist the carnivorous feeding typical of the group. A simple circumoral ciliature is associated with nematodesmata which support a simple rhabdos-type cytopharynx. Haptorids are also carnivorous gymnostomes with an apical or subapical oval or slit-like mouth. Toxicysts with paralytic and proteolytic properties are often abundant in the oral area. An array of coronal circumoral cilia contribute nematodesmata forming a complex rhabdos. These ciliates may be large, e.g. *Dileptus* (Fig. 7.7b), extensible, e.g. *Lacrymaria* (Fig. 7.7c) or specialised predators like *Didinium* (Figs. 3.6, 7.8).

The pleurostomatids are usually long, laterally-flattened ciliates whose slit-like cytostome lies along the (presumed) ventral, usually concave, edge of the anterior part of the body. The ciliature is often largely confined to the ciliate's right surface, and the circumoral ciliature and rhabdos are simple. These ciliates are often large and carnivorous, eating other ciliates and even small metazoans. *Loxophyllum* (Fig. 7.7d), *Amphileptus* (Fig. 7.7e) and *Lionotus* (Fig. 9.12f) are familiar examples.

The cytostome of vestibuliferans lies within an apical, subapical, or occasio-nally posterior, vestibular depression containing modified somatic ciliature associated with a rhabdos-type cytopharynx. Typical groups of this assemblage are the trichostomes and entodiniomorphs, but the position of colpodids is less clear. Trichostomes are free-living or symbiotic in animals, both vertebrate and invertebrate; they are holotrichous with simple vestibular ciliature and include the parasitic or endocommensal *Balantidium* (Fig. 7.9a), and endocommensals like *Isotricha* or *Dasytricha* found in the rumen of ruminant mammals (p. 280) or *Blepharocorys* in the caecum of horses, elephants etc. Entodiniomorphs are found in similar habitats in the digestive tracts of herbivorous mammals, e.g. ophryoscolecids like *Entodinium* (Fig. 7.9b) in the rumen of ruminants and cycloposthiids like *Cycloposthium* (Fig. 7.9c) in the caecum of horses etc. Members of this group are recognised by the restriction of somatic and oral

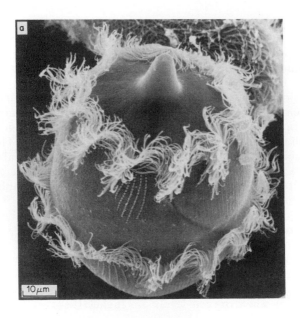

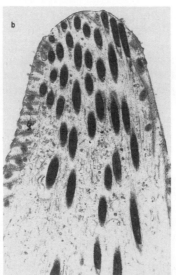

Fig. 7.8　Electron micrographs of the haptorid gymnostome *Didinium nasutum*. **a**, Scanning electron micrograph of the whole ciliate, length about 100 μm) Micrograph by D.I. Barlow. **b**, Transmission electron micrograph of a longitudinal section through the proboscis showing the bands of fibrils at the sides and the giant toxicysts at the centre.

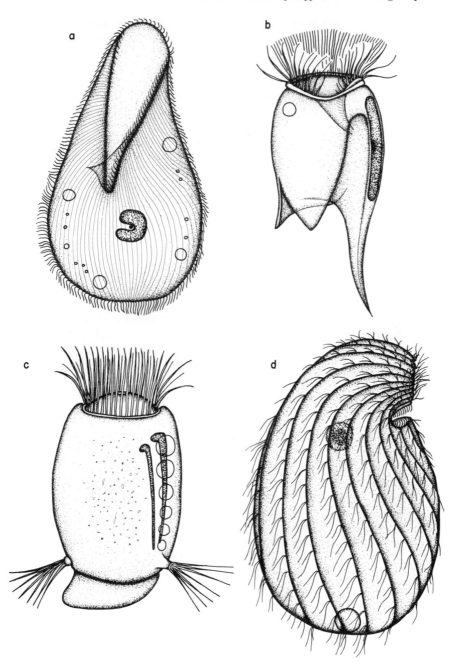

Fig. 7.9 Vestibulifera and a colpodid; **a**, the trichostome *Balantidium entozoon* (70 μm) from the frog rectum: **b**, the ophryoscolecid entodiniomorph *Entodinium caudatum* (70 μm) from the cow rumen: **c**, the cycloposthiid entodiniomorph *Cycloposthium bipalmatum* (100 m) from the horse large intestine: **d**, the colpodid *Colpoda cucullus* (75 μm) from soil.

ciliature to tufts or bands of apparently fused syncilia, the rest of the surface being covered by a very rigid naked pellicle, often with pronounced ridges or spines. The vestibulum with its group of syncilia is often retractile by contraction of cytoplasmic filaments. A rhabdos-type cytopharynx leads into the cytoplasm, and allows ingestion of fragments of vegetation (p. 66); the cytoproct is also quite elaborate and large skeletal plates often occur in the cytoplasm (Furness and Butler, 1985). The roles of vestibuliferan ciliates in the rumen is discussed on p. 280.

Colpodids have a more organised oral ciliature in a depression which has for some time been regarded as a vestibulum. However, the cytopharynx is lined by ribbons of postciliary microtubules, and hence should be classified as a cyrtos. Oral cilia appear to be rather like membranelles, suggesting that the oral depression should be regarded as a buccal cavity. These are holotrichous ciliates with kineties of dikinetids bearing ribbons and transverse microtubules. They are free-living forms typically found in soil and, as may be expected for edaphic forms, they frequently form cysts. *Colpoda* (Fig. 7.9d) is a very common genus and *Tillina* is also widespread.

The ciliature in members of the hypostome assemblage (Phyllopharyngea) is generally rather simple; the somatic ciliature is often reduced and the oral ciliature, which may lie within an atrial depression, surrounds a ventral cytostome which leads to a conspicuous cyrtos-type cytopharynx. This assemblage includes the nassulids, synhymeniids, cyrtophorids, chonotrichs, rhynchodids and apostomes. Nassulids and synhymeniids are characterised by a prominent band of perioral cilia, the hypostomial frange, running obliquely across the anterior part of the ventral surface just posterior to the oral area. In synhymeniids the frange is extensive and more or less fused, while in nassulids the frange is limited to the (ciliate's) left side and may be reduced to a few pseudomembranelles. Members of both groups are typically cylindrical holotrichous ciliates, but some nassulids like *Pseudomicrothorax* are laterally flattened. While *Nassula* (p. 66) and *Pseudomicrothorax* (p. 67) (Fig. 3.7) ingest algal filaments through their cytopharynges, smaller nassulids are bacterivores.

Cyrtophorids lack a frange, and care should be taken to avoid confusion of the frange with a prominent preoral suture at the left (Fig. 7.10a) in many of these ciliates. They are usually dorso-ventrally flattened, with cilia only ventrally, and have a large, often prominently-curved, cyrtos formed from very few nematodesmata. Most are able to attach to surfaces by thigmotactic ventral ciliature or an adhesive 'podite'. Some are bacterivorous and many live as epibionts on various invertebrates, where they may be histophagous. *Chilodonella* (Fig. 7.10a) is a common free-living form, as is *Dysteria*, which has a beak-shaped podite, whilst hypocomids are epibionts that feed through a protruding cytopharyngeal tube, and may be more correctly placed with rhynchodids, since they lack nematodesmata (Grell and Meister, 1983).

The sessile vase-shaped chonotrichs are almost all found as epibionts on crustaceans and are unlikely to be confused with any other ciliates. Many have a stalk, a rigid pellicle without somatic ciliature, and reproduce by budding to release a migratory ciliated tomite (larva), formed externally or internally within a brood pouch. A deep spiral funnel encloses the atrium with its perioral ciliature and a simple cytopharynx lacking nematodesmata. *Spirochona* (Figs. 7.10b, 9.13b) is a common freshwater form, although most chonotrichs are marine.

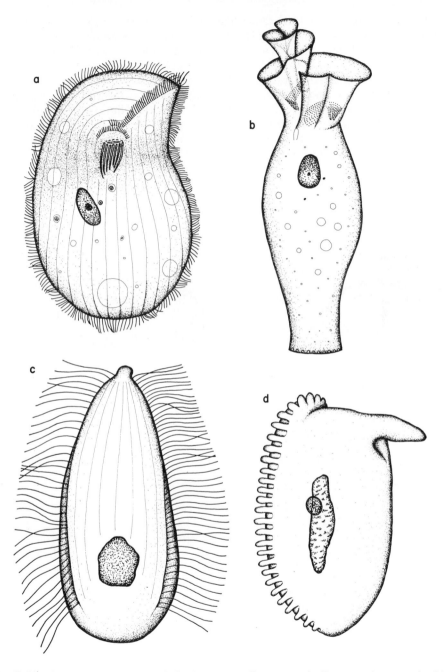

Fig. 7.10 Some representative phyllopharyngean (hypostome) ciliates; **a** the cyrtophorid *Chilodonella cucullulus* (140 μm), seen from the ventral side: **b**, the chonotrich *Spirochona gemmipara* (100 μm) from the gills of the crustacean *Gammarus*: **c**, the rhynchodid *Ancistrocoma pelseneeri* (60 μm) from the gills of the bivalve mollusc *Mya*: **d**, the rhynchodid *Gagarius gagarius* (40 μm) from the gills of the bivalve mollusc *Mytilus*.

Rhynchodids are small ciliates with an anterior, tubular, sucker-like tentacle, which bears toxicysts, and by which they attach to and feed upon the tissues of molluscs, usually marine bivalves, and sometimes marine worms. Some have an anterior ventral patch of thigmotactic cilia, e.g. *Ancistrocoma* (Fig. 7.10c) and others lack cilia altogether, except in the larval stage, e.g. *Gagarius* (Fig. 7.10d). The apostomes are also parasites on a range of marine and freshwater invertebrates, and even on other apostomes. Their somatic ciliature is complete but sparse, in spiral rows, and often with an anterior thigmotactic area, but there is only fragmentary oral ciliature. The cytostome is either absent or small and ventral, and is often accompanied by a 'glandular' rosette (e.g. *Foettingeria* Fig. 7.11a). The life cycles of apostomes are complex, with a series of polymorphic stages, and were the subject of a classic study by Chatton and Lwoff (1935). A trophont grows throughout the feeding stage without dividing (Fig. 7.11b); then, in the tomont stage, there is a period of division (palintomy),

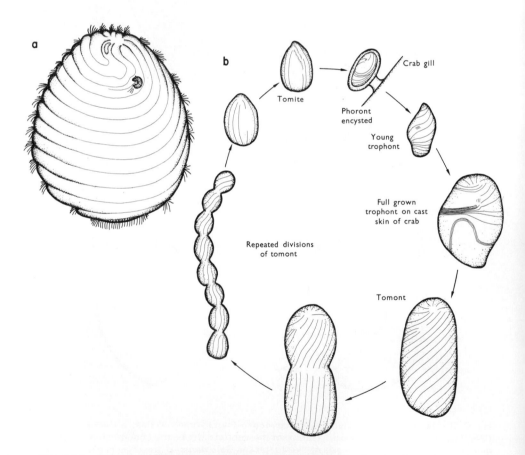

Crab gill

Tomite

Phoront encysted

Young trophont

Full grown trophont on cast skin of crab

Repeated divisions of tomont

Tomont

Fig. 7.11 Apostome ciliates; **a**, the trophont stage of *Foettingeria acinetiarum* (up to 1 mm long) from the gastrovascular cavity of sea anemones: **b**, life cycle of *Polyspira delagei* found as an encysted phoront on the gills, or as a trophont on the cast skin, of the crab *Eupagurus bernhardus*. (**b** Redrawn from Lwoff, 1950.)

usually within a cyst, resulting in the liberation of numerous tomites, which swim away and encyst upon the surface of a host as a phoront stage. The phoront of *Foettingeria* is found on the skin of a crustacean and hatches to release a trophont when an anthozoan eats the crustacean; the tomont is found on the outside of the anthozoan. Many feed by osmotrophy or pinocytosis of host fluid, whilst others are phagotrophic.

The suctorians are distinctive ciliates because of the presence of multiple (usually) suctorial tentacles with haptocysts (p. 64) and the lack of cilia in the sessile adult stage. The adults are generally found attached to the surface of a marine or freshwater invertebrate, or occasionally to an inanimate substratum, by a non-contractile stalk. Their ciliate nature is confirmed by the hetero-karyotic nuclear complement and sexual conjugation process. They reproduce by budding, which may be external, evaginative or in a brood pouch, and results in the release of a ciliated migratory larval stage whose ciliary basal bodies are derived from non-ciliated kinetosomes of the adult. The larval kinetids and infraciliature are similar to those in cyrtophorids. The larvae settle on an appro-priate surface and secrete a stalk from a 'glandular' scopula region. Most suctorians feed on other ciliates captured on simple tentacles, e.g *Discophrya* (Fig. 7.12a) and *Acineta* (Fig. 7.12b), whilst *Dendrocometes* has branched tentacles and is found on gills of freshwater gammarid crustaceans (Figs. 7.13a,b, 9.13c), and *Ephelota* has suctorial tentacles interspersed with prehensile tentacles supported by a complex axial array of microtubules (Figs. 7.12c, 7.13c). *Tachyblaston* is a suctorian that is parasitic on *Ephelota*. In many cases the tentacles are retractile (Al-Khazzar *et al.*, 1984); this is especially evident in the single, very extensible tentacle of *Rhyncheta*, used to capture sessile flagellates (Hitchen and Butler, 1974) or the funnel-like tentacle of *Choa-nophrya*, which is used to feed on organic debris (Hitchen and Butler, 1973).

The remaining groups have more elaborate oral ciliature, consisting in poly-hymenophora of an extensive array of adoral membranelles with one or several rows of paroral cilia, and in oligohymenophora of a small number of adoral polykinetids and a paroral membrane formed of a single row of dikinetids. In the latter group the cytostome lies in an anterior to ventral buccal cavity which also contains at least some of the oral ciliature (e.g. Fig. 7.6). Hymenostomes, scuticociliates, astomes and peritrichs may be grouped in an oligohy-menophoran assemblage.

Several very familiar ciliates like *Tetrahymena*, *Colpidium* and *Glaucoma* are included in the hymenostome group. *Tetrahymena* (Fig. 1.4) is a most impor-tant ciliate, which has probably been the subject of more scientific papers than any other protist, primarily because it has been maintained in axenic culture for some decades and has been found to be excellent material for biochemical studies (Corliss, 1973; Elliott, 1973); references to such studies abound in this book. Some authors also include *Paramecium* (Fig. 7.14) and related peniculine ciliates in this group, but this classification is disputed (p. 220); this genus is also very well known and has been the subject of more books than any other among protists (e.g. Beale, 1954; Jurand and Selman, 1969; van Wagtendonk, 1974; Wichterman, 1953, 1986). Hymenostomes have dense holotrichous somatic ciliature, with kinetodesmata that are often conspicuous, and a ventral buccal cavity, typically containing a paroral (undulating) membrane and three adoral membranellar polykinetids without nematodesmata. They are generally bacteri-vorous freshwater ciliates, but the Ophryoglenina (Canella and Rocchi-Canella,

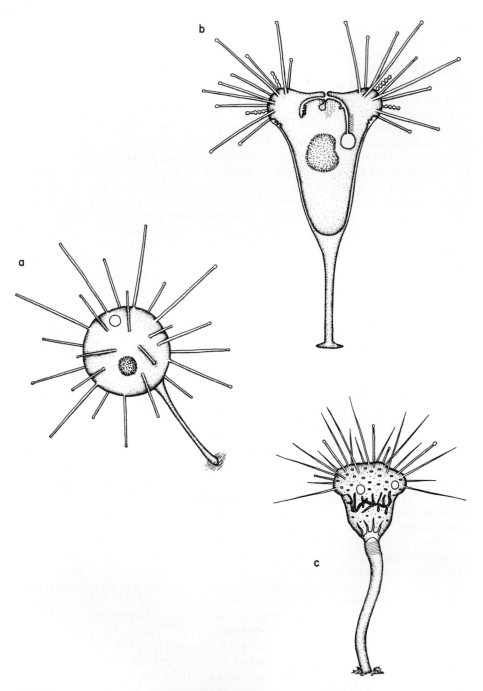

Fig. 7.12 Suctorian ciliates; **a**, *Discophrya collini* (50 μm) with suctorial tentacles and a secreted stalk, but no lorica: **b**, *Acineta tuberosa* (100 μm) showing the stalked lorica, suctorial tentacles and an early stage in the endogenous budding of a larva within a brood chamber: **c**, *Ephelota gemmipara* (200 μm), a stalked marine form with pointed prehensile tentacles and capitate suctorial tentacles.

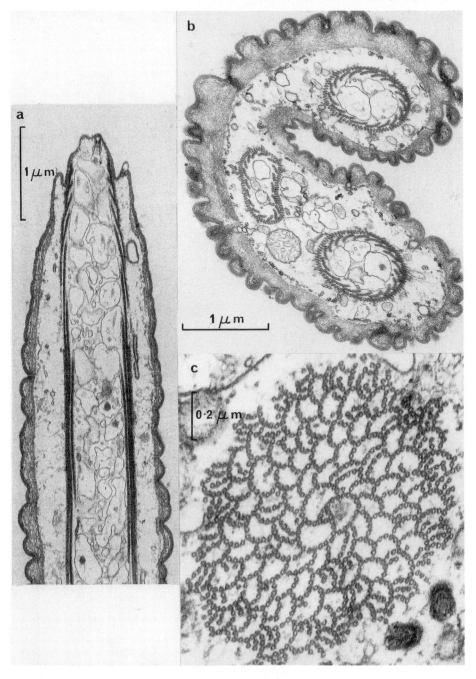

Fig. 7.13 Electron micrographs of sections through suctorian tentacles; **a**, a longitudinal section through the tip of a tentacle of *Dendrocometes paradoxus*, showing the microtubule cylinder and a haptocyst at the apex: **b**, a transverse section through a main branch of a tentacle of *Dendrocometes* showing three cylinders of microtubules: **c**, a transverse section through the microtubular axis of a prehensile tentacle of *Ephelota gemmipara*. Micrographs by C.F. Bardele, **a** and **b** from Bardele (1972).

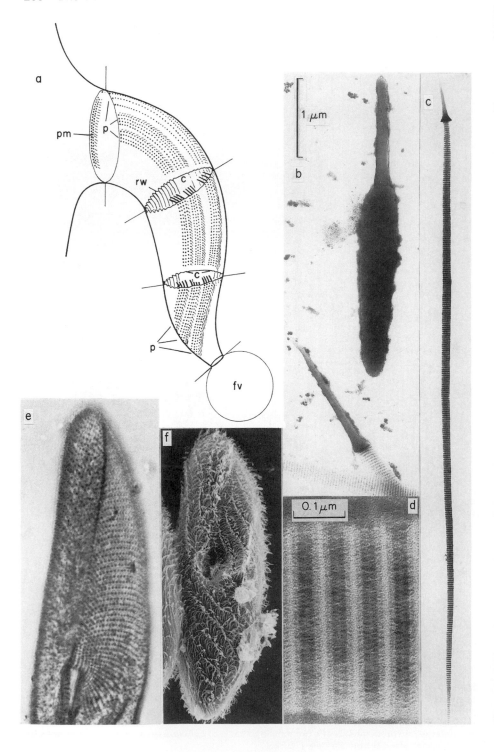

1976) include histophagous forms with a more complex life cycle; following massive growth of a parasitic trophont, they encyst and undergo palintomy to release numerous small migrant ciliates; important forms include *Ichthyophthirius* on freshwater fish and *Cryptocaryon* on marine fish. Members of the peniculine group have a deep buccal cavity with long polykinetids called peniculi (usually of four rows of cilia) at the left within it, and a prominent ribbed wall at the right of the cytostome, supported by postciliary microtubules of kinetids of the paroral membrane, which is at the right margin of the buccal cavity. The oral cilia, unlike those of hymenostomes in general, carry nematodesmata, and explosive trichocysts line much of the pellicle. These peniculine ciliates differ from 'true' hymenostomes in the mode of stomatogenesis, which is buccokinetal in peniculines and parakinetal in tetrahymenines, as well as in the form of the buccal cavity.

The scuticociliates are close to the hymenostomes and sometimes classed within them. They are generally small forms with sparse somatic ciliature, but some of their cilia may be long, especially the one or more caudal cilia, and some cilia have a thigmotactic role. The paroral membrane is normally the most prominent part of the oral ciliature, and may be subdivided into three sections, usually accompanied by three polykinetids at the left around a shallow buccal cavity; nemadesmata are absent. Most scuticociliates are marine, as endosymbionts of various invertebrates and associated with detrital particles in pelagic and interstitial habitats (Figs. 9.11b, 9.12g), and they are also found in freshwater and soil. The paroral membrane usually serves to filter out bacteria for food (Fenchel, 1980), e.g. *Cyclidium* (Fig. 7.15a), but other forms are histophagous, algivorous or carnivorous. The thigmotrich forms commonly live within the mantle cavities of molluscs; they have a prominent anterior (sometimes left or dorsal) thigmotactic ciliated area, and often the cytostome is posterior, e.g. *Ancistrum* (Fig. 7.15b) and *Boveria* (Fig. 7.15c).

Large, and often long, mouthless ciliates constitute the astome group, whose members are usually found as parasites living osmotrophically in the gut of annelids (de Puytorac, 1954). They have dense uniform somatic ciliature with complex infraciliature in which kinetodesmata dominate and may form the basis of a substantial endoskeleton. There is often an anterior holdfast organelle in the form of a sucker, hook or spines, or sometimes thigmotactic ciliature. Reproduction may involve incomplete fission, resulting in the formation of chains of individuals. *Anoplophrya* is common in earthworms and *Durchoniella* (Fig. 7.15d) is from a marine worm.

Fig. 7.14 *Paramecium.* **a**, Detail of buccal cavity structure seem from the left and in a series of sections: the ciliature comprises a paroral (or endoral) membrane (*pm*), based on dikinetids that give rise to postciliary microtubules supporting a ribbed wall (*rw*) as in *Tetrahymena*, three long peniculi (*p*), each of four rows of cilia, which curve down the left side opposite to the ribbed wall, and a cytostomial surface (*c*), which widens towards the bottom as the ribbed wall (ventrally) and a naked ribbed wall (dorsally) become narrower; food vacuoles (*fv*) form terminally at the dorsal side. **b–d**, Electron micrographs of trichocysts, undischarged (**b**) showing the tip and dense matrix of the main body, discharged (**c**) showing the tip and extended shaft, and with the detail of banding of the shaft after discharge (**d**) (Micrographs by L.H. Bannister). **e**, Light micrograph of part of *Paramecium caudatum* stained with silver to show the arrangement of kinetosomes in the oral groove leading to the entrance to the buccal cavity. **f**, Scanning electron micrograph showing metachronal waves of the general body surface and oral groove anterior to the buccal overture (Micrograph by D.I. Barlow).

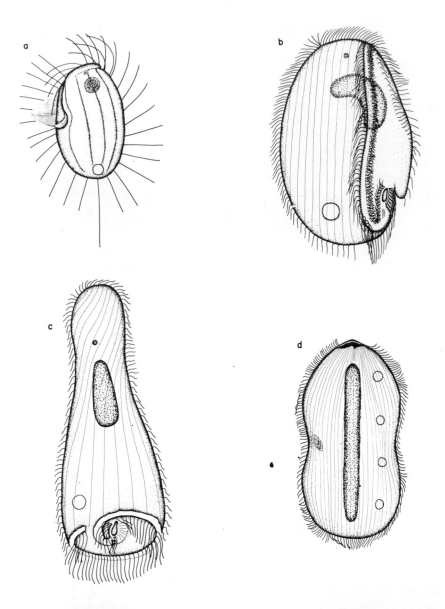

Fig. 7.15 Scuticociliates and an astome ciliate; **a**, *Cyclidium glaucoma* (20 μm), a free-living bacterivore: **b**, the thigmotrich *Ancistrum mytili* (40 μm) from the gills of a bivalve mollusc: **c**, the thigmotrich *Boveria subcylindrica* (50 μm) from the respiratory tree of an echinoderm: **d**, the astome *Durchoniella brasili* (150 μm) from the polychaete worm *Audouinia*, showing the anterior skeletal bar which is developed into a prominent hook in some other genera.

Fig. 7.16 (see *opposite page*) Representative peritrich ciliates; **a**, *Vorticella convallaria* (80 μm body), showing an extended zooid and a partially contracted one: **b**, the mobiline peritrich *Trichodina pediculus* (50 μm) found on Hydra: **c**, *Opercularia ramosa* (cell body 140 μm) has a non-contractile stalk: **d**, *Ophrydium sessile* (300 μm) forms gelatinous clusters of stalkless zooids with green symbionts: **e**, *Cothurnia imberbis* (80 μm) secretes an erect lorica within which both daughters may remain temporarily after division until one emigrates to form its own lorica. *bc*, Buccal cavity: *cv*, contractile vacuole: *h*, haplokinety: *mn*, macronucleus: *p*, polykinety: *sh*, sheath of stalk and *sp*, spasmoneme (myoneme) of stalk.

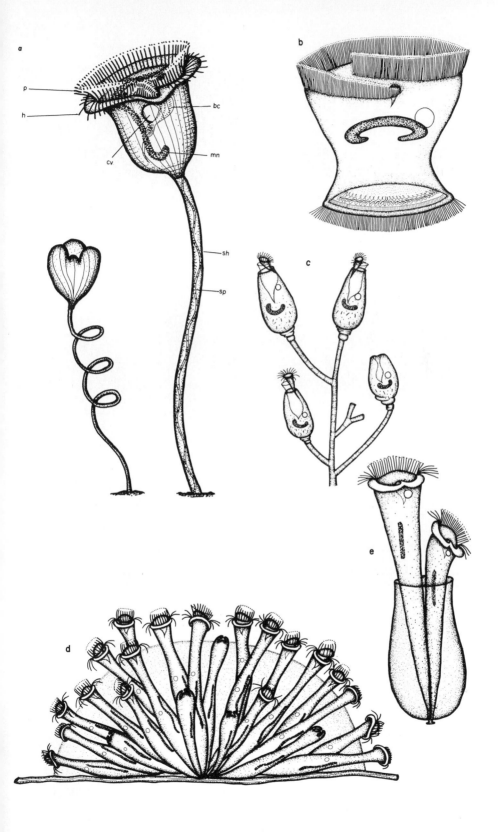

a

p

h

cv

bc

mn

sh

sp

b

c

d

e

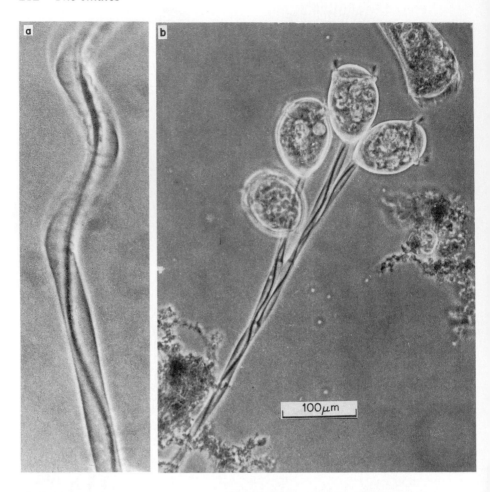

Fig. 7.17 Photomicrographs of peritrichs; **a**, a stalk of ***Vorticella*** re-extending following contraction and showing the distortion of the sheath caused by spasmoneme shortening: **b**, a small colony of ***Carchesium*** in which the stalk is branched but a separate spasmoneme occurs in each branch (cf. ***Zoothamnium*** in which the spasmoneme is continuous throughout the colony).

Vorticella (Fig. 7.16a) is probably the most widely-recognised ciliate, and displays typical peritrich features. In this very numerous group the oral ciliature, in the form of a paroral membrane ('haplokinety') and adoral polykinetids ('polykinety'), extend out of the deep buccal cavity ('infundibulum') and run for one or more turns around the apical (= ventral ?) end of the body; typically this ciliature is used to filter bacteria from the water, e.g. in polluted water or sewage (p. 273). The form of the oral ciliature suggests an evolutionary sequence from hymenostome scuticociliates to thigmotrichs to peritrichs; there are no nematodesmata, but a complex filamentous network surrounds the buccal cavity. Many peritrichs are sessile, usually stalked, and then lack somatic ciliature except as a temporary aboral band of locomotor cilia in the migratory

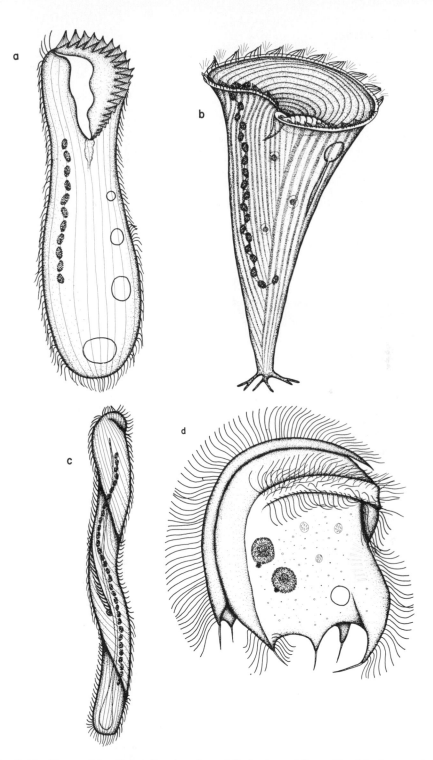

Fig. 7.18 Heterotrich ciliates (**a–c**) and an odontostome (**d**); **a**, *Condylostoma patens* (400 μm): **b**, *Stentor coeruleus* (1.5 mm) with blue pigment granules: **c**, *Spirostomum ambiguum* (2 mm): **d**, *Saprodinium dentatum* (70 μm).

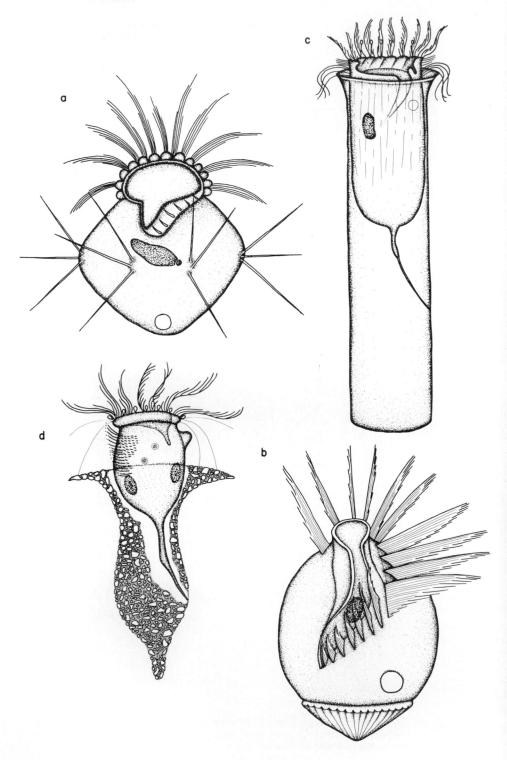

larval stage (telotroch). However, the mobiline peritrichs have a permanent girdle of cilia around the flattened aboral pole of the body (Fig. 7.16b), which assist their movement over the surfaces of host animals, vertebrate or inverte-brate, marine or freshwater; the aboral pole often also has a characteristic ring of denticles and radial myonemes, which may aid the thigmotactic adhesion to the host. Some species of *Trichodina* are pathogens of fish. Among sessile forms the body and the stalk possess highly contractile filamentous organelles (myonemes and spasmoneme, respectively, p. 29). The stalk is secreted from a 'glandular' scopula region rich in kinetosomes. The sessile species are frequently colonial (Figs. 7.16c,d. 7.17, 9.16a,e), or may be loricate (Fig. 7.16e); a few forms swim freely in the adult stage, e.g. *Telotrochidium*. Conjugation is unusual in that it involves complete fusion of a microconjugant with a macroconjugant, and binary fission appears unusual among ciliates in having a division plane nearly parallel to the body axis.

Polyhymenophora are perhaps better known as spirotrich ciliates, typified by the presence of a long curved adoral zone of membranelles (AZM), which often extends out of the buccal cavity onto the body surface. This assemblage includes the heterotrichs, odontostomes, oligotrichs, tintinnids and hypotrichs.

Several familiar ciliates are classed as heterotrichs, which have a dense, holo-trichous ciliature whose somatic dikinetids usually lack kinetodesmata but bear prominent postciliodesmata. The AZM usually dominates the oral ciliature, but in *Blepharisma* (Fig. 3.4) (Giese, 1973) and *Condylostoma* (Fig. 7.18a) the paroral (undulating) membrane is also prominent. These are large ciliates, as are the strongly contractile and transiently sessile *Stentor* (Fig. 7.18b) (Tartar, 1961), the free-swimming *Spirostomum* (Fig. 7.18c) and the sessile loricate *Folliculina* (Fig. 9.11g). In some anaerobic polysaprobic forms the somatic ciliature is reduced, as in *Metopus*, or almost absent, as in *Caenomorpha*; members of both genera are classed as sulphide ciliates and carry ecto- and endo-symbiotic bacteria (p. 281) (Fenchel *et al.*, 1977). Most heterotrichs are free-living, but the group includes parasites such as *Nyctotherus*, found commonly in the frog rectum. In the same polysaprobic habitats as *Metopus* are found bacterivorous odontostome ciliates like *Saprodinium* (Fig. 7.18d). These odontostomes are small, laterally-compressed ciliates with a ridged pellicle and few somatic cilia. There are also few, small membranelles in the AZM, and no paroral membrane, in members of this small group.

The oligotrichs, as their name suggests, also have relatively few somatic cilia, although they do have a prominent AZM around the anterior end of the body and usually a small paroral membrane based on a row of monokinetids. The AZM forms an incomplete circle and may be subdivided into an inner sector, within the buccal cavity, and concerned with feeding, and an outer sector of larger membranelles on the body surface, concerned with locomotion. In some naked forms the somatic cilia are reduced to a few, long, cirrus-like bristles, e.g. *Halteria* (Fig. 7.19a) or are apparently absent, e.g. *Strombidium* (Fig. 7.19b).

Fig. 7.19 Naked oligotrich and loricate tintinnid ciliates; **a**, *Halteria grandinella* (30 μm) a freshwater oligotrich: **b**, *Strombidium lagenula* (50 μm) a marine oligotrich: **c**, *Eutintinnus latus* (350 μm) a marine tintinnid with totally secreted lorica: **d**, *Stenosemella campanula* (150 μm) a freshwater tintinnid in whose lorica sand grains are incorporated.

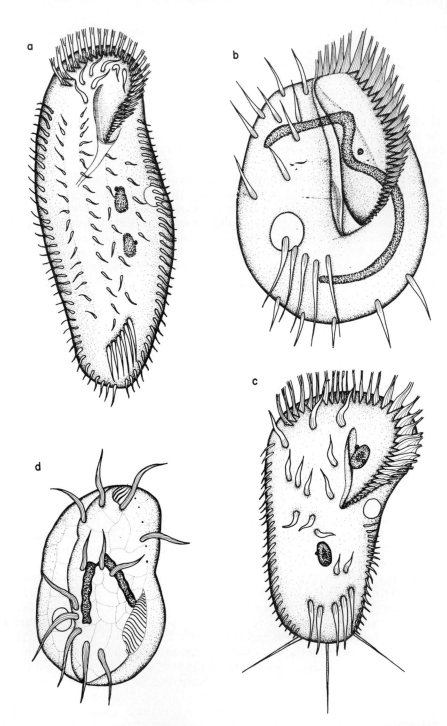

Fig. 7.20 Representative hypotrich ciliates; **a**, the stichotrich *Paraurostyla weissei* (250 μm): **b**, the familiar sporadotrich *Euplotes patella* (100 μm) which commonly has only four cirri at the caudal margin; **c**, *Stylonychia mytilus* (200 μm) a sporadotrich with numerous cirri: **d**, *Aspidisca major* (70 μm) a large member of a sporadotrich genus in which the number of compound cilia is reduced.

In tintinnids, all of which are loricate, the AZM forms a closed circle and the somatic ciliature persists as a few, short, specialised rows near the oral end of the body, e.g. *Eutintinnus* (Fig. 7.19c) and *Stenosemella* (Fig. 7.19d). *Strobilidium* is a genus of aloricate forms that may be more closely related to tintinnids than to oligotrichs. A perilemma (p. 32) often covers the membranelles and some-times the whole cell in oligotrichs and tintinnids. Most species of both groups are pelagic marine forms and some are of substantial ecological importance; a few naked species occur as symbionts in echinoids, and a few are found in mosses and soils, whilst there are several important naked and tintinnid species in freshwater. The loricae of tintinnids are of diverse shapes and may incor-porate agglutinated particles, but the composition of the 'chitinous' secreted material does not seem to be certain. Because of their largely marine habitat, tin-tinnids are often less familiar to protistologists than other ciliates, yet they form the largest order of ciliates, with some 1200 species (Corliss, 1979).

Some of the hypotrichs are very well known, particularly *Euplotes*. These ciliates are typically dorso-ventrally flattened and in addition to the prominent AZM most species have one or more paroral membranes. The somatic ciliature is concentrated into cirri, based upon polykinetids, which are essentially confined to the ventral surface of the flattened body, whilst on the dorsal surface dikinetids in rows bear short, non-motile, (sensory ?) bristle cilia. The cirri are used by these ciliates to walk or run forward and dart back over surfaces in a very characteristic manner, under a system of ciliary control studied in detail by H. Machemer and colleagues in *Stylonychia* (Machemer, 1986) (p. 41), and which does not require the 'neuromotorium' proposed by earlier authors, and erroneously perpetuated in so many textbooks. At least two types of hypo-trichs are recognised, the stichotrichs are usually more elongate forms that bear numerous cirri in several to many ventral or spiral rows (e.g. *Paraurostyla*, Fig. 7.20a), while in the sporadotrichs the cirri, except marginal ones, if present, occur in clumps on the ventral surface rather than in clear longitudinal rows (e.g. *Euplotes*, Fig. 7.20b; *Stylonychia*, Fig. 7.20c; *Aspidisca*, Figs. 7.20d, 9.11f, 9.16f; *Diophrys* Fig. 9.11e). There is a tendency in some of these sporadotrichs for the AZM to become less conspicuous and the cirri more prominent, notably in *Aspidisca*, which also lacks dorsal cilia. The compound cilia of some hypotrichs may be enclosed in a perilemma, but this appears not to be present in euplotids (Bardele, 1981). In addition to locomotion over surfaces using cirri, many of these ciliates swim very effectively using membranelles and cirri. The membranelles are also used to create water currents and to filter food from these currents (p. 62). Hypotrichs are abundant freshwater and marine ciliates, most commonly found in interstitial habitats, but frequent on any surfaces, including sewage flocs.

Ciliate classification and phylogeny

As in so many areas of Biology, the progressive changes here reflect the introduction and exploitation of new techniques. Earlier classifications depended only on the presence and arrangement of external cilia, separating off suctorians from ciliates among the 'Infusoria', and permitting the recog-nition of most of the groups mentioned in the previous pages. Thus they dis-tinguished between holotrichs, spirotrichs, peritrichs and chonotrichs, and allowed subdivision of holotrichs into gymnostomes, trichostomes, astomes,

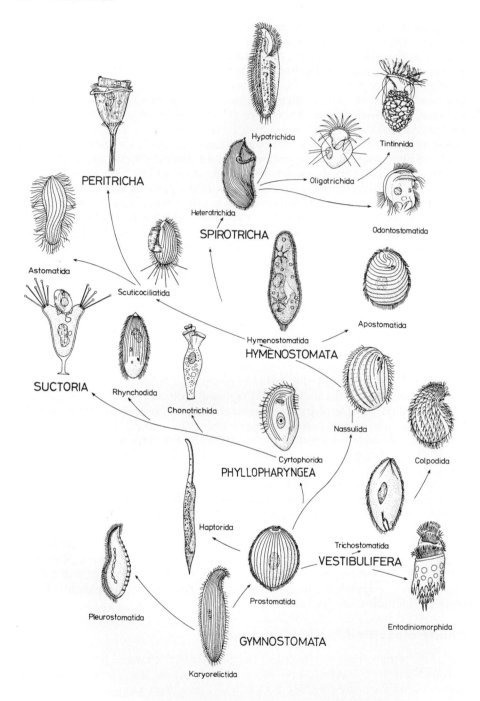

Fig. 7.21   A scheme showing possible evolutionary relationships between the ciliate groups mentioned in this chapter in the form of a phylogenetic tree. The diagram also indicates the groups included in the three classes of de Puytorac *et al.* (1974). Diagram kindly provided by C.F. Bardele.

hymenostomes, etc., and spirotrichs into heterotrichs, oligotrichs, hypotrichs etc. Methods of staining that allowed discrimination of patterns of infraciliature in the definition of the groups led to recognition that the suctorians, chonotrichs and peritrichs were more closely related to holotrich ciliates than previously thought, so that in the classification published in 1961 by Corliss the sub-phylum Ciliophora contained a single class, the Ciliata, which was sub-divided into the two sub-classes Holotricha and Spirotricha, following the proposal of Faure–Fremiet (1950). Various other authors, particularly Jankowski (1967, 1973) have produced modifications of these earlier schemes.

The steady accumulation of evidence about an ever wider range of ciliate species, largely by electron microscopy, has provided much better means of recognising the homology of or difference between comparable structures in different groups, and this evidence has been used by several groups of workers to propose profound changes in the arrangement of groups within the Ciliophora, which has come to be regarded as a phylum. The first major revision of this type was published by de Puytorac *et al.*, (1974), who, influenced by some of Jankowski's ideas, divided the phylum into three classes according to the organisation of oral structures: the Kinetofragminophorea, with oral ciliature

Table 7.1 Classification of the Phylum Ciliophora as used by de Puytorac *et al.* (1974), Corliss (1979) and Levine *et al.* (1980).

Phylum Ciliophora
  Class Kinetofragminophorea
    Sub-class Gymnostomata
      Order Karyorelictida, e.g. *Trachelocerca, Loxodes, Geleia.*
      Order Prostomatida, e.g. *Coleps, Holophrya, Prorodon.*
      Order Haptorida, e.g. *Lacrymaria, Dileptus, Didinium, Mesodinium.*
      Order Pleurostomatida, e.g. *Amphileptus, Litonotus, Loxophyllum.*
    Sub-class Vestibulifera
      Order Trichostomatida, e.g. *Sonderia, Balantidium, Isotricha.*
      Order Entodiniomorphida, e.g. *Diplodinium, Cycloposthium.*
      Order Colpodida, e.g. *Colpoda, Tillina, Cyrtolophosis*.
    Sub-class Hypostomata
      Order Synhymeniida, e.g. *Nassulopsis.*
      Order Nassulida, e.g. *Nassula, Furgasonia, Pseudomicrothorax.*
      Order Cyrtophorida, e.g. *Chilodonella, Brooklynella, Dysteria.*
      Order Chonotrichida, e.g. *Spirochona, Stylochona.*
      Order Rhynchodida, e.g. *Ancistrocoma, Gargarius, Sphenophrya.*
      Order Apostomatida, e.g. *Foettingeria, Polyspira.*
    Sub-class Suctoria
      Order Suctorida, e.g. *Ephelota, Acineta, Discophrya, Dendrocometes.*
  Class Oligohymenophorea
    Sub-class Hymenostomata
      Order Hymenostomatida, e.g. *Tetrahymena, Ophryoglena, Paramecium.*
      Order Scuticociliatida, e.g. *Uronema, Cyclidium, Ancistrum, Boveria.*
      Order Astomatida, e.g. *Anoplophrya, Durchoniella.*
    Sub-class Peritricha
      Order Peritrichida, e.g. *Vorticella, Cothurnia, Trichodina.*
  Class Polyhymenophorea
    Sub-class Spirotricha
      Order Heterotrichida, e.g. *Stentor, Metopus, Bursaria, Folliculina.*
      Order Odontostomatida, e.g. *Saprodinium.*
      Order Oligotrichida, e.g. *Halteria, Strombidium, Eutintinnus.*
      Order Hypotrichida – stichotrichs e.g. *Urostyla* and sporadotrichs e.g. *Oxytricha, Stylonychia, Aspidisca, Euplotes.*

based upon modified somatic ciliature, the Oligohymenophorea, with a small number of specialised compound oral cilia, and the Polyhymenophorea, with many compound oral cilia (Table 7.1). The same three classes were adopted by Corliss (1979) and, with slight changes in spelling, by Levine *et al.* (1980). Within these classes there were some differences between these three schemes in the status and relationships of some groups, which are at too detailed a level to pursue here, so that the widely-available scheme of Levine *et al.*, (1980) is the one outlined in Table 7.1, and used to provide the basis of Fig. 7.21.

Meanwhile, more attention was being paid to the ultrastructure of somatic kinetids, and, noting the two major fibre patterns associated with somatic kineties, one based upon kinetodesmata and the other on bundles of postciliary microtubules (Sleigh, 1966), Gerassimova and Seravin (1976, in Russian, but described in English by Seravin and Gerassimova, 1978) proposed the division of ciliates into two Classes, the Kinetodesmatophora, including hymenostomes, astomes, apostomes, thigmotrichs and peritrichs, and the Postciliodesmatophora, including gymnostomes, trichostomes, rhynchodines, spirotrichs, chonotrichs, suctorians and entodiniomorphs. Further comparisons of the organisation of the infraciliature and its morphogenesis led Small and Lynn (1981) to propose a different alignment of groups, dividing the Ciliophora into three Sub-phyla (Table 7.2), the Postciliodesmatophora, with postciliodesmata originating from dikinetids, the Rhabdophora, in which transverse micro-

Table 7.2   Classification of the Phylum Ciliophora used by Small and Lynn (1985) to the level of subclasses, with some colloquial names and examples from the text that belong to the groups named.

Phylum Ciliophora
  Sub-phylum Postciliodesmatophora
    Class Karyorelictea, e.g. ***Trachelocerca, Loxodes, Geleia***.
    Class Spirotrichea
      Sub-class Heterotrichia, e.g. ***Blepharisma, Stentor***, odontostomes.
      Sub-class Choreotrichia – tintinnids and oligotrichs, e.g. ***Halteria***.
      Sub-class Stichotrichia, e.g. ***Urostyla, Oxytricha, Stylonychia***.
  Sub-phylum Rhabdophora
    Class Prostomatea, e.g. ***Coleps***.
    Class Litostomatea
      Sub-class Haptoria – haptorids e.g. ***Didinium***, amphileptids, ***Dileptus***.
      Sub-class Trichostomatia – entodiniomorphs, trichostomes, ***Isotricha***.
  Sub-phylum Cyrtophora
    Class Phyllopharyngea
      Sub-class Phyllopharyngia, e.g. ***Chilodonella, Dysteria***, rhynchodids.
      Sub-class Chonotrichia, e.g. ***Spirochona***.
      Sub-class Suctoria, e.g. ***Ephelota, Acineta, Discophrya***.
    Class Nassophorea
      Sub-class Nassophoria, e.g. ***Nassula***, peniculines, e.g. ***Paramecium***.
      Sub-class Hypotrichia, e.g. ***Aspidisca, Eupiotes, Diophrys***.
    Class Oligohymenophorea
      Sub-class Hymenostomatia – hymenostomes, scuticociliates.
      Sub-class Peritrichia, e.g. ***Vorticella, Trichodina***.
      Sub-class Astomatia, e.g. ***Anoplophrya, Durchoniella***.
      Sub-class Apostomatia, e.g. ***Foettingeria, Polyspira***.
      Sub-class Plagiopylia, e.g. ***Sonderia***.
    Class Colpodea, e.g. ***Colpoda, Tillina, Bursaria***.

tubules from oral dikinetids support the cytopharynx, and the Cyrtophora, in which postciliary microtubules of oral dikinetids support the cytopharynx. This scheme, with minor alterations, is used in the new *Illustrated Guide* (Small and Lynn, 1985). Space is not available here for detailed comment on the arrangement of classes and sub-classes within this scheme, but the position of colpodids and peniculines, and the separation of euplotid hypotrichs from the stichotrich and other sporadotrich hypotrichs are of special interest. The authors emphasise that they expect their scheme to be revised as new information becomes available. It will be interesting to see how well the two main present schemes stand the test of time. Certainly many new sources of information can contribute to the discussion; one such is the distribution of different patterns of intramembranous particles in ciliary membranes (Bardele, 1981), which supports some of the conclusions of Small and Lynn. However, in spite of the great increase in the range and detail of information available, it is still often limited to very few species in many of the taxa, and confidence in the proposed relationships will grow when the correlations can be improved by availability of more detailed data. The outlines of a phylogenetic tree corresponding to the classification of Small and Lynn is included for comparison in Fig. 7.22.

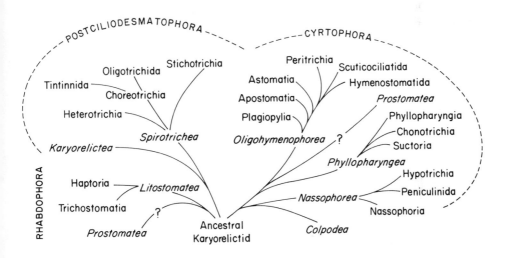

Fig. 7.22 A phylogenetic tree of Ciliophora based on the classification in Table 7.2, using information from Small and Lynn (1985) and additional personal comments from E.B. Small, who suggests that recent evidence would put the Prostomatea among the Cyrtophora.

8

Spore-forming parasitic protozoa

As mentioned in chapter 1 (pp. 5 and 8) the protists considered in this chapter all belong to phyla whose members are parasites in animals or protists and which usually produce infective spores. The form of the spores (Fig. 1.5), the form of the infective cells that emerge from the spores and the manner of their emergence are among features that distinguish the different phyla. The groups probably have little in common other than their mode of life and the possession of spores; their cells are very different and there is no evidence of any close phylogenetic relationships between them.

Phylum Sporozoa (= Apicomplexa)

All members of this phylum are endo-parasites with more or less complex life-cycles. They have long been regarded as distinct from other protozoa because of the presence of spores containing characteristic infective sporozoites, hence the name. However, in a number of groups the sporozoites are not enclosed in spores, and the character of these infective stages, which are elongate cells that move by gliding or body flexion, and lack either flagella or pseudopodia, is more universal than the possession of spores. Studies of fine structure have revealed several features, notably in the apical complex, that appear to be the important diagnostic characteristics of the group (Vivier, 1979). These features occur widely in all major groups of Sporozoa, and have led to wide use of the name Apicomplexa for this phylum (Levine *et al.*, 1980); however, since the name Sporozoa is familiar and well understood, change seems unnecessary (Corliss, 1984).

The apical complex (Fig. 8.1) is present in the infective stages of the life cycle, sporozoites and merozoites, and sometimes in the trophozoites. It comprises a conoid (occasionally absent) in the form of a truncated cone of short spiral microtubules, through the centre of which run extensions of the rhoptries, presumed secretory organelles whose main parts lie deeper in the cell. Around the conoid is an apical ring which links the anterior terminations of an array of longitudinal sub-pellicular microtubules, and associated with the rhoptries are a number of smaller dense inclusions usually called micronemes. The apical complex appears to function in attachment to or penetration of host cells; the rhoptries in particular, and perhaps also the micronemes, are supposed to release products that cause invagination of host cell membranes as a prelude to entry of the parasite into a host cell vacuole, e.g. in *Plasmodium* (Bannister 1977).

In extracellular stages of the life cycle the cell membrane is typically underlain

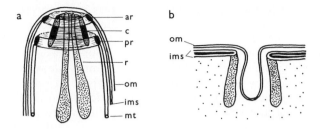

Fig. 8.1 Some characteristic structures of sporozoa; **a**, polar organelles include the anterior apical ring *ar*, the conoid *c*, the posterior apical ring *pr*, rhoptries *r*, the outer membrane *om*, inner membranes *ims* and microtubules *mt*: **b**, a micropore, with outer membrane invaginated within the cylindrical dense collar and inner membranes interrupted at the outer end of the collar.

by a double membrane layer, apparently formed from the flattening of pellicular alveoli or vesicles, as found in dinoflagellates (p. 110) and ciliates (p. 190), giving a triple-layered structure. One or several micropores (Fig. 8.1) are present, and where these occur there is a break in the inner membranes; these micropores are thought to represent cytostomal sites of endocytosis. Nuclear division is a closed pleuromitosis (p. 47), sometimes with a cytoplasmic centriole composed of 9 singlet microtubules, and sometimes with only a dense plaque on the nuclear envelope at the poles of the half spindles. Meiosis, which apparently takes place in a single division (p. 71), is zygotic and the main stages of the life cycle are haploid. Mitochondria usually have few, rather vesicular or ampullar cristae, but in forms parasitic in blood cells the cristae or even the mitochondria tend to be lost. Golgi bodies and food reserves in the form of lipid globules and paraglycogen granules are often present. Flagella are absent, except on the male gametes of some species.

The life cycle typically shows a sequence of stages (Fig. 8.2) characterised by episodes of multiple fission (schizogony) that result in great increases in

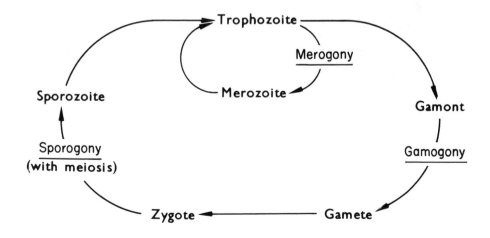

Fig. 8.2 The sequence of stages in the life cycle of a typical sporozoan.

numbers. The sporozoite that enters the host grows as a trophozoite in an extra-cellular (most gregarines) or intracellular position. It may then undergo schizo-gony (merogony) producing numerous merozoites, which grow once more as trophozoites. Merogony often occurs repeatedly, but is absent from most grega-rines. Eventually the trophozoites transform into gamonts within which gametes are formed, with or without a further phase of schizogonic division (gamogony); division is usual in formation of male gametes, which may some-times be flagellate and sometimes amoeboid. The zygote formed by gamete fusion undergoes meiosis in its first division, and may undergo further schizo-gonies in a process of sporogony producing few to many sporozoites. These may be enclosed in a spore coat (gregarines and coccidians) or transmitted naked to the next host by invertebrate vectors (haemosporidians and piroplasms).

The phylum contains four clearly distinct groups of organisms, the gregarines, coccidians, haemosporidians and piroplasms, and these will be discussed separately here. Whilst authorities agree that the gregarines form a distinct group, some place the haemosporidians within the coccidian group and class piroplasms separately (Levine *et al.*, 1980; Levine, 1985), whilst Vivier (1982) proposed grouping haemosporidians and piroplasms as the Haema-tozoa, with a growth stage in blood cells and no spores, and separated the spore-forming coccidians. The infective 'zoospore stage' of *Dermocystidium* (= *Perkinsus*), a parasite of oysters, has features of the apical complex, including a conoid-like structure, rhoptries, micronemes and a polar ring, as well as micro-pores, additional membranes and microtubules in the pellicle; however there is no trace of sex or schizogony and the zoospore is biflagellate with a posterior naked flagellum and an anterior flagellum with unilateral mastigonemes (Perkins, 1976). Some authors regard it as a sporozoan (Levine *et al.*, 1980), whilst others suggest it is a flagellate (Vivier, 1982). It is clearly not a member of any of the groups described below, and, while its features may suggest a connec-tion between sporozoa and heterokont flagellates, more information is required to classify it correctly.

The gregarines

These are distinguished from most other sporozoa by having extracellular trophozoites, which are often large and mobile; their gametogenesis is preceded by fusion and encystment of gamonts, with large numbers of gametes produced by each female gamont, and each oocyst forming a spore. They share with coccidians the presence of a typical conoid and typical mitochondria, binary fission of the nucleus, the production of spores and the presence of amylopectin food stores. In most gregarines multiplication is restricted to the phases of gamogony and sporogony; in the absence of merogony each sporozoite can produce only one trophozoite, so large infections are less likely to occur than in forms with merogony.

The most familiar members of the class Gregarinea are species of *Monocystis* or related genera commonly found in the seminal vesicles of earthworms. A typical life cycle is shown in Fig. 8.3. It is assumed that the earthworms eat the sporocysts of *Monocystis* and that the sporozoites escape in the gut and find their way to the seminal vesicles. Here the trophozoite stage grows amongst the developing sperm, probably without any intracellular stage, eventually reaching

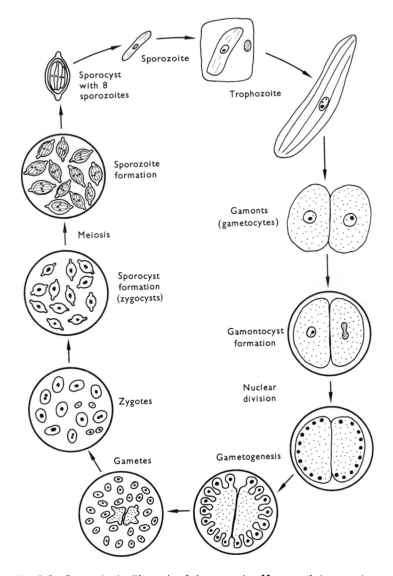

Fig. 8.3   Stages in the life cycle of the gregarine *Monocystis* (*see* text).

a length of several hundred μm (or even 1 mm in *Nematocystis*). Two full-grown trophozoites associate in pairs as gamonts, and a common gamontocyst wall is secreted around them. Large numbers of amoeboid gametes are formed by each gamont, and these subsequently fuse in pairs within the gamontocyst. Following the secretion of a sporocyst wall around each zygote, these diploid cells divide three times forming eight haploid sporozoites within each sporocyst. Gamonto-cysts containing numerous sporocysts are frequently found in the seminal vesicles of earthworms, but it is not clear how these find their way to the soil to infect other worms, except at the death of the host.

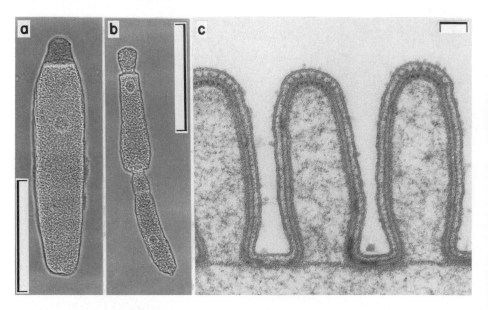

Fig. 8.4 Micrographs of trophozoites of *Gregarina*. *a,b*, Photomicrographs of *Gregarina* from the gut of a mealworm showing in **a** the division of the body into protomerite and deutomerite regions and in **b** two individuals attached together in syzygy; **c**, an electron micrograph of a section through epicytic folds of *Gregarina garnhami* from the locust *Schistocerca*, showing the three membrane layers and filaments associated with inner and middle membranes at the crests of the folds (Micrograph from Walker *et al.*, 1979). Scales in **a** and **b**, 100 μm and in **c**, 0.1 μm.

Other gregarines may commonly be found in the body cavities and gut of other worms and in insects. A good source of material for laboratory study is the gut of larvae of the meal-worm (*Tenebrio*), where several species of *Gregarina* may be found. In these forms the body has three parts, the epimerite (with apical complex) by which the young parasite anchors itself to a host cell, a clearer protomerite region and a more dense deutomerite region which contains the nucleus. Older trophozoites lose their attachment to host cells (Fig. 8.4a), and may be found gliding unidirectionally over surfaces in a remarkable fashion (King, 1981); the surface of the body is formed into complex longitudinal folds within which are longitudinal filaments (Fig. 8.4c) (Walker *et al.*, 1979), but the mechanism of gliding has so far eluded explanation. These trophozoites associate in pairs (in syzygy) at an early stage in growth (Fig. 8.4b), but continue to glide actively together. The first layers of the gamontocyst are not formed until growth is completed, and the cysts may be passed out with the faeces before gametogenesis and spore formation have taken place inside them.

The archigregarines, regarded as the most primitive gregarines, are exemplified by *Selenidium* (Fig. 8.5), species of which are found in the gut of marine polychaetes. In these forms the trophozoite looks very like a large, longitudinally or spirally ridged, sporozoite, which retains an apical complex and performs pendular movements. Each zygote forms a spore with four sporozoites. Reports of merogony in the life cycle of these forms are doubtful (Vivier,

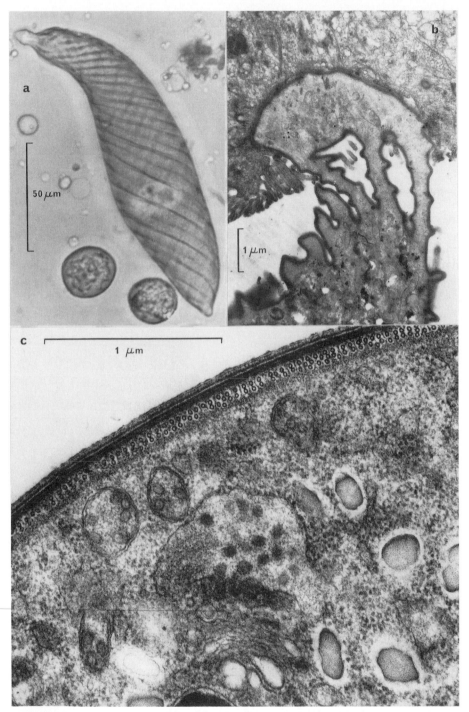

Fig. 8.5 The archigregarine *Selenidium terebellae*. **a**, Photomicrograph of a trophozoite from the gut of the marine worm *Terebella lapidaria*: **b**, electron micrograph showing the attachment of the anterior end of a trophozoite to the gut wall of the host: **c**, electron micrograph of a transverse section through a trophozoite showing the three membrane layers underlain by a layer of microtubules in an epiplasmic zone and mitochondria and a dictyosome in the cytoplasm. Micrographs by A.G.H. Dorey.

1982). **Monocystis** and **Gregarina** are eugregarines, as are the majority of gregarine sporozoans; generally eight sporozoites are formed from each zygote in these forms. However, there is a third group, the neogregarines (or schizogregarines), which are rather poorly known, but are distinguished by the occurrence of merogony during the trophic stage, which occurs in the malpighian tubules, intestine or haemocoel of insects, e.g. **Mattesia**.

The Coccidea

As considered here these are typically spore-forming parasites (i.e. excluding haemosporidians). Their trophozoites are mostly non-motile and usually intracellular, so that they are smaller than gregarines, and often show a rapid loss of the apical complex and the two inner membranes after entering host cells, where they reside within a parasitophorous vacuole. This appears identical with a phagocytic vacuole, but the parasite apparently evades digestion (p. 24). Each female gamont produces a single female gamete, and gamonts of the two sexes undergo gametogenesis without encystment, and usually before meeting together; they are usually oogamous with small flagellated male gametes. Most

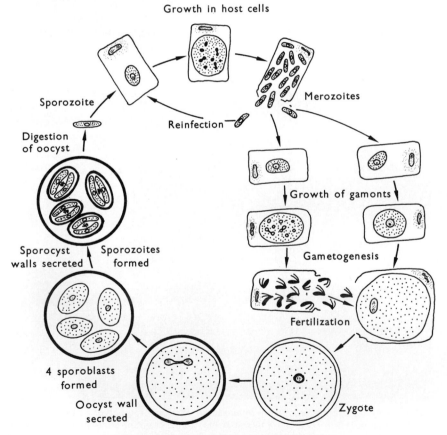

Fig. 8.6 Stages in the life cycle of the coccidian *Eimeria steidae*, a parasite in the liver cells of rabbits (see text).

of these features they share with haemosporidians, and they also share some features with gregarines (p. 224).

These common features of coccidians can be illustrated by describing the life cycle of *Eimeria steidae* (order Eimeriida), which is parasitic in rabbits (Fig. 8.6). The oocyst ingested by the rabbit hatches in the small intestine of the host, where the cyst wall is probably digested by trypsin. The sporozoites which emerge enter the gut wall and pass via the hepatic portal blood system to the liver where they enter the epithelial cells of the bile ductules. Within these cells the parasite grows as a trophozoite, apparently feeding phagocytically, before undergoing merogony to produce numerous merozoites which break out of the cell and enter other host cells as trophozoites. After several cycles of merogony the trophozoites develop into gamonts instead of schizonts. Two forms of gamont can be distinguished: macrogamonts which form a single macrogamete that remains within the host cell, and microgamonts which divide to produce numerous microgametes, each with two free flagella and one recurrent attached flagellum, that swim away to fuse with macrogametes. The zygote secretes a thick oocyst wall, divides meiotically (supposedly by a one-division meiosis, p. 71) and mitotically to form four sporoblast cells, each of which secretes a sporocyst wall and divides once more to form two sporozoites. The oocysts leave the host cells early in sporogony and this process is completed outside the host within a few days of the cysts being voided with the faeces. Contamination of food is presumed to be the means of infection of new hosts, and the coprophilous habits of rabbits make infection more likely. This parasite causes severe hepatitis of the rabbits and is sometimes fatal.

Other species of *Eimeria* are pathogenic in various domestic animals (Davies *et al.*, 1963; Baker, 1969; Kreier, 1977), including *E. necatrix* and *E. tenella*, which are found in the small intestine and caeca of chickens and are often fatal, species causing diseases in other farmed birds, and several species which cause various forms of 'coccidiosis' in cattle. Oocysts of the related species *Isospora* can be distinguished from those of *Eimeria* by the presence of four sporozoites in each of two sporocysts within the oocyst (Fig. 8.7); in the same way other

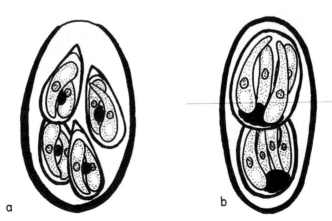

Fig. 8.7 Oocysts of two common genera of coccidians, **a**, *Eimeria* (35 × 20 μm) and **b**, *Isospora* (30 × 15 μm) (*see* text).

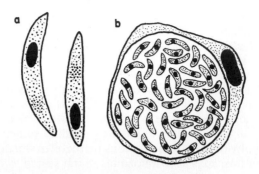

Fig. 8.8 *Toxoplasma gondii*; **a**, two isolated zoites (5 μm long) and **b**, a cyst within a host cell, which may be reduced to a wall around the parasites.

genera of eimeriids may be distinguished by the number and arrangement of sporozoites in the oocyst (Levine, 1985). Several species of *Isospora* are found in domestic animals, including the pathogenic *I. bigemina*, which is found in the small intesetine of dogs. Related species found in man are only mildly pathogenic.

Most species of *Eimeria* and *Isospora* are quite specific in their hosts; by contrast the parasite long known as *Toxoplasma gondii* has been found in a very wide range of birds and mammals, including man, from all parts of the world. The rate of infection with this parasite is estimated to exceed 25% in man and to exceed 50% in some domestic animals in Britain and the USA, and may reach even higher levels in some other lands. The very small parasites appear in macrophage cells, where they multiply by a special form of binary fission called endodyogeny (in which two daugher cells form within the parent cell cytoplasm) until they eventually fill the cell and burst it. Repeated cycles of this activity produce enormous numbers of small parasitic cells. Some of these 'zoites' find their way to other tissues – nervc, liver, kidney, muscle and lung – where they form resistant cysts containing numerous *Toxoplasma* cells (Fig. 8.8). For many years the taxonomic relationships of these forms were obscure and they were classed separately until typical apical complexes were found in electron micrographs. Then in 1970 it was shown that suitable infected hosts produced characteristic oocysts in the faeces, and careful study of the disease in cats revealed that ingestion of *Toxoplasma* cysts is followed by profuse merogony and gametogony of a coccidian type in the small intestine (Hutchison *et al.*, 1970). These observations led to the conclusion that *Toxoplasma* is related to *Isospora*, and indeed undergoes a normal eimeriid cycle in the cat, but if it is eaten by certain alternative hosts it is unusual in forming cysts containing trophic cells (typical sporozoan cysts are formed following zygote formation). Toxoplasmosis is normally a mild disease in human adults, except in cases of immune deficiency, but is a more severe pre-natal disease through which the parasite may damage the liver, retina and brain, causing jaundice, blindness or death.

A parasite with some similar features is *Sarcocystis*, which is found encysted in the muscles of warm-blooded vertebrates, including man. The cysts may be 1 or 2 mm long and contain numerous parasite cells in various stages of develop-

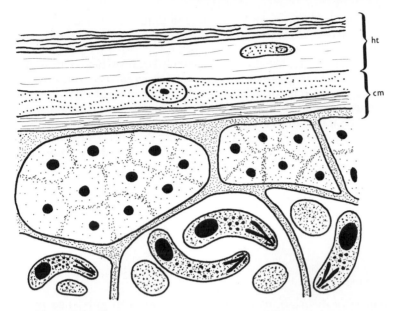

Fig. 8.9 *Sarcocystis tenella* from sheep muscle. A portion of the cyst is shown surrounded connective tissue of the host (*ht*) and enclosed within a three-layered cyst membrane (*cm*), the innermost layer of which is continous with trabeculae which isolate compartments containing trophozoites of *Sarcocystis* at different stages of growth – some are immature and of irregular shapes, while others are crescentic with polar organelles and about 10 μm long.

ment (Fig. 8.9). It was found early in the 1970s that infective merozoites from cysts of *S. tenella* in sheep would enter cells of the small intestine of a dog, transform to gamonts, then gametes, the zygotes from which form oocysts typical of *Isospora* (see Markus *et al.*, 1974). If these oocysts are eaten by a sheep the sporozoites released enter blood vessels and undergo several cycles of merogony in endothelial cells before settling in muscles, where they grow and multiply extensively to produce a special type of infective merozoite in the resting cyst. This life cycle is a typical eimeriid one except that it is heteroxenous (involving more than one host), rather than the homoxenous (single host) cycle found in *Eimeria*. The parasite has been renamed *S. ovo-canis*.

Another eimeriid that has a heteroxenous life cycle is *Aggregata eberthi*, whose merogony occurs in a crustacean and whose gamogony and sporogony in cephalopod molluscs results in the release of numerous resistant sporocysts each with three sporozoites. Blood parasites of the genus *Lankesterella* are eimeriids found in birds, reptiles and amphibia. They differ from malarial parasites in that multiplication phases occur in tissue cells of the host; e.g. growth stages of *L. minima*, including the oocysts, occur in frog endothelial cells – these oocysts release sporozoites that enter erythrocytes which are taken up by leeches and pass without development to another frog when it eats the leech.

In members of the order Adeleida, gamonts associate together before gametogony. *Klossia* is an adeleid parasite found in the kidney of gastropod molluscs, including terrestrial snails and slugs. It undergoes merogony, gamont formation, mating of gamonts, gametogony, fertilisation and sporogony in the epithelium of the kidney, and oocysts containing sporozoites in thin-walled sporocyst envelopes are passed out in the urine. A new host is infected by ingestion of cysts.

Some blood parasites are classed as adeleids because their gamonts pair in the gut of the vector before gametogenesis. *Haemogregarina* trophozoites live and undergo merogony in blood cells (and some other cells) of vertebrates; gamonts formed in blood cells are ingested by a blood-sucking invertebrate vector, leech, insect, tick or mite, and emerge from blood cells to associate in pairs, form gametes and motile zygotes in the host's gut. Zygotes complete sporogony to produce sporozoites which may either be injected into a vertebrate host by the vector when it feeds, or in some forms may be taken in by a vertebrate when it eats the vector.

The small order Protococcida (= Coelotrophida) includes parasites that grow extracellularly in the gut or coelom of annelids. They lack merogony but show gametogony with biflagellate male gametes, e.g. *Grellia* (= *Eucoccidium*).

The Haemosporidea

Comprehensive descriptions of this group have been published by Garnham (1966) and Kreier, (1977). Haemosporidia share with piroplasms a reduction of the mitochondria and of the golgi system, the lack of a conoid (usually), an absence of amylopecten food reserves, and a multiple nuclear division. In both groups the gamonts, like those of coccidia, develop into gametes without encystment, each macrogamont producing a single macrogamete (except perhaps in piroplasms). The zygote is a motile ookinete which undergoes sporogony by direct division to form many sporozoites without spore coats. The trophozoites grow intracellularly, mostly in blood cells, where merogony occurs; in this stage the groups differ in that the cell membrane of haemoporidians is very closely enclosed by the membrane of the host vacuole, while piroplasms lie directly in the cytoplasm, with only one membrane between the cytoplasm of the host cell and the parasite. Because of this difference, haemosporidians feed by endocytosis of host cytoplasm while piroplasms feed by direct absorption without phagocytosis. Parasites in both groups are heteroxenous, with essentially similar life cycles.

Haemosporidia have a life cycle which involves gamogony and sporogony in dipteran insects and merogony in vertebrates. Species in several haemosporidian genera (including *Plasmodium*) whose members enter vertebrate erythrocytes metabolise haemoglobin with the production of characteristic pigment masses within the parasite cells, but these pigment masses are not found in other parasites, including piroplasms and all members of other groups, found within erythrocytes.

Several species of the malarial parasite *Plasmodium* are found in man, all with a similar life cycle (Fig. 8.10). Naked fusiform sporozoites are injected into the skin with the saliva of an infected *Anopheles* mosquito as it commences to feed. Sporozoites pass in the blood to the liver, where they enter parenchyma cells and grow. After between 5 and 15 days they undergo a stage of merogony, known as exoerythrocytic schizogony, which results in the production of 10–30 000 merozoites. Upon release these merozoites enter erythrocytes and

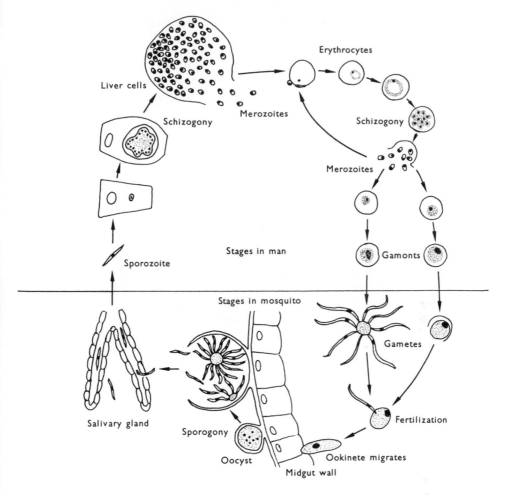

Fig. 8.10 Stages in the life cycle of the malaria parasite *Plasmodium* (*see* text).

become trophozoites which are first seen as a ring stage, having engulfed masses of host cytoplasm, and subsequently grow to fill much of the cell before commencing schizogony. The 6–24 merozoites that are released at lysis of the erythrocyte enter other erythrocytes and repeat the growth and merogony cycle. Later some of the trophozoites become gamonts instead of schizonts, and differentiate to form macrogamonts and microgamonts within the erythrocytes. These gamonts remain dormant in the erythrocyte unless this is ingested by a mosquito. Development of parasite gamonts quickly recommences in the stomach of the insect and both forms of gamont emerge from the red cells. The macrogamont becomes a spherical macrogamete without division. The microgamont undergoes three nuclear divisions, and develops eight flagellar projections; one nucleus enters each projection, and this then breaks away as a motile microgamete. Fertilisation takes place in the stomach of the mosquito. The zygote is an active ookinete which burrows through the walls of the stomach and comes to lie on the outer surface of the stomach, where it secretes a thin oocyst wall. Growth of the zygote is followed by sporogony, which results in the production of very large numbers of sporozoites within each oocyst. The sporozoites break out of the oocyst into the haemocoel of the insect and many of them pass to the salivary glands, whence they may be injected into a new vertebrate host at the next feed of the mosquito.

The most pathogenic species of malaria parasite found in man is *P. falciparum* which probably causes more human deaths in tropical countries than any other disease organism. The other species, *P. vivax, P. malariae* and *P. ovale*, do not usually cause fatal disease, although they cause debilitating recurrent fevers. When the erythrocytes burst and the merozoites are released, other materials are liberated into the blood and among these are toxic substances that cause fever. Merogony tends to become synchronised so that vast numbers of erythrocytes burst at the same time, giving rise to fever every 48 hours in three of the species, or every 72 hours in *P. malariae*. The erythrocytes containing parasites may normally be found circulating freely in the blood, but in *P. falciparum* only the ring stages and mature gamonts are common in blood smears, since erythrocytes containing trophozoites of this species tend to stick in the blood capillaries; this occurs because knobs that develop on the eythrocyte membranes form focal junctions with endothelial membranes (Aikawa and Miller, 1983). The ensuing capillary blockage results in oxygen shortage in tissues and often rupture of the blood vessels. If capillary blockage occurs in the brain, death usually follows, and if it occurs elsewhere it may also cause serious damage. The production of antibodies by the host (Deans and Cohen, 1983) or treatment with drugs (Peters, 1970; Fitch, 1983) can enable the host to destroy erythrocytic stages, but in *P. malariae* and *P. vivax* the disease organism may persist for many years as a latent infection in exoerythrocytic sites, from which merozoites may emerge to cause further outbursts of fever ('relapses').

Other species of *Plasmodium* are found in other mammals, birds and reptiles. Two species that have been widely studied because they can be maintained in laboratory animals are *P. berghei* in rats and mice and *P. gallinaceum* in chickens; these are used among other things for the testing of antimalarial drugs. Species of *Haemoproteus*, which are common in birds and reptiles, do not have an erythrocytic schizogony stage, so that the only parasites found within red blood corpuscles are the gamonts; their schizogony takes place in endothelial cells of the blood vessels and the parasites are transmitted by midges and other dipterans. *Leucocytozoon*, which is common in domestic and wild birds, and is usually transmitted by blackflies (*Simulium*), is again only found in blood cells (red or white) às gamont stages, merogony occuring in other tissues.

The piroplasms

These are small parasites found in the erythrocytes of vertebrates; they have many of the features of haemosporidia, as mentioned on p. 232 (Baker, 1969; Kreier, 1977), but they do not contain pigment masses, are surrounded by only a single membrane within the host cell and are transmitted by ticks. Piroplasms have been found in vertebrates of all classes, and occasional infections have been recorded in man, particularly in cases of immune deficiency or removal of

Fig. 8.11 Two cells of **Babesia bigemina** in an erythrocyte cell.

the spleen. Several species of **Babesia** and **Theileria** are pathogenic in domestic animals, and are probably responsible for the most serious diseases of domestic animals throughout much of the world. Red-water fever in cattle is the best known of these, but others in dogs and horses are widespread.

In the vertebrate host sporozoites of **Theileria** first enter lymphocytes where they undergo repeated schizogonies and cause hypertrophy of lymph nodes; those merozoites that enter erythrocytes seem not to multiply there, but probably become gametocytes available to infect another tick. **Theileria** only passes from one tick stage to the next, and does not pass transovarially. By contrast, **Babesia** seems to occur only in red cells, and sporozoites may enter them directly; within the blood cells the small parasites, only 1–5 μm across (Fig. 8.11), divide to produce a small number (often 2 or 4) merozoites, which break out of the cell and release fever-inducing toxic products and blood pigments which are excreted – giving the name red-water fever. Infected animals may show anaemia and fever, accompanied by blockage of capillaries and anoxia of the tissues, as well as other manifestations of disease.

The discovery in 1893 by Smith and Kilborne of the transmission of red-water fever by ticks was the first demonstration of the transmission of a disease by an arthropod vector. However, only recently has information been obtained about the details of the parasite life-cycle in the tick. In the case of **Babesia**, for example, Rudzinska *et al.*, (1984) followed the fate of parasites taken up with a blood meal by larval ticks using electron microscopy, and demonstrated a series of events comparable with those seen when **Plasmodium** is ingested by a mosquito, but with differences of detail. Most erythrocytes ingested by the tick are lysed by enzymes, but a few escape intact, and some of these contain parasites. Maturation of gametocytes involves first the appearance of cytostomes displaying active phagocytosis (not seen within vertebrate hosts), and then the development of protruding tails and an unique arrowhead structure. These gametocytes divide once or twice, before or after emerging from the erythrocytes, and fuse in pairs – apparently only one member of each pair having an arrowhead. The zygote, with a single arrowhead, penetrates a cell of the gut wall, emerges into the haemocoel and migrates directly or indirectly to the salivary gland, where sporogony takes place to produce the sporozoites. The active zygote (ookinete) appears to use the arrowhead and a coiled membranous structure in place of (missing) components of the apical complex in release of substances involved in penetration of and migration through host cells. Ticks take few, large, blood meals, and they may only feed once in each instar of growth. Parasites within the tick survive the moult of the host to be injected with saliva at the next feed. In some instances **Babesia** enters the eggs of the tick and survives to the next generation to be injected into the first vertebrate from which the young tick feeds.

Dactylosomatid piroplasms are found in the erythrocytes of cold-blooded vertebrates, but their life cycle and mode of transmission are unknown.

Phylum Microsporidia (= Microspora)

These are intracellular parasites found in animals and protists (Vivier, 1979; Weiser, 1985a; Canning and Lom, 1986; Larsson, 1986). They form unicellular spores which contain a single sporoplasm, with no limiting membrane separating it from other spore structures, a coiled tubular filament, which is not within a polar capsule, a membranous polaroplast, often with lamellate and vesicular regions, and often a spongy 'posterior vacuole', which, with the polar filament, is derived from a Golgi-type structure. The membrane of the trophic stage is usually in direct contact with host cytoplasm, although later some parasite cells occupy parasitophorous vacuoles; molecular nutrients are absorbed directly without phagocytosis or pinocytosis. Flagella and mitochondria are absent, and there is doubt about the presence of a true Golgi system. During the closed acentriolar pleuromitosis the intranuclear spindle microtubules converge upon polar plaques on the nuclear envelope. Microsporidian ribosomes are of prokaryotic size (70S), and examination of the ribosomal RNAs of *Vairimorpha* show them to be unusual in both the large subunit (Vossbrinck and Woese, 1986) and especially in the apparently primitive 18S RNA of the small subunit (Vossbrinck *et al.*, 1987), suggesting a very early divergence from the eukaryote stock (p. 5). Multiplication (merogony) takes place within host cells either by binary fission or schizogony, and is followed by sporogony in which two spore wall layers, one with chitin, and the internal structures of the spore (Fig. 8.12a) are formed. Meiotic nuclear division is thought to occur during sporogony (at least in some species); autogamic nuclear fusion has been reported to occur late in merogony in some microsporidian parasites of invertebrates. Many species possess diplokaryons – pairs of closely appressed nuclei – which may be present for a greater or lesser part of the life cycle, including the spore stage, e.g. in *Nosema*; the ploidy level of these nuclei is uncertain, and may vary.

Two species of the genus *Nosema* cause epidemic diseases of economic importance. Spores of *Nosema apis* may be found in the cells of the midgut wall and malpighian tubules of adult honey bees, in which this parasite causes the destructive nosema disease. Mature spores are about 5 μm long and 3 μm wide and have a thickened wall, at one pole of which is attached an inverted tubular filament that is coiled around the single binucleate sporoplasm (Fig. 8.12a). Following ingestion of a spore by the host, the polar filament is everted, probably by expansion of the polaroplast or posterior vacuole, and the amoeboid sporoplasm, which gains a single membrane upon emergence, is injected directly through the tubular filament into the cytoplasm of a cell in the host gut wall. As the parasite grows and multiplies the host cell becomes distended and hypertrophied. Each sporont produced in the host cell develops into a single spore. These spores may be shed into the gut and passed out with the faeces to spread the infection to new hosts, or may germinate in the same insect. Another important species of the same genus, *Nosema bombycis*, is responsible for

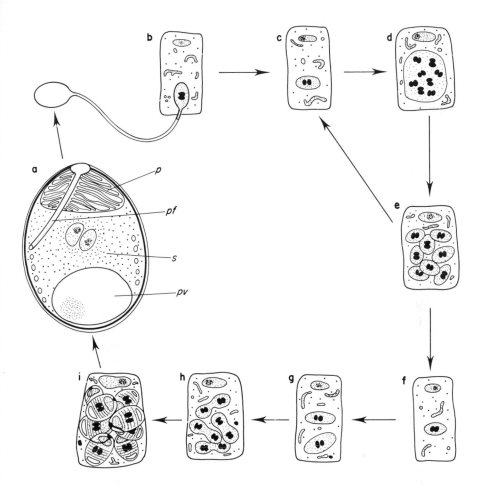

Fig. 8.12 The life cycle of the microsporidian **Nosema**; **a**, the mature spore with binucleate sporoplasm *s*, polar filament *pf*, polaroplast *p* and posterior vacuole *pv*: **b**, the sporoplasm emerges from the everted polar filament directly into the cytoplasm of a host cell, grows **c**, to form a plasmodium **d** which divides by merogony to produce merozoites **e** that enter other host cells **c** to repeat merogony or **f** to undergo sporogony: the latter parasites divide by binary fission **g**,**h** to produce numerous sporoblasts which develop into spores **i**.

pebrine disease of silkworms, in which tissue cells of any type may be infected at any stage of growth, so that even eggs may be infected and the larvae may die without spinning silk coccoons.

Microsporidia generally invade the muscles, intestinal epithelium, lymphocytes and adipose tissue of invertebrates, especially insects, but almost all animal groups are hosts to them. Microsporidia parasitic in vertebrates (Canning and Lom, 1986) are presumed to hatch in the gut, but their mode of transfer to remote sites of multiplication is uncertain. *Encephalitozoon cuniculi* has been found in the kidney and connective tissues of rodents and other mammals of many groups, always in parasitophorous vacuoles; it may be carried in infected macrophages from primary sites to many other tissues, especially the kidney. This is the commonest microsporidian to have been reported from man, generally only from people with deficient immunity, and has been reported to be capable of transplacental transmission. A species of **Enterocytozoon** also occurs in man.

A common disease of sticklebacks (**Gasterosteus**) and some other fish from freshwaters is caused by species of **Glugea** which cause the formation of white cysts, up to 5 mm across, in the skin or muscles of the fish. Other species of this genus and of other genera are common in fish, amphibians and reptiles.

These microsporidia belong to the class Microsporea. The spores of members of the class Rudimicrosporea have a simpler polar tube; these forms parasitise gregarines. It is likely that many species of microsporidia remain to be studied and described. Spore characters are important in classification (Fig. 8.12 and 8.13).

Phylum Haplosporidia (= Ascetospora in part)

These are parasites in marine invertebrates, notably annelids and molluscs (Sprague, 1979; Desportes and Nashed, 1983). Their unicellular spores contain a single, uninucleate, amoeboid sporoplasm but no polar filament. The sporoplasm in **Minchinia**, a parasite in the mollusc **Dentalium**, develops as an extracellular plasmodium with diplokaryotic nuclei. Division of these multinucleate plasmodia, probably associated with meiosis, results in the formation of uninucleate cells that appear to fuse in pairs to form binucleate sporoblasts. The two nuclei in these cells fuse together and then a major part of the cytoplasm partially separates from the nucleated cell and grows around the latter, first surrounding it like a cup and then enclosing it as a complete spore wall, except for an apical opening that is closed by an operculum, e.g. **Haplosporidium**, (Fig. 8.13b). At the apex of the inner sporoplasm

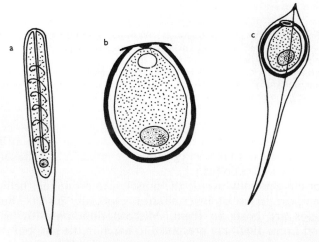

Fig. 8.13 Spores of a microsporidian **a** and two haplosporidia **b**,**c**; a, *Mrazekia caudata* (20 μm) from the worm **Tubifex**: **b**, *Haplosporidium limnodrili* (10 μm) from the worm **Limnodrilus**: **c**, *Urosporidium fuliginosum* (10 μm, excluding projection) from the worm **Syllis**.

is a globular mass of vesicular material, derived from endoplasmic reticulum, which is thought to expand to displace the operculum when the spore germinates.

These parasites possess mitochondria with few, rather vesicular, cristae, but lack flagella. In the closed nuclear mitoses the spindle microtubules converge on polar plaques well inside the nuclear envelope. Dense, membrane-bound granules named 'haplosporosomes' are present in the spores of these cells. The spores in some species have long projections, e.g. *Urosporidium* (Fig. 8.13c). There are a number of similarities between these parasites and the Microsporidia, but also some marked differences.

Phylum Myxosporidia (= Myxospora, = Myxozoa)

Myxosporidia are parasites found in the body cavities or tissues of vertebrates and invertebrates (Kudo, 1966; Poisson, 1953; Weiser, 1985b). Their spores are complex multicellular structures constructed usually of two or more valves enclosing one or more infective amoeboid sporoplasms and one to several polar capsules containing coiled filaments. These polar capsules with extrusible polar filaments are closely similar in their structure and morphogenesis to cnidarian nematocysts. Because of these features, and the existence of multicellular stages in which some cells are purely somatic and have no reproductive future, the Myxosporidia are now usually regarded as having been derived from multicellular animals, possibly members of the cnidarian stock. They are mentioned here because they have been traditionally regarded as protozoa. Notable cytological features are the presence of plate-like cristae in the mitochondria, a total absence of flagella, closed acentriolar nuclear division with internal spindles centred on plaques at the nuclear envelope, and the occurrence of endogenous divisions in which one cell is formed totally within another (a feature also reported by light microscopy in some cnidarians, but requiring ultrastructural confirmation).

Myxobolus pfeifferi, a member of the class Myxosporea, causes a 'boil' disease of cyprinid fish, as a result of which vast numbers of spores escape from burst cysts in the surface tissues of the fish. At one end of each spore are two polar capsules, each containing a coiled polar filament, and in the remaining space between the two shell valves is a single sporoplasm that is initially binucleate, but later appears to have only a single nucleus (Fig. 8.14). When a spore is swallowed by a suitable host fish, the polar filaments are everted and presumably serve for anchorage in the gut of the host. The amoeboid sporoplasm emerges when the spore wall breaks, passes through the gut wall and migrates, possibly in the blood vessels, to the muscles and connective tissues of the body wall. Here the parasite grows and its nucleus divides to form a syncytial mass which may break up and spread in a multiplicative trophozoite stage. The host tissue around the parasite becomes modified to form a thick envelope and a cyst is formed that may be easily visible to the naked eye. Certain nuclei of the syncytium become surrounded by dense cytoplasm and a cell membrane to form generative cells. Two of these cells come together with one enclosing the other (in some other species the nucleus of a single generative cell divides and one daughter cell is formed endogenously within the other). Each cell divides, the outer ones forming envelope cells of the pansporoblast, and each inner cell

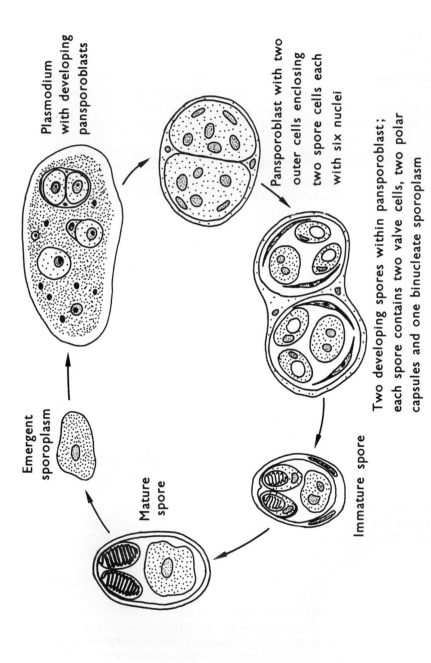

Plasmodium with developing pansporoblasts

Pansporoblast with two outer cells enclosing two spore cells each with six nuclei

Two developing spores within pansporoblast; each spore contains two valve cells, two polar capsules and one binucleate sporoplasm

Immature spore

Mature spore

Emergent sporoplasm

Fig. 8.14   The life cycle of the myxosporean *Myxobolus pfefferi* (see text).

forming a sporoblast. Two complete spores are formed within each pansporoblast, one from each of these inner cells. Division of a sporoblast cell forms one cell which will divide twice more to produce two valve cells and two capsule cells, and a second sporoplasm cell, which does not divide again although its nucleus divides (Fig. 8.14). Most nuclei of the spore are diploid, but the nuclei of the sporoplasm cell appear to be haploid (Uspenskaya, 1976), and are believed to undergo autogamous fusion later. Some reports suggest that formation of gametic nuclei and nuclear fusion takes place during the plasmodial stage of the cycle in myxospora. After maturation spores remain dormant, but, following release by rupture of the cyst, they may be swallowed by a new host and commence a new cycle of growth.

Other members of the class Myxosporea are also parasites of cold-blooded vertebrates, principally fish. They are best identified by spore characters, since they differ in the number and shape of shell valves and in the number and arrangement of polar capsules (Fig. 8.15a–d). Some have been seen to show intracellular developmental stages with characteristic endogenous division, but these forms did not produce mature spores (Lom and Dykova, 1985). Others, like *Sphaerospora renicola*, may develop, and presumably multiply, extracellularly in sites like renal tubules, where they produce only small pseudoplasmodia, each equivalent to the pansporoblast of *Myxobolus* (Lom, Dykova and Lhotakova, 1982); two uninucleate sporoplasms are found in spores of *S. renicola*. Generally these extracellular forms, whether they are in tissues

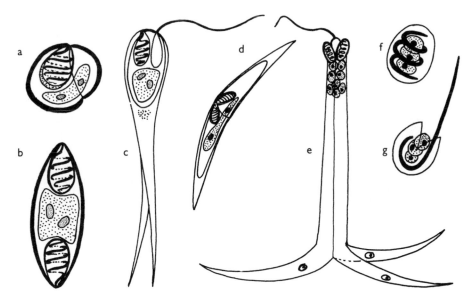

Fig. 8.15   Spores of myxosporidia and Helicosporidium; **a**, the myxosporean *Unicapsulina muscularis* (6 μm) from halibut: **b**, the myxosporean *Myxidium lieberkuhni* (20 μm long) from pike: **c**, the myxosporean *Henneguya* sp. (50 μm long) from freshwater fish, with one capsule everted: **d**, the myxosporean *Ceratomyxa arcuata* (25 μm long) from gall bladders of marine fish: **e**, the actinosporean *Triactinomyxon ignotum* (150 μm long) from gut epithelium of the worm *Tubifex*: **f,g**, *Helicosporidium parasiticum* from dipteran larvae, showing complete spore (5 μm) and a hatching spore with filamentous cell and three sporoplasms.

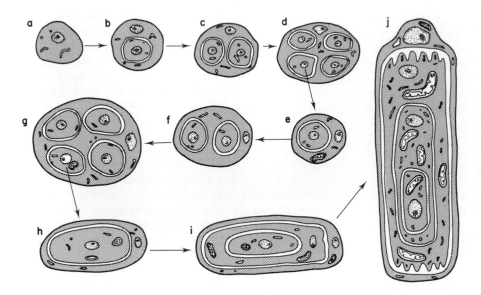

Fig. 8.16 Sporogenesis in *Paramyxa paradoxa*, from the larva of a polychaete worm. The amoeboid parasite **a** develops among host cells but does not enter them. This stem cell grows and first undergoes an endogenous division **b**, and the inner cell divides, usually twice producing four secondary cells or sporonts **d**. Each sporont undergoes an endogenous division **e**, and the inner cell divides twice to produce four tertaiary cells **g**. Each tertiary (or sporal) cell undergoes three successive endogenous divisions, **h,i** and **j**. The final full grown spore is about 20 μm long and consists of three concentric sporal cells surrounding a central sporoplasm. Four sporal cells develop in each sporont and four or more sporonts within each stem cell. The dense bodies in some sporal cells appear to be haplosporosomes. (Information from Desportes, 1981, 1984.)

like muscle or liver, or whether they are in body cavities like bile ducts or kidney ducts, seem to be fluid feeders that show active pinocytosis rather than phagocytosis. Some forms are of economic importance since they parasitise fish that are commercially valuable; they may reduce the growth of the fish, spoil their flesh or kill them, and sometimes the diseases may become epidemic.

A class of organisms named the Actinosporea (= Actinomyxida) has been described as parasites in annelids and sipunculids, e.g. ***Triactinomyxon ignotum*** in ***Tubifex***. Their multicellular spores usually have several uninucleate sporoplasms enclosed by three valves and with three polar capsules. Sporoplasms that emerge from spores in the host gut penetrate the gut wall cells or pass to the body cavity and enter lymphocytes. Within these cells the parasites grow, multiply, undergo meiosis and nuclear fusion to form zygotes, each of which forms a spore after a series of mitotic divisions. A recent report that an actinomyxid stage occurs in the life cycle of ***Myxobolus cerebralis*** (a myxosporean) (Wolf and Markiw, 1984) throws doubt upon the status of this taxonomic group (Corliss, 1985).

Helicosporidium, a single species parasitic in arthropods, may also be a myxosporidian. The spores of this species contain three uninucleate sporoplasms surrounded by an elongate cell formed into a single coiled filament, and are enclosed in a single membrane (Fig. 8.15f,g).

Members of the Class Paramyxea are parasites in marine molluscs, polychaetes and crustaceans (Desportes, 1984). Because their spores lack polar capsules, and simple centrioles composed of nine singlet microtubules occupy the poles of the spindle of the open mitotic nucleus, this Class may come to be recognised as a separate phylum, although these organisms share with the myxosporidia a characteristic production of multicellular spores during which endogenous divisions produce one cell inside another, and may have had a metazoan origin.

Developing spores of *Paramyxa* have been found between cells of the gut epithelium of polychaete larvae, where presumably multiplication of trophic cells had taken place (Desportes, 1981). Separate stem cells undergo a succession of endogenous and lateral divisions (Fig. 8.16) that result in the formation of multicellular spores in which the sporoplasm is enveloped successively by three other cells as well as sheathing remains of outer envelope cells. The extrusion of a small cell (like a polar body in animal oogenesis) in the later stages of sporogenesis is suggestive of a meiotic process. In other genera, e.g. *Marteilia*, the number of spores and number of enveloping cells varies (Desportes, 1984).

Table 8.1   A list of the groups mentioned in this chapter, with examples.

Phylum Sporozoa (Apicomplexa): about 5000 species in four classes:
    Class Gregarinea: usually divided into three orders:
        Order Archigregarinida, e.g. ***Selenidioides, Selenidium*** (disputed)
        Order Eugregarinida, e.g. ***Gregarina, Monocystis, Selenidium*** (?)
        Order Neogregarinida, e.g. ***Mattesia, Lipotropha***
    Class Coccidea: may be divided into three orders:
        Order Protococcida (Coelotrophida), e.g. ***Grellia, Coelotropha***
        Order Adeleida, e.g. ***Klossia, Haemogregarina, Adelea***
        Order Eimeriida, e.g. ***Eimeria, Isospora, Aggregata, Lankesterella***
    Class Haemosporidea: one order:
        Order Haemosporida, e.g. ***Plasmodium, Haemoproteus, Leucocytozoon***
    Class Piroplasmea: two orders:
        Order Piroplasmida, e.g. ***Babesia, Theileria***
        Order Dactylosomatida, e.g. ***Dactylosoma***
The genus Perkinsus may be the sole member of a Class Perkinsea within this phylum, or may be closer to flagellates.

Phylum Microsporidia (Microspora): about 800 species in two classes:
    Class Rudimicrosporea: one order
        Order Metchnikovellida, e.g. ***Metchnikovella***
    Class Microsporea: two orders:
        Order Minisporida, e.g. ***Hessea, Chytridiopsis***
        Order Microsporida, e.g. ***Nosema, Vairimorpha, Glugea, Encephalitozoon***

Phylum Haplosporidia: a few tens of species in a single order:
    Order Balanosporida, e.g. ***Minchinia, Haplosporidium***

Phylum Myxosporidia (Myxospora): about 1200 species in two classes:
    Class Myxosporea: two orders:
        Order Bivalvulida, e.g. ***Myxobolus, Ceratomyxa, Henneguya, Myxosoma***
        Order Multivalvulida, e.g. ***Kudoa, Trilospora, Hexacapsula***
    Class Actinosporea: a single order:
        Order Actinomyxida, e.g. ***Triactinomyxon***
The Class Paramyxea, with a few species, e.g. ***Paramyxa, Marteilia***, may belong here, or may be a separate phylum.

The outer cells of these spores often contain dense bodies like haplosporosomes (p. 239). It is not known how sporoplasms escape from these complex spores.

Most authorities now recognise the main groups mentioned in this chapter as phyla, but differ in their views about the minor groups, and about subdivisions within the main phyla. The listing given in Table 8.1 is taken from various sources.

9
Ecology

The lives of free-living and of the various forms of endobiotic protists are quite different. Feeding relationships and physico-chemical factors dominate the lives of free-living protists, while parasites must be concerned with migration from host to host and interaction with host defences. The ecology of protists exploiting these different modes of life will therefore be considered separately. The main emphasis of the discussion will concern protozoan ecology, but the roles of photosynthetic protists will be considered in outline; more detailed accounts of the ecology of these autotrophs are available elsewhere (e.g. Round, 1981).

Protists in ecosystems

The energy that maintains the cycling of organic carbon through almost all eco-systems is obtained from the sun by means of the process of photosynthesis; a possible exception is the sulphur-based ecosystem associated with thermal vents in the deep ocean (Tuttle *et al.*, 1983). The organic material built up by photosynthesis in photoautotrophs is either eaten by herbivores or is broken down by decomposer organisms (Fig. 9.1). The herbivores may in turn be eaten by predatory carnivores, and other carnivores may eat these; any organic matter that is not assimilated by these animals (i.e. dead bodies, faeces, excreted or secreted matter) will provide food for decomposers. The decomposers themselves form the food of animals which in turn may fall prey to carnivores. The carbon and other 'nutrient' elements incorporated by anabolic processes into the bodies of autotrophs are released to a nutrient pool as CO_2 and various salts by catabolic processes of organisms of all types. The cycling of these elements, notably nitrogen, phosphorus and sulphur, but also many trace elements like magnesium, silicon etc. as well as carbon, is driven by the flux of solar energy passing through the trophic system as chemical energy and being released as heat.

Although the role of photosynthetic protists has long been recognised, the contribution made by protozoans to energy flux and carbon cycling in the major ecosystems is only now beginning to be understood in quantitative terms, and their role in the 'microbial loop' (through decomposers) in particular is now gaining recognition. The ecological roles of free-living phagotrophic protists have recently been expertly reviewed by T. Fenchel (1987). The biomass of protists is often small, but their high rates of growth and reproduction, their efficient incorporation of food material and easy availability as food for higher animals all contribute to their ecological importance. Favourable conditions for protists are often restricted in both space and time; transient microhabitats (e.g.

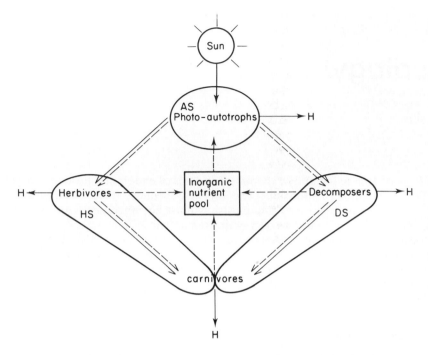

Fig. 9.1 Main trophic pathways in a generalised ecosystem. Energy from the sun is captured by members of the autotroph sub-system (*AS*), whose products provide food for both the herbivore sub-system (*HS*) and the decomposer sub-system (*DS*). This energy drives a circulation of nutrients, and is eventually lost as heat (*H*). Decomposers also break down the organic products and dead bodies of uneaten heterotrophs as well as autotrophs.

around a detritus particle) may provide ideal conditions for growth in a wider habitat that is less hospitable. They may be nearly as dependent as bacteria upon specialised niches or consortia of organisms with mutually dependent metabolism (Sieburth, 1987), but this does not reduce the global importance of either group. The roles played by protists in carbon cycling in the major ecosystems will be considered first as a background to discussions of energetics, population biology and cycling of other inorganic nutrients.

Roles of protists in carbon cycling

Protists contribute substantially to the photosynthetic primary production of organic matter in aquatic ecosystems. The preponderance of small, rapidly-reproducing plants in planktonic environments is reflected in the fact that the plant biomass of the oceans is only about 4×10^9 t (dry wt.) in comparison with a total world plant biomass of some 1850×10^9 t, while the annual net primary production of oceanic phytoplankton is some 50×10^9 t (dry wt.), compared with about 170×10^9 t for the whole world (Anderson, 1981). Phytoplanktonic organisms in several groups are responsible for this primary production in the oceans. The proportions of annual production contributed by different categories in a temperate sea might be as follows: diatoms 31%, dinoflagellates 28%, other flagellates and coccoid eukaryotes 25% and cyanobacteria 16%. The amount and composition of marine phytoplankton will vary with time and place, depending upon the intensity of illumination and the availability of essential nutrients, and its net productivity ranges from about 25 g C m^{-2} yr^{-1} in oligotrophic waters of the central parts of subtropical oceans to about 350 g C m^{-2} yr^{-1} in shallow seas over the continental shelf (Eppley and Peterson, 1979) (production is often expressed in terms of the weight of organic carbon, which constitutes about half of the dry weight biomass, or as its energy content (in joules), synthesised in unit time); polar and coastal seas in general have higher nutrient concentrations and productivity there is more often limited by light. In temperate waters production may be limited by light in the winter and by nutrients in the summer, so that the phytoplankton flourishes (blooms) most abundantly during transition periods of spring and autumn. Photosynthetic members of the picoplankton (diameter 0.2–2.0µm), which are cyanobacteria and the smallest flagellate and coccoid eukaryotes (chlorophytes, chrysophytes and haptophytes), appear to grow at lower light intensities and less stringent nutrient levels than larger forms and are now recognised to provide an important component (20–80%) of primary production in most seas much of the time. Diatoms (10–50µm) and dinoflagellates (10–40µm), which occur in both nanoplankton (2–20µm) and microplankton (20–200µm) categories, both tend to form blooms in spring and autumn, diatoms responding preferentially to nitrate and dinoflagellates to ammonium salts as sources of nitrogen. The other nanoplanktonic plants are from a diversity of mostly flagellate groups, varying in different locations and whose contributions to primary production are less well known.

Planktonic plants also abound in freshwater lakes, where diatoms may dominate, with freshwater flagellates perhaps less abundant but showing similar diversity to marine forms. Cyanobacteria usually form filamentous or gelatinous aggregates in freshwater plankton and so provide a different size of food resource compared with the (usually) separate cells of marine cyanobacteria. In freshwater ecosystems as well as coastal marine habitats the primary production is often dominated by macrophytes, giving net primary productivities of 100 g to 1 kg (dry wt) m^{-2} yr^{-1}, or more, in lakes and streams and as high as 4 kg m^{-2} yr^{-1} in kelp beds (similar to the highest productivity of terrestrial habitats). The brown, green and red seaweeds of littoral and shallow marine environments and the macroalgae and higher plants growing in lakes and streams provide physical support for protists but little in the way of direct nutrients; they do however contribute much organic matter to decomposer pathways (see below) (Stuart *et al.*, 1981). Many of the protists that live attached to these plants as well as non-living surfaces in shallow water habitats are also photosynthetic forms, belonging to various groups of flagellate and coccoid protists, including diatoms. The primary production by these attached microscopic plants is often substantial; amounting for example to about 100 g dry wt m^{-2} yr^{-1} for epilithic algae in a shallow stream (Marker, 1976) and 35 g C m^{-2} yr^{-1} (70 g dry wt m^{-2} yr^{-1}) for algae epiphytic on macrophytes in a lake, where it was 31% of the total littoral production (Allen, 1971).

Substantial quantities of organic compounds are released (secreted ?) by aquatic plants into the water. In addition, a large part of the biomass of the phytoplankton in the oceans and of both microscopic and macroscopic plants of shallow marine and freshwater habitats dies and is broken down to particulate organic matter (POM) and dissolved organic matter (DOM) by physical action, aided by various decomposer organisms, particularly fungi and bacteria. In freshwater ponds and streams and some coastal marine habitats the autochthonous plant material (that formed within the habitat) is joined by allochthonous plant material (that formed in another habitat) which is mainly terrestrial and is washed into the habitat from the land or from inflow streams, and therefore also contributes to the POM and DOM pools of these ecosystems. The annual input of allochthonous biomass into a stream may equal or exceed the autochthonous production. All this POM derived from plant sources, together with POM from the dead bodies and faeces of animals will either be eaten by detritivores or degraded to DOM.

Significant amounts of DOM derived from POM or released by animals and plants provide a very substantial organic resource in many ecosystems (Fig. 9.2). Indeed, commonly 80–90% of the organic carbon in the sea is present as DOM (ranging on average from 0.3 mg C l^{-1} at 300 m depth to 2.0 mg C l^{-1} at the surface). This DOM may be taken up to some extent by many organisms, but the vast bulk of it seems to be absorbed by bacteria which have by far the largest surface/volume ratio; it provides a nutrient resource that maintains heterotrophic bacterial concentrations of 10^4–10^6 ml^{-1} (representing a biomass of about 10 mg C m^{-3}) in the euphotic zones of the oceans. The concentration of DOM in coastal marine and freshwater habitats, and particularly in their sediments (as well as in damp soils), is often much higher and can maintain much more dense populations of bacteria in the range 10^6–10^9 bacteria ml^{-1}.

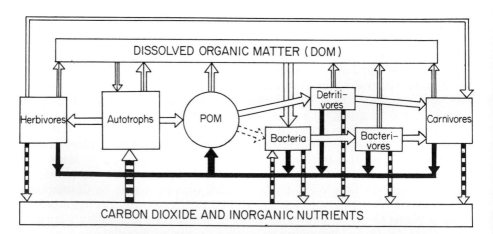

Fig. 9.2 The flow of organic matter and nutrients in an ecosystem. Autotrophs use CO_2 and salts, and sometimes dissolved organic matter (DOM), in the synthesis of organic compounds which are either eaten by herbivores, broken down as particulate organic matter (POM) or released as DOM. Other members of both the herbivore (left) and decomposer (right) subsystems also release DOM and produce POM as faeces or on death, as well as releasing inorganic salts and CO_2. Bacteria absorb dissolved organic matter, some of it derived by external digestion of POM, and also use nutrient salts and CO_2 in synthetic reactions (see text). The direct use of POM by detritivores is questionable, for many detritivores may derive nutrients principally from the bacteria associated with the detritus they eat.

Many protists, including some desmids (Vincent, 1980) and diatoms (Hellebust and Lewin, 1977) are osmotrophic heterotrophs, dependent upon DOM (*see* Chapter 3). Decomposition processes dependent upon bacteria are very important in almost all ecosystems, even in the open sea where life might seem sparse (Fig. 9.3), and are promoted in sewage treatment processes discussed below.

The microscopic plants and heterotrophic bacteria of these autotrophic and decomposer sub-systems provide the food of heterotrophic protists of various sorts. The relative size of these protists and their food sources determine trophic relationships and feeding mechanisms. Small organisms in suspension can be collected by protists (or multicellular animals) by the types of microphagous filtration techniques mentioned in Chapter 3. Thus bacteria and picophyto-plankton can be caught by active filtration by flagellates and some ciliates and by passive interception by amoeboid forms, and larger suspended phytoplank-ton and other protists are caught in the pseudopod systems of amoeboid forms or the filters of larger ciliates, crustaceans and other larval and adult inverte-brates of benthos and plankton. Bacteria and photosynthetic protists are often most numerous at surfaces, either as neuston in the thin surface skin at the air/water interface, or on solid substrata, and these organisms may also be collected by microphagous techniques of sweeping the surface with a ciliary or flagellar suction current, or with a mobile pseudopodial net. Macrophagous techniques are more specialised (Chapter 3), and may even be limited to one or a few species of prey. These allow small amoebae or smaller flagellates (e.g.

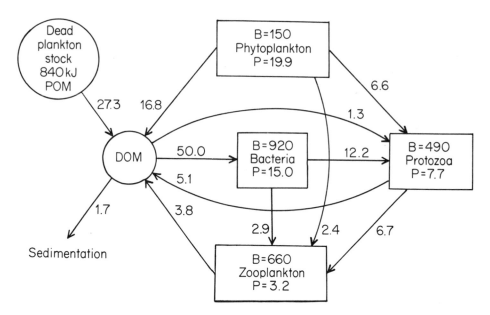

Fig. 9.3 Energy flow in the 0–100 m depth zone of a plankton ecosystem in the sea of Japan during a period after the spring phytoplankton bloom when heterotrophic activity is intense (data for 12 June 1972 extracted from Sorokin, 1977). Figures for biomass (*B*) are in mg C m^{-2} and for production (*P*) in kJ m^{-2} day^{-1}. Numbers by arrows indicate the energy content (in kJ m^{-2} day^{-1}) of organic material passing between components of the system.

bodonids) to capture individual bacteria, larger flagellates to eat diatoms or other small protists, and ciliates and amoebae to eat cyanobacterial filaments, diatoms, flagellates, other ciliates or even small metazoans and reportedly detritus (dead organic debris). Each ecosystem therefore contains a diversity of heterotrophic protists which are involved in the energy flux of both herbivore and decomposer sub-sytems.

Ecological energetics

The biomass built up in autotrophs by photosynthesis provides the energy required by organisms in both herbivore and decomposer sub-systems in the form of chemical bonds of organic compounds. Clearly part of this energy will be lost to entropy at each stage, and the efficiency of transfer of energy from one stage to the next varies widely. Of the available photosynthetic active radiation (PAR) falling on a water surface, only 0.5% or less is converted into plant biomass by phytoplankton communities as gross primary production (GPP), and 30–40% of this is lost by respiration in the plants, so that 60–70% of GPP is available as net primary production (NPP) to other sub-systems of the planktonic ecosystem.

Any consumer (predator or decomposer) assimilates organic matter from its food, but some organic components in the food consumed are not utilised in building the body of the consumer and are rejected in faeces and urine (i.e. Consumption = Assimilation + rejecta (faeces and urine)). The assimilation efficiency (A/C) is highest in carnivores ($\sim$ 80%) and lowest in saprotrophs ($\sim$ 20%), largely reflecting the proportion of the diet that is not suitable for assimilation (Table 9.1). Of the assimilated material, a part must be used in respiration to provide energy for growth, body maintenance, movement etc. so that the eventual production of consumer biomass in growth and reproduction (Production = Assimilation – Respiration) will vary with these metabolic costs; the net production efficiency (or net growth efficiency) (P/A) is of course very low in adult homoiothermic vertebrates ($\sim$ 2%), but is high in protozoans and

Table 9.1   Growth efficiencies of various protists growing on food of different types.

Species	Food	Temp.	Assimilation efficiency	Production efficiency Net	Gross	Basis	Reference
Amoeba	ciliates	20°C	55%	73%	40%	Energy	Rogerson
proteus	ciliates	10°C	59%	49%	29%	Energy	(1981)
Acanthamoeba	yeast	25°C	58%	63%	37%	Energy	Heal (1967)
Stentor coeruleus	ciliates	20°C	75%	96%	72%	Energy	Laybourn (1976)
Tetrahymena pyriformis	peptone bacteria	25°C 25°C	23%	37%	9% 50%	Carbon Weight	Curds and Cockburn (1968)
Ochromonas sp.	bacteria	20°C		59%	34%	Carbon	Fenchel (1982b)
Pleuromonas jaculans	bacteria	20°C		60%	43%	Carbon	Fenchel (1982b)

growing invertebrates ($\sim 60\%$ or more). Clearly only a minor part of the food consumed eventually appears as production in the consumer, and examples of gross production efficiency (gross growth efficiency or yield (P/C) of protozoa are given in Table 9.1. As a general rule of thumb, about 30% of the food consumed by an actively growing protozoan is used in respiration, 30% is egested or excreted and 40% contributes to growth.

It is interesting that the gross production efficiency of protozoans (or bacteria and yeasts) changes little with growth rate (Curds and Cockburn, 1968; Fenchel, 1982b) in spite of the energy losses by respiration. This suggests that the respiratory energy lost is predominantly used for growth, with very little used for maintenance and movement. Support for this belief comes from measurements of the rate of respiration of protozoans under different nutritional conditions (Fenchel and Finlay, 1983). Actively growing protozoa have respiratory rates that are proportional to body weight$^{0.75}$, and which fall on a line extrapolated downwards from that relating respiratory rate and body weight for poikilothermic metazoans. Under starvation conditions the rate of respiration per cell falls away rapidly for a range of protozoans, sometimes decreasing over a few days to as little as 2% of the rate measured in growing cells. This is essential if a starving protist is to survive, for in active growth about half of the food assimilated is used in respiration, and without a severe reduction in respiration rate the starving cell would quickly need to resort to cytoplasmic breakdown to supply its energy needs. A reduction in size partly accounts for the reduction in respiration; in starved *Tetrahymena*, for example, a reduction to 11.3% of the original cell volume accompanied by a reduction in respiration to 2.6% of the original rate/cell, produced a reduction of the respiration per unit volume to 23% of the original level. The reduction in respiratory rate is particularly marked in small cells (eg by 90% on a respiration per unit volume basis for the small flagellate *Ochromonas*), which suggests that this is a strategy for survival on an absolute time scale, smallest cells having the least reserves. Only under known growth conditions can one make meaningful predictions about the respiratory rate of a protozoan. Changes in structure and patterns of synthesis occur in such protists upon starvation (Fenchel, 1987); a similar response to starvation occurs in bacteria, where it is shown to be an 'active' phenomenon associated with production of 'starvation proteins' (Kjelleberg, 1987). Some species take the response a stage further by transforming to resting cells (cysts) with a resistant cell coat (see below), but organisms that retain a low level of activity are able to respond more quickly to the return of good conditions than encysted organisms and are therefore well equipped to expoit transient microhabitats.

In addition to the losses which occur in secondary production within the consumer, there is the question of the efficiency with which energy flow takes place from one trophic level to another. The ratio of the production of a consumer to the production of its food organism (a 'gross ecological efficiency'), which could be studied in a laboratory situation of for example a ciliate eating bacteria (p. 253), has less meaning in a field situation where several consumers may compete for the same food, or large fractions of the available food may pass to decomposers.

Protozoan populations and feeding relationships

The types of heterotrophic protist present in a community will clearly depend upon the types of food present, and will relate to feeding preferences and mechanisms (see Chapter 3), which restrict the food resources that can be exploited. Some predatory ciliates, for example, *Didinium nasutum* (eating *Paramecium*) and *Actinobolina radians* (eating *Halteria*), are rather specific in the prey they will eat. Some prey are distasteful and are avoided, for example certain pigmented bacteria such as *Chromobacterium violaceum* are toxic and may be rejected by bacterivores (Curds and Vandyke, 1966), though the mechanism of avoidance is unknown.

The sizes of populations of protozoa will depend upon the abundance of appropriate food. Laboratory studies show that protozoan populations follow a typical growth curve (Fig. 9.4). The presence of the lag phase (I) depends on the condition of the organisms used. The slope of the exponential phase (II) as well as the population density in the stationary phase (IV) depend primarily on food concentration, but may vary with such factors as temperature and with the availability of any limiting factor (e.g. a vitamin) in the diet; if some factor other than energy supply is limiting in the stationary phase, then this phase may be much prolonged. If the food supply is maintained and a proportion of the protozoans are continuously removed, the population will continue to multiply exponentially; this maximal growth, which requires a balance between food supply and cropping of the population, may seldom occur in nature, so that for example populations may vary cyclically and/or the protozoans may compete with each other (see below).

Fenchel (1980a,b, 1982b, 1987) has provided useful discussions on parameters affecting suspension feeding in protists. He points out that the food requirements of a microphagous flagellate or ciliate can be met if the volume of

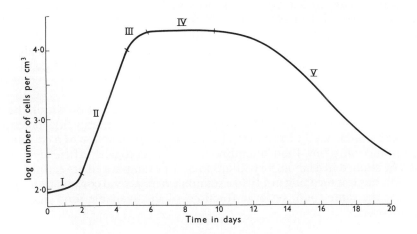

Fig. 9.4 Phases in the growth of a population of *Tetrahymena* in culture.

water it can filter in unit time (F) and the concentration of nutritious particles (Cp) that it can extract from the water are together above the minimum required for growth, i.e. FCp must exceed a certain value. However, the rate of food uptake may be limited by the 'handling time' (t) required to take in each unit (or 'mouthful') of food, and less food may be ingested than is filtered, so the actual uptake (U) $= FCp$ $(1\text{-}tU)$. Clearly t will determine the maximum rate of food uptake (Um), ($= 1/t$), and therefore the maximum growth rate. The value of F gives an indication of the minimum concentration of particles at which the organism can sustain itself, and the value of ($Um/F = K$) is a half-saturation constant; this K is the particle concentration at 50% of maximum food uptake and gives a guide to the food particle concentration at which the organism is adapted to live.

Some of these features can be illustrated by results of growing **Tetrahymena** monoxenically on the bacterium **Klebsiella aerogenes** (Figs. 9.5 and 9.6). The feeding rate of individual ciliates is seen to increase with bacterial concentration ($U \propto Cp$); each ciliate is estimated to clear the bacteria from 10^{-5} ml of medium in an hour, and the half maximal growth rate occurs at a bacterial concentration of 11.7 mg dry wt. l^{-1}, which is equivalent to about 10^8 cells ml^{-1} (Curds and Cockburn, 1968). As the concentration of bacteria is further increased (at a particular ciliate size and concentration), both the individual feeding rates of the ciliates and their growth rates level off, presumably because t now limits uptake and the feeding mechanism of the **Tetrahymena** becomes saturated at a maximum level (Um). At a particular bacterial concentration the feeding rate of individual ciliates decreases as ciliate concentration increases to a high level,

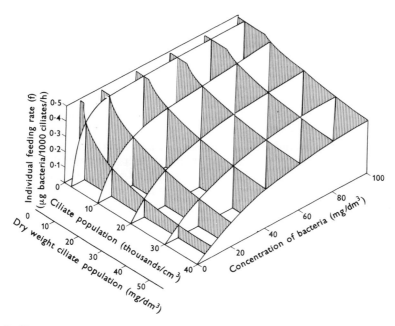

Fig. 9.5 The effect of bacteria concentration and population density of ciliates on the individual feeding rate of **Tetrahymena pyriformis** on the bacterium **Klebsiella aerogenes**. Data by Curds and Cockburn (1968), with permission of HMSO.

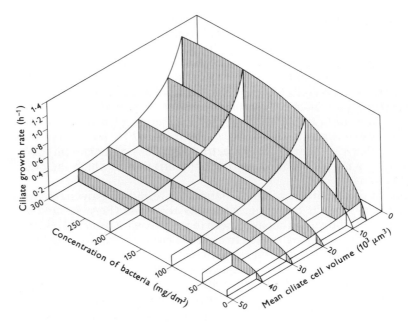

Fig. 9.6   The effect of bacteria concentration and the mean cell volume of the ciliates on the specific growth rate of **Tetrahymena pyriformis**. Data by Curds and Cockburn (1968), with permission of HMSO.

probably as a result of interference between the feeding currents of ciliates (which may then perhaps ingest fewer bacteria per food vacuole). Smaller ciliates reproduce more quickly than larger ones, at a particular bacterial concentration (Fig. 9.6), though the change in production of biomass will be less marked; ciliates also tend to become smaller during the later stages of exponential growth when reproductive rate falls.

The requirements of suspension feeders exploiting food particles of different sizes have also been considered by Fenchel (1980b,c, 1982b) (*see also* Chapter 3). Microflagellates and some smaller ciliates have filters fine enough to catch any bacteria, but the required level of Cp may be fairly high because the drag forces in the fine filter restrict the value of F. Ciliates with a coarser filter have higher clearance (F) rates, but cannot catch bacteria, and survive by catching smaller numbers of larger particles, which are present at much lower densities but have much higher biomass (*see* Table 9.2). In fact maximum clearance rates are approximately proportional to the preferred particle size. A bacterivorous ciliate will filter about 10^4 times its body volume per hour, and requires a bacterial concentration of 10^7 or 10^8 ml^{-1} for growth and values for K for these ciliates will be of this order. Such high concentrations of bacteria are found only locally in aquatic habitats, so while bacterivorous ciliates may control bacterial numbers in polluted waters and around sediments rich in decomposers, they otherwise act as opportunists exploiting temporary rich resources. The bacterivorous microflagellates may clear 10^5 or 10^6 body volumes per hour, and require between 10^6 and 10^7 bacteria ml^{-1} for growth; such flagellates are probably the principal bacterivores of the marine plankton, and are thus the

Table 9.2 Characteristics of suspension feeding processes in different categories of feeding relationships. Data from Fenchel (1980b, c, 1982b)

Type of Protist	Microflagellate	Ciliate	Ciliate	Ciliate
Food	bacteria	bacteria	microflagellate	ciliate
Clearance rate (range) body volumes hr^{-1}	$5 \times 10^4 - 10^6$	$3 \times 10^3 - 10^4$	$10^4 + 5 \times 10^4$	$5 \times 10^4 - 5 \times 10^5$
Clearance rate (range) ml hr^{-1}	$2 \times 10^{-6} - 10^{-5}$	$10^{-6} - 10^{-4}$	$10^{-3} - 10^{-2}$	$10^{-2} - 5 \times 10^{-1}$
Specific example	*Ochromonas* sp.	*Glaucoma scintillans*	*Paramecium caudatum*	*Bursaria truncatella*
Clearance rate, ml hr^{-1}	10^{-5}	1.3×10^{-5}	10^{-3}	0.43
Minimum filter mesh, μm	–	0.2	0.4	8
Food particle size range, μm	0.2–2 (–5)	0.2–1	1–5	20–60
Minimum food particle concentration mg dry weight l^{-1}	0.5	0.8–8	0.08–1.3	0.002–0.15
Body length, μm	7	35	200	800
Body volume, μm^3	200	4×10^3	10^5	5×10^7

grazers which provide the necessary link to maintain bacterial concentrations at about 10^6 ml^{-1} in many such habitats, channelling carbon from the microbial loop on to zooplankton. Ciliates, along with metazoan filter feeders, play a role at the next trophic level in these habitats by feeding on the microflagellates. Since metabolic rate decreases with increase in body size (*see* discussion above on respiratory rates), one might expect smaller cells to require relatively higher feeding rates; in one study small ciliates ingested 80–120% of their body volume per hour and divided once every 2–4 hours whilst large ciliates ingested only 10–30% of their body volume per hour and divided once every 10–30 hours (Fenchel, 1980b).

In the Limfjord Fenchel (1982d) found bacterivorous flagellates at populations of about 10^3 ml^{-1} and bacteria at about 10^6 ml^{-1}. This number of flagellates could filter about 20% of the Limfjord water each day and if bacterial division rates lie in a normal range (of once every one or two days, in this situation) the flagellates could be the principal consumers of the pelagic bacteria and be responsible for limiting their numbers. That this is the case is indicated by the coupled oscillations of bacterial and flagellate numbers, in which increases and decreases of bacterial numbers are closely followed by rise and fall of flagellate populations with a cycle time of about 16 days, in a classical, Volterra-type, prey/predator pattern. A similar cycling of nanoplanktonic flagellates and the oligotrich ciliates that feed upon them showed a 30-day periodicity in a mediterranean bay (Ibanez and Rassoulzadegan, 1977). These are field samples of the laboratory results of Gause (1935) (Fig. 9.7), in which increases in numbers of *Paramecium* reduced yeast populations, depriving themselves of food so that their population decreased, allowing the yeast numbers to build up again and provide food for another multiplication of *Paramecium*.

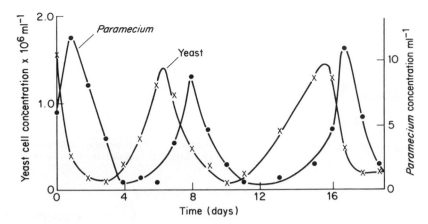

Fig. 9.7 Prey/predator oscillations in populations of the yeast *Schizosaccharomyces exiguus* and the ciliate *Paramecium aurelia*. (Data from Gause, 1935.)

Search-and-capture feeding (p. 66) may be compared with suspension feeding. *Didinium*, for example, searches (a process equivalent to filtration) for *Paramecium* that may be present at low concentrations, and because of the large size of prey in comparison with the predator's mouth (see p. 65), a 'handling time' of up to two hours is required to reform the mouth before another *Paramecium* can be eaten. As in the case of *Tetrahymena* eating *Klebsiella*, the feeding rate of *Didinium* is limited by Cp at low densities of *Paramecium* and by t at high densities of prey (Fig. 9.8). One would expect a similar relationship in other examples, e.g. *Pseudomicrothorax* (p. 67) must search for suitable filaments of blue-green algae to eat, and t may be determined by the availability of membrane to form food vacuoles. Laboratory experiments show that when a predator like *Didinium* is confined in a restricted space with *Paramecium*, it will multiply and eat all of the prey, but if the habitat is made spatially (or temporally) complex, then both predator and prey populations survive (Maly, 1978), and predator/prey cycles like those seen in Fig. 9.7 may occur.

It is generally assumed that no two species of organism occupy precisely the same ecological niche but will compete to some extent with others in 'adjoining' niches. For example, Fenchel (1980) has demonstrated that filter feeding ciliates display a range of (often fairly narrow) filter ranges which imply occupancy of different niches. Present information is inadequate to compare the full specifications of niches occupied by two organisms, particularly protozoa, but some indication of feeding interactions between protozoa in similar niches can be gained from the result of some experiments by Gause (1934). *Paramecium aurelia* and *P. caudatum* will each grow separately on either the bacterium *Pseudomonas pyocyanea* or the yeast *Saccharomyces exiguus*; when both species of *Paramecium* are grown together on the bacterium, *P. aurelia* outgrows *P. caudatum*, but when both are grown together on the yeast, *P. caudatum* outgrows *P. aurelia*. In nature habitats are complex and provide an enormous diversity of niches in which different protozoa can thrive; inevitably physico-chemical conditions will change, emigration and immigration will occur and a considerable variety of interactions must take place within and

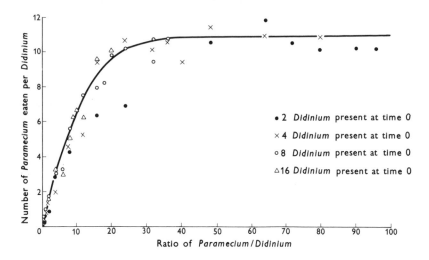

Fig. 9.8 The number of *Paramecium* eaten in a 20 hour period by each Didinium present at different initial ratios of *Paramecium: Didinium* for four initial population densities of *Didinium*. In calculating the feeding rate, the number of *Paramecium* consumed is divided by the average number of *Didinium* present during the 20 hour period to take account of the increase in number of *Didinium*. The *Paramecium* had been starved, and control experiments showed that very few of them would divide during the 20 hour period. Unpublished data by P. Brobyn.

between protozan populations – in general the most important of them concerned with feeding relationships. In spite of temporary, sometimes violent, fluctuations, particular habitats tend towards stability and averaged over longer periods of time contain a similar range of species in similar proportions.

Nutrient cycling

An outline of the circulation of nutrients in ecosystems has already been given in Figs. 9.1 and 9.2. The composition of the nutrient pool must be expected to vary from time to time, according to the release of C, N and P, and other nutrient elements not considered here, from organisms within the biosphere (including any nitrogen-fixers) and entry of N or P from geological, atmospheric or pollutant sources. The proportions of C, N and P in living organisms of different types are likely to be different, according for example to the relative proportions of carbohydrates, lipids and protein in their bodies. The food of an organism is therefore unlikely to contain C, N and P in the concentrations it needs, so that it must make up nutrient deficits either by absorbing the nutrient from the environmental pool or by ingesting additional food and rejecting surplus nutrients of other types to the nutrient pool or storing them.

It was noticed some years ago that phytoplankton samples had an atomic ratio near to $C_{106} N_{16} P$ – the 'Redfield Ratio', while the ratio of these atoms available in the oceans was about $C_{1000} N_{15} P$ (Redfield, 1958); C is therefore never limiting, but shortage of either N or P could limit carbon assimilation. In freshwater the proportion of phosphorus to nitrogen in the environment is often lower, and ratios around $C_{174} N_{31} P$ have been reported (Schindler, 1977), so production in the average freshwater system will be limited by P rather than N. Bacteria however often have five or more times as much P per unit of dry weight as the tissues of plants, and the relative N content of bacteria is also higher (Pomeroy, 1980). When bacteria obtain organic matter from degrading plants, they obtain the additional P and N required from the nutrient pool, in competition with autotrophs (Fig. 9.2). However, the phosphorus content of protozoans is nearer to that of plants than of bacteria, so bacterivorous protozoans may be expected to release the excess P (and N) to the nutrient pool; indeed ciliates have been found to release P (Taylor and Lean, 1982) and flagellates to release N (Sherr *et al.*, 1983). Bacteria, protozoa and other heterotrophs will also tend to release C, N and P to the pool because something like one third of the organic matter assimilated is usually broken down in respiration at each trophic level. Bacteria, and at least autotrophic protists, as well as higher plants, have mechanisms for the very active uptake of orthophosphate from water, and often much of this may be surplus to requirements for the recycling of P is apparently very rapid (a matter of minutes), particularly in euphotic waters. N cycles more slowly, and may be more easily stored. Nutrient cycling may be especially tightly coupled between symbiotic algal protists and their hosts (p. 284).

Autotrophic protists clearly assist cycling by taking nutrients from the pool and building them into organic compounds. Heterotrophic protists catabolise some of these organic compounds and return N and P to the pool, in the same way as members of any other trophic level, but do they have any additional role? It appears that bacterivorous protozoans stimulate the bacterial turnover of N and P (by up to 10 times) and accelerate detritus breakdown by cropping bacteria and maintaining their exponential multiplication; experimental addition of PO_4^{3-} and NO_3^- stimulates these processes still further (Fenchel and Harrison, 1976). Others believe that the additional recycled P and N may be excreted by the protozoans as surplus from their bacterial food (Johannes, 1965). It is currently believed that heterotrophic protists enhance nutrient cycling (how remains controversial), stimulate bacterial growth and provide a nutritional link between bacteria and zooplankton, acting to keep energy and nutrients available and passing them up the food chain (Taylor, 1982). Not only do bacterivorous protozoa accelerate decomposition, they also stimulate autotrophic growth, in both aquatic habitats and soils (see p. 275).

The influence of ecological factors on the lives of protists

Protists with the ecological roles outlined above occur in almost any body of water, salt or fresh, permanent or temporary, as well as in damp soil, damp moss, snow and within the bodies of other animals and plants. These habitats provide conditions within the range of ecological tolerance of protists, the more stringent habitats requiring more particular specializations of the protistan inhabitants. Ecological factors of particular importance in the life of various protists are water, temperature, oxygen, light, pH and salinity; if these factors are within favourable limits for a species of protist, then its occurrence and abundance will depend upon the availability of suitable food and the extent of predation.

Active phagotrophic protists need water because an area of unprotected cell membrane is necessary for purposes of feeding and such active protists are therefore absolutely limited to damp environments, although some coccoid green protists are a little more tolerant because of their cell wall. Protists are most abundant in aquatic habitats, but are also characteristic inhabitants of soils and polar regions where free water is only present for short periods; protists from these habitats are able to escape shortage of water by encystment or the formation of spores or sclerotia, which will be considered in more detail later. Protists inhabiting soils, moss, temporary pools or snow have the ability to excyst rapidly on the return of favourable conditions as one may see in the occurrence of **Chlamydomonas nivalis** – ('red snow') in thin films of water in part-melted snow (Hoham, 1980).

Dormant stages of protists, including cysts, spores and sclerotia, survive extreme temperatures; for example, the cysts of the soil ciliate **Colpoda** have been found to germinate after immersion in liquid air for seven days or heating to 100°C for three hours. The lower temperature limit for the active life of protists is provided by the freezing point of the surrounding water, and many protists will grow and reproduce at temperatures down to almost 0°C in fresh waters and about -2°C in sea water. The optimum temperature for the life of a strain of protists seems to be at least partly a matter of acclimation; the same species of organism may be found in cold streams at below 5°C and in hot springs at 40–50°C or more, but 'cold strains' generally have a lower optimum temperature for growth and division than 'hot strains'. Some species, by contrast, are characteristic of more restricted temperature ranges; for example, the snow algae often have temperature optima in the 1–5°C range (Hoham, 1980), and it is common experience that different foraminiferans, diatoms and dino-flagellates characterise polar and tropical seas.

The optimum temperature for the growth of a protist in laboratory culture may vary with the food provided, e.g. **Euglena gracilis** grown in the dark in a casein proteose medium without acetate had a maximal division rate at 10°C while identical cultures grown with acetate had a maximal division rate at 23°C. In an experimental study of the growth and division of **Tetrahymena pyriformis** it was found that although growth was possible over the whole experimental range from 5°C to 35°C, division (i.e. population increase) only occurred at temperatures between 7.5°C and 32.5°C (Fig. 9.9). The volume of the ciliates remained low at about 20°C (the normal culture temperature), but volumes of 2–3 times the normal were reached at extreme temperatures. Growth and respiration both commonly have Q_{10} values of between 2 and 3 in the temperature range between 10 and 20°C. E. Zeuthen and his colleagues have used temperature changes to control the time of division in **Tetrahymena** and thereby obtained cultures in which the stages of the cell cycle were synchronized (see p. 92). Similar division-synchrony may be achieved with the flagellate **Astasia** using heat treatment and with photosynthetic flagellates using alternate periods of dark and light.

An important feature of the aquatic environment that varies with temperature is the amount of dissolved oxygen; water saturated with oxygen contains 14 mg oxygen l^{-1} at 0°C and only 9 mg l^{-1} at 20°C. Since the metabolic rate of organisms, and hence their utilization of oxygen, also tends to increase with rise in temperature, oxygen is more likely to limit the abundance of aerobic organisms in warmer waters. The oxygen content of fresh waters varies seasonally and diurnally, depending on the extent of photosynthetic and respiratory

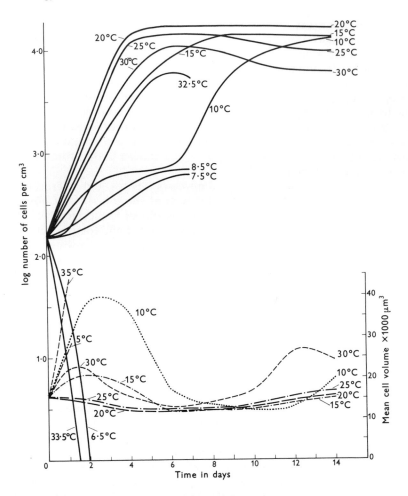

Fig. 9.9 The effect of temperature on the population density and cell size in cultures of *Tetrahymena pyriformis*. The continuous lines indicate the changes in population density at different temperatures following the introduction of ciliates from a culture maintained at 20°C into tubes of proteose peptone. The broken lines indicate the changes in mean cell volume at different temperatures. Unpublished data by J. Bullock.

activites of the organisms present. Few free-living protists seem to be capable of living in the complete absence of oxygen, but many tolerate very low levels of oxygen, provided other substances common in anoxic situations such as H_2S and CO_2 are not present at high concentrations. A few species including *Trepomonas agilis*, *Caenomorpha medusula*, *Pelomyxa palustris* and *Saprodinium putrinum*, are characteristically found in anaerobic freshwater habitats, and *Trepomonas* soon dies if the water is aerated. Some of these forms depend upon electron acceptors in symbionts for their anaerobic respiration (p. 282). A number of other forms such as *Spirostomum ambiguum*, *Stentor coeruleus*, *Loxodes striatus*, *Amoeba proteus*, *Actinosphaerium*, *Difflugia*,

Peranema and many more, are often found in regions of low oxygen content, and are at least capable of enduring oxygen lack if not of thriving without oxygen (see p. 18). While some gut parasites may be facultative anaerobes like many of these freshwater forms, it is believed that the flagellates in the gut of termites and the ciliates in the rumen contents of cattle and sheep are obligate anaerobes. It is clear that some protozoa make use of anaerobic pathways of respiratory metabolism, and it is likely that many more need very little oxygen although they cannot rely exclusively on anaerobic respiration.

The carbon dioxide content of water tends to vary inversely with the dissolved oxygen. Most protists are fairly tolerant of low CO_2 concentrations, but high concentrations are toxic to many species, particularly those from situations that are normally well oxygenated. Thus, forms from oligosaprobic lakes, such as **Codonella, Ceratium hirundinella** and **Synura uvella** were found to be very sensitive to CO_2, while **Paramecium putrinum** and **Polytoma uvella**, characteristic of polysaprobic conditions, were tolerant of high CO_2 concentrations. The CO_2 content of water is vital in that it provides the carbon source of autotrophic organisms, but it is also important because it combines with water to form carbonic acid and tends to lower the pH of the water. Because of the diurnal fluctuations of photosynthetic activity it is commonly observed that the pH of water rich in algae rises considerably in the day and falls at night. In correlation with this pH change, it is interesting that the pH-sensitive species **Spirostomum ambiguum** was found to be free-swimming by night but to retreat into the layer of decaying leaves by day when the pH rose.

The pH of the environment may vary for other reasons. Certain waters may become very acid because of the release of humic acids in the absence of buffering salts which would otherwise permit bacterial action on these organic compounds; such water contains characteristic protists – desmids and a few flagellates. Even more extreme acid conditions are found in some cases where industrial effluents or mine drainage enter streams. Records of protists from such habitats indicate great tolerance by some species, extending to a pH of 1.8 in the case of **Euglena mutabilis, Chlamydomonas** sp., **Urotricha** sp. and **Oxytricha** sp. In acid soils testaceous amoebae may be the most common organisms.

Protists belonging to some groups are found only in the sea and those of others only in fresh water, but there are few species that are able to live in both habitats. The difference in salinity between sea and fresh water involves a substantial change in osmotic pressure and in the concentrations of various ions, and it is therefore a considerable feat for a protist to survive the transition from fresh water to sea water or the reverse (see p. 21). While it has been found to be possible to transfer **Cyclidium glaucoma** directly from fresh water to sea water, and to grow the freshwater **Amoeba lacerata** in salinities of up to 4.4%, a majority of forms probably have a much more restricted tolerance, e.g. **Paramecium caudatum** would live at a salinity of about 1.5% · but not above, while **Cryptomonas ovata** var. **palustris** would not even tolerate a rise in salinity to 0.03%. Very high salinities occur in certain situations in salt lakes and salt marshes and in such habitats the ciliate **Fabrea salina** was found at a salinity of 20%, **Prorodon utahensis** and **Uroleptus packi** are ciliates that were found in water with a salt content of 23% and the flagellate **Dunaliella salina** has been recorded from waters with a salt content above 250 g l^{-1} in Lake Elton.

All protists, particularly pigmented ones, absorb light energy. A particular pigment absorbs light energy of particular wavelengths so different organisms have different spectral sensitivities Thus the efficiency of use of different parts of the spectrum for photosynthesis varies in different groups according to the balance of pigments present (see also p. 98). Many protists, notably those with 'eyespots' (p. 99) show phototactic migrations, both in water and more solid media, as exemplified by the swimming of *Volvox* and *Euglena* towards light of moderate intensity, and the diurnal migrations of marine dinoflagellates in the plankton, of *Amphidinium* and various diatoms in sand and of *Euglena deses* in mud. Interestingly enough, a phototactic response is also shown by some green ciliates – ciliates containing symbiotic zoochlorellae. Exceedingly strong light is damaging; many protists may respond to high light intensities by negative phototaxis, as in *Volvox* and some species of *Euglena*, while in some other species of *Euglena* (*E. sanguinea* and *E. haematodes*) a protective red pigment is found to migrate to the body surface in bright light and to migrate inwards in weak light, and a number of other flagellates that live at high altitudes or latitudes tend to carry red pigmentation e.g. *Chlamydomonas nivalis* in 'red snow'.

Encystment and Excystment

Protists appear to produce cysts, spores or sclerotia for a variety of reasons (Corliss and Esser, 1974). Many protists secrete resistant cell walls within which they can remain dormant through periods of adverse conditions, including dispersal as dried cysts. Thus the cysts of soil protists, of some parasites and of protists from temporary bodies of water, among others, are able to survive drought and other severe conditions; the walls of such cysts are highly resistant and rather impermeable and within such cysts the respiration rate falls to a low level. In other cases the cyst wall is a relatively permeable layer and the encysted organism cannot survive drying out, e.g. the cyst of *Didinium*, and thin-walled reproductive cysts, in which one or many divisions may take place, are common in some ciliates. Flagellates and some other protists may encyst following sexual fusion and this cyst may be resistant or thin-walled.

The factors which cause encystment vary in different species. Usually these factors are adverse environmental features like shortage of food, extreme pH, increase in salt concentrations or high temperatures, all of which might signal the drying up of the environment of a soil amoeba like *Acanthamoeba* or a soil ciliate like *Colpoda*. Such organisms may also encyst as a result of overcrowding, lack of oxygen or accumulation of metabolic water following rapid growth. In the amoebo-flagellate *Naegleria* the response of the amoeba to changed conditions includes encystment on drying and the development of flagella with dilution of the medium (p. 166).

In an experimental study of the life cycle and encystment of *Acanthamoeba* by Neff and Neff (1969) (Fig. 9.10), encystment was promoted by transfer from a nutrient growth medium to an inorganic salt medium containing calcium and magnesium ions. Failure to complete the S phase of the growth/division cycle is supposed to have triggered encystment. It is likely that all adverse conditions causing encystment result in an interference with DNA synthesis, and hence block the cell cycle at the S phase. In the experiments the induction process was found to lead to a rounding up of the cell and a massive breakdown of cell com-

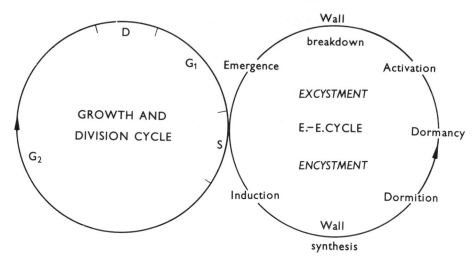

Fig. 9.10 The relationship between the growth-division cycle and the encystment-excystment cycle of **Acanthamoeba**. Modified from a figure by Neff and Neff (1969)

ponents, including many mitochondria and storage granules. It was followed by two waves of wall synthesis, the first produced an outer phospho-protein layer and the second an inner cellulose layer; these synthetic reactions involved specific RNA synthesis and the formation of new enzymes. All detectable metabolism ceased within a few days after the completion of the cyst wall, there was no longer any cytoplasmic movement within the cyst, and the organism entered a period of dormancy in which it could survive for years. In the dormant state such cysts can endure high and low temperatures, drying, low and high pressures and absence of oxygen. Different components, e.g. silica, lipids, proteins, polysaccharides, are found in the cyst walls of different protists and they are formed from 'molecular secretions' or products released from mucocysts or more directly from Golgi vesicles.

Comparable detailed information on the induction of excystment of *Acanthamoeba* is not available, but, following an initial stage of activation that is dependent upon RNA synthesis and protein synthesis, there is some digestion of the cellulose wall and a progressive increase in activity of the organism, and within 12 to 30 hours the amoeba emerges. A number of studies of excystment of the soil ciliate *Colpoda* indicate that soluble substances present in hay infusions, principally organic acids and sugars, provide the necessary stimulus for induction of excystment in this organism, so that it is presumed that the resistant cyst wall is permeable to such substances. Induction and increased activity is often followed by osmotic uptake of water and rupture of the cyst, though in some cases an emergence pore is present.

Cyst stages clearly have ecological importance in that they provide for survival through periods of adverse conditions, such as temporary or seasonal lack of food, low temperature or drought, and allow an important means of dispersal.

Features of the life of protists in various habitats

According to estimates by J.O. Corliss (1984) about 120 000 species of protists have been described, rather more than 50 000 of them being fossil species (largely foraminifera and diatoms). Of the 60 000 plus living species more than half belong to groups that are entirely heterotrophic, one third to groups that are entirely photosynthetic and rather less than 10 000 to those protistan phyla which contain a fair proportion of both autotrophs and heterotrophs. There are about 10 000 living species of parasitic protists, largely sporozoa and parasitic flagellates. About two thirds of the remaining 50 000 living species are marine forms, diatoms, foraminifera, radiolaria, dinoflagellates, ciliates, red and brown algae being important groups. Diatoms, dinoflagellates and ciliates are also important in fresh water, and most green algae, euglenids and heliozoa occur in freshwater.

Protists from marine habitats

The most extensive marine habitats are those of the open ocean. The majority of oceanic protists known at present are planktonic forms found within 100 m or so of the surface of the sea, in the region where light energy powers photosynthesis and abundant food is produced for heterotrophic forms. The primary producers and some other inhabitants of this zone have already been mentioned (p. 247). Planktonic protozoa also occur in deeper waters, even at extreme depths, but the abyssal forms known are largely phaeodarian radiolaria. The various groups of radiolaria and a few species of foraminifera are the characteristic planktonic amoebae, with long spines and often oil droplets to aid flotation, and long slender pseudopods for food capture; often they depend heavily on symbiotic algae for their nutrition and in some waters the dinoflagellates symbiotic in radiolaria contribute a major proportion of the primary production, with acantharia being particularly important in tropical waters, but less numerous in colder waters. Several groups of bacterivorous flagellates consume picoplankton, and tintinnids and other oligotrich ciliates as well as dinoflagellates are other important phagotrophs, many of these larger forms eating smaller protists. Local concentrations of protozoa, principally bacterivorous flagellates and ciliates, are found associated with 'marine snow' (the rain of organic debris, dead plants and animals and the dense faecal pellets of planktonic crustaceans) in the open oceans (Caron *et al.*, 1982).

A wider range of protist species is found where there are more substrata for attached forms. These are occasionally present in oceanic regions where there are numbers of epizoic protozoa such as peritrich and chonotrich ciliates attached to crustaceans (see p. 268). In coastal regions there are abundant surfaces, inanimate as well as living, upon which protists may settle. The majority of species of foraminifera are benthic species, many of them from shallow waters, and there are characteristic species of flagellates and ciliates which occur in close association with or attached to various forms of substratum; these include dinoflagellates like *Amphidinium* and *Oxyrrhis*, and ciliates like vorticellid peritrichs, hypotrichs (e.g. *Euplotes* and *Diophrys*), gymnostomes (e.g. *Trachelocerca*), suctorians (e.g. *Ephelota*) and heterotrichs like *Condylostoma* and the loricate *Folliculina* (*see also* Fig. 9.11).

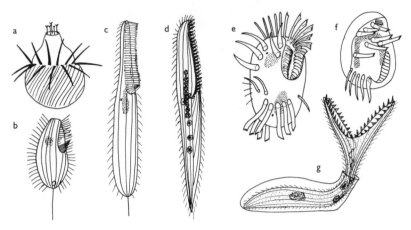

Fig. 9.11 Some ciliate protozoa common in shallow benthic marine habitats; **a**, the haptorid gymnostome *Mesodinium pulex* (25 μm): **b**, the scuticociliate *Uronema marinum* (40 μm): **c**, the scuticocilate *Cohnilembus verminus* (100 μm): **d**, the heterotrich *Gruberia lanceolata* (500 μm): **e**, the sporadotrich hypotrich *Diophrys appendiculata* (80 μm): **f**, the sporadotrich *Aspidisca steini* (30 μm): **g**, the loricate heterotrich *Folliculina aculeata* (350 μm), and further examples in Fig. 9.12.

The colonisation of aquatic sediments by protists depends on their chemical and mechanical features; the ecology of such microbenthos communities has been extensively studied by T. Fenchel (1969). In sediments with grain sizes below about 100μm all of the larger protists are excluded, and even a small proportion (<5%) of fine clay particles tends to exclude protists altogether by filling the interstices between larger grains. Some specialized ciliates inhabit interstitial habitats of marine sands (Fig. 9.12), especially those with grain sizes between 150 and 250μm. These psammophilic forms are either small species, or extremely flattened and often elongate ciliates, showing marked thigmotactic attachments to the sand grains over which they creep while collecting bacteria, other protists or organic detritus (Dragesco 1963a,b). In these situations the microorganisms form distinct communities (Fenchel, 1969, 1987). Beneath the oxidised surface layer of marine (or freshwater) sandy sediments is an anaerobic sulphide zone, marked by reducing conditions associated with plentiful deposits of organic matter but no free oxygen for oxidation processes. The organic matter is broken down by fermenting, methanogenic and sulphate reducing bacteria; the activity of the latter is particularly evident in H_2S production and the black sulphides of the sediment. A number of other bacterial species living nearer the surface oxidise reduced compounds like H_2S using CO_2 in anaerobic photosynthesis or in chemoautotrophic processes of microaerophilic forms. Many protists migrate into or live within the sulphide layer; these include potentially autotrophic diatoms, euglenids and dinoflagellates as well as amoebae and ciliates, particularly obligately anaerobic and mainly bacterivorous ciliates like **Caenomorpha**, **Metopus** and **Saprodinium** in the lower regions and herbivorous forms like **Condylostoma**, **Frontonia**, **Strobilidium** and **Tracheloraphis** in the surface regions, with predatory ciliates also specific to certain layers. A restricted range of invertebrates, mostly flatworms, nematodes and sometimes gastrotrichs live in the sulphide zone with a wider range of groups in the surface

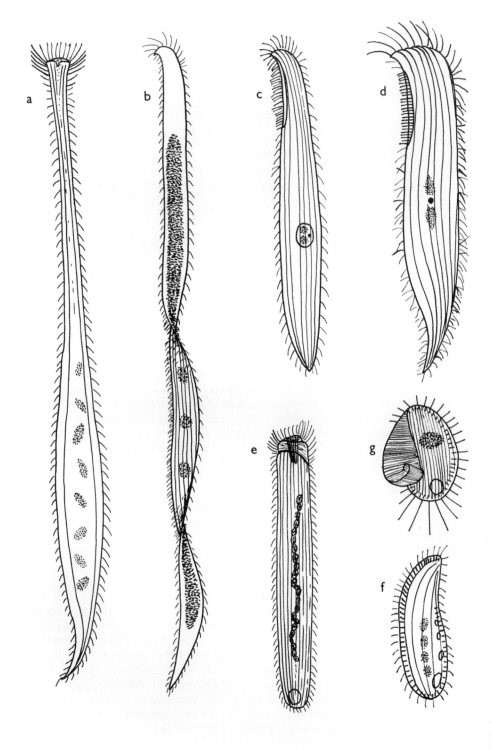

zone; many of these will include protists in·their diet (Fenchel and Riedl, 1970).

Freshwater Protists

Freshwater habitats are somewhat more diverse and include puddles and tempo-rary bodies of water as well as lakes and rivers. The larger lakes have a true open-water plankton, including among the autotrophic forms dinoflagellates like *Ceratium* and *Peridinium*, chlorophytes like the flagellates *Pandorina* and *Volvox* and the desmids *Staurastrum* and *Arthrodesmus*, diatoms like *Asterionella* and *Fragillaria* and smaller chrysophytes; in freshwater phyto-plankton these eukaryotes may be outnumbered on occasions by cyanobacteria. The availability of nutrient salts required for the growth of plants influences the abundance of phytoplankton in lakes. Waters with few salts are referred to as oligotrophic, and the rather sparse phytoplankton is dominated by characteris-tic diatoms and desmids and such chrysophycean flagellates as *Dinobryon*, *Mallomonas* and *Synura*; eutrophic waters with a higher content of nutrient salts support a rich and diverse flora, often dominated by cyanobacteria and with a different range of diatoms and desmids, and often including dinoflagel-lates, although *Ceratium* and some species of *Peridinium* may also be found in oligotrophic waters. Among the zooplankton of lakes there are a few freshwater representatives of the tintinnid ciliates, e.g. *Codonella*, and some other ciliates, but generally it is found that as in the sea the coastal (neritic) regions have a richer fauna and flora. Protists that are truly planktonic seldom survive in flowing waters, except for lake-like plankton in the largest rivers, but consider-able numbers of protists that have been washed away from habitats near surfaces are common in the flowing water column.

Many of the protists found in streams and rivers as well as around the margin of lakes and ponds live in close contact with surfaces, either as attached forms or as creeping forms maintaining contact with the surface with pseudopodia or thigmotactic cilia or flagella. In stationary water and even in quite strong currents there are sedentary epiphytic and epilithic diatoms and bacteria that provide food for protozoans on those surfaces. Not only is food more abundant near surfaces, but in moving water habitats the organisms that live on surfaces find themselves in the more slowly moving fluid of the hydrodynamic boundary layer (Silvester and Sleigh, 1985).

Many other protozoans, especially bacterivorous species, live epizoically on animals of various phyla, some in specific associations, others less specific. Often the protozoans are sessile forms, as illustrated in Fig. 9.13, where the water currents maintained by the crustacean may provide an abundance of different types of food for different protozoans occupying different regions of

---

Fig. 9.12 Some ciliates from marine sand; **a**, the karyorelictid *Tracheloraphis phoenicopterus* (1 mm): **b**, the ribbon-shaped karyorelictid *Kentrophoros lanceolata* (500 μm), with a dense mat of sulphur bacteria on the dorsal surface: **c**, the karyorelictid *Geleia fossata* (350 μm): **d**, the karyorelictid *Remanella rugosa* (200 μm), seen from the dorsal surface, but showing the ciliary meridians of the ventral surface: **e**, the haptorid *Chaenia gigas* (500 μm): **f**, the pleuro-stomatid *Litonotus* sp. (60 μm): **g**, the scuticociliate *Pleuronema coronatum* (80 μm); some of the examples shown in Fig. 9.11 are also found in marine sands.

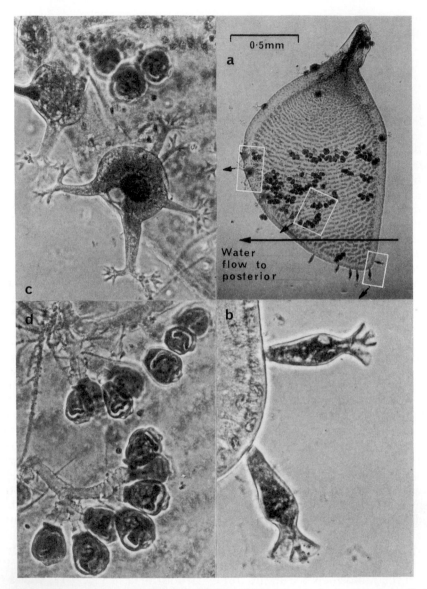

Fig. 9.13   Epizoic ciliate protozoans on a gill plate of *Gammarus pulex*; **a**, the whole gill plate stained to show up the protozoa and labelled to indicate the orientation of the gill plate in the water current created by limb movements of the crustacean: **b**, the chonotrich *Spirochona gemmipara*: **c**, the suctorian *Dendrocometes paradoxus*: **d**, the peritrich *Epistylis* sp.

the gill. Similar forms may be present on insect larvae, crustaceans and annelids, depending on the habitat. Motile epizoic forms include such hypotrichs as *Kerona* on *Hydra* and mobiline peritrichs like *Trichodina*, *Urceolaria* and *Cyclochaeta*, common on the skin and gills of fish, polychaetes and many other animals, where they may harm the host.

The sediments deposited from flowing waters in regions of slower flow have very similar characteristics to the benthic regions of lakes and marine habitats mentioned above, with a strong dependence on the amount of organic matter present and the availability of oxygen. The anaerobic conditions that prevail in sediments may extend upwards into the water above the floor of a lake during periods of thermal stratification and poor oxygen supply to the deeper water; under these conditions large populations of microaerophilic ciliates like *Loxodes*, *Spirostomum* and *Prorodon* follow the oxycline up into the water, migrating periodically between the bacteria-rich anaerobic water below the oxycline and the less nutritious aerobic water above (Bark, 1981; Finlay, 1982; Finlay and Fenchel, 1986). In flowing waters the oxygen levels remain high (except in polluted conditions) and normally the surface layer of sediments will remain oxidised, with a range of (often) motile epipelic diatoms and autotrophic flagellates as well as herbivorous and bacterivorous protozoa. A number of workers have recognised quite characteristic associations or communities (microbiocoenoses) of protists and prokaryotes, usually in sediments or around filamentous plants. For example, E. Faure-Fremiet (1951) described the part played by ciliates in the life of a mat of filaments of the sulphur bacterium *Beggiatoa*. At first the mat of filaments was fairly clean and there were a few *Colpidium* eating free-swimming bacteria around the *Beggiatoa*. Increase in the number of these ciliates was followed by the arrival of predatory ciliates like the gymnostome *Lionotus* and the hymenostome *Leucophrys* (= *Tetrahymena*), and a reduction in the numbers of *Colpidium* led to an increase in the numbers of *Glaucoma* eating *Beggiatoa* and *Chilodonella* eating associated blue-green algae or filamentous bacteria. A rapid multiplication of these last two ciliates led to the breakdown of the mat of *Beggiatoa* and the production of detritus which forms the food of such ciliates as *Paramecium* and the prostomatid *Coleps*. At all stages the number of ciliates closely associated with the mat of filaments was much higher than that in the surrounding water or on the adjacent mud.

The close association between protozoa, even many of the more active swimming ciliates, with filamentous growths of algae and bacteria, had been emphasised in an earlier study by L.E.R. Picken (1937). In waters of lower organic content the filamentous growths of cyanobacteria, particularly *Oscillatoria*, formed the basis of the community of organisms, and in waters of high organic content the organisms were associated with growths of 'sewage fungus' (a complex association in which the filamentous bacterium *Sphaerotilus* and the fungus *Leptomitus* are usually dominant). In both communities there were protozoa feeding on bacteria and on diatoms, but the numbers of different protozoa varied with the relative quantities of these two major forms of food, diatoms being more abundant in cleaner water and bacteria being abundant in polluted water. Both communities included forms consuming detritus and carnivorous protozoa as well as some omnivorous species; some relationships between the organisms involved are shown in Fig. 9.14. It is interesting that the protozoa maintained their close contact with the filamentous surfaces even when the diatoms were also common on the surrounding mud, so that it seems as if the specific association may depend upon mechanical and chemical characteristics of the substratum, the filaments of algae and bacteria providing a preferred micro-environment. Comparable microbiocoenoses have been described

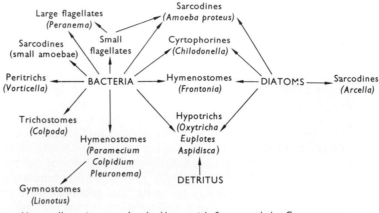

Almost all species eaten by the Heterotrich *Stentor* and the Gymnostome
Dileptus eats *Stentor* as well as most other species

Fig. 9.14   The major protozoan components of a freshwater food web where the sources of
food are bacteria, diatoms and detritus. Data from Picken (1937).

from estuarine and marine sediments (Webb, 1956; Fenchel 1969); diatoms and
bacteria again provide the main food and filamentous sulphur bacteria or
cyanobacteria provide mechanical support.

The species of organisms found in these freshwater habitats vary according to
the amount of organic matter present, as well as according to such features as
the oxygen content, calcium, nitrate and phosphate concentrations and pH.
Water with a high organic content is described as polysaprobic, and waters with
medium and low organic contents are referred to as mesosaprobic and oligosaprobic respectively, while extremely pure water is described as katharobic. It is
possible to encounter any of these categories of organic content in waters that
are classed as eutrophic on the basis of mineral salt levels, but oligotrophic
waters will generally have rather low levels of dead organic matter. The largest
numbers of protistan individuals occur in polysabrobic conditions, but
normally rather few species are represented, principally bacterivorous forms
tolerant of very low oxygen levels, such as the flagellates **Oikomonas** and **Bodo**,
the ciliates **Paramecium putrinum** and **Vorticella microstoma**, forms like **Polytoma uvella** that depend upon dissolved organic matter, and a few photosynthetic forms like **Euglena**. A richer and more diverse range of protists occur in
mesosaprobic conditions, including some pigmented flagellates like **Chlamydomonas** and **Cryptomonas** and a characteristic flora of desmids and diatoms as
well as such colourless flagellates as **Peranema, Anthophysa** and **Chilomonas**,
ciliates like **Colpidium, Carchesium, Vorticella, Stentor, Euplotes** and
Aspidisca, the heliozoans **Actinophrys** and **Actinosphaerium** and such amoebae
as the testacean **Arcella**. Autotrophs tend to predominate in oligosaprobic
conditions, including characteristic diatoms and desmids and such flagellates as
Dinobryon, Gonium, Volvox and **Ceratium**, accompanied by a rather sparse
fauna of herbivorous and detritus-eating ciliates and amoebae.

Protozoa, pollution and sewage treatment

A variety of forms of pollution occur in natural waters as a result of human activity. Organic pollution is probably most widespread, but toxic chemicals, including heavy metals, and also extreme pH, heated effluents, particulate effluents and radionuclides all damage ecosystems. Protists as well as other organisms have been used to monitor such pollution, and various protists have been used as indicator species in 'saprobicity indices' based upon the scheme of Kolkwitz and Marsson (1908) (Fjerdingstad, 1971). The communities of protists are rather complex, and vary with substratum, so that attempts to standardise pollution monitoring procedures have used artificial substrata such as glass slides or small blocks of polyurethane foam, the latter being favoured because they provide a three-dimensional microhabitat (Henebry and Cairns, 1980). Such methods using protists are valuable because of the short response time of protistan communities, and because in general protists tend to show similar responses to macroscopic organisms, e.g. in respect of the production of metallothionein proteins that bind heavy metals (Albergoni and Piccinni, 1983).

Most streams and rivers show variations in the content of organic matter along their length because of the inflow of organically polluted water from tributaries (Hynes, 1960). Such organic compounds are oxidised by the organisms living in the stream as the water flows along, and at a distance below the site of pollution the organic materials will have disappeared leaving water enriched in mineral salts. Close to the site of pollution there will be intense bacterial activity associated with a high utilisation of oxygen in the breakdown of organic compounds; if the pollution is extensive, the oxygen content may be reduced to zero and the water may contain hydrogen sulphide and ammonia. In such anaerobic conditions few eukaryote organisms occur, including such bacterivorous flagellates as **Bodo**, along with the abundant swimming bacteria and carpets of sewage fungus. As the organic matter is oxidised, and the consumption of oxygen falls, aeration and the reappearance of photosynthetic plants cause the oxygen content to rise again, so that the general character of the stream reverts from the polysaprobic condition towards the mesosaprobic and eventually to the oligosaprobic state. Bacterivorous protists will be most abundant in more polluted places, but with the appearance of algae, herbivorous ciliates and amoebae become common. Laboratory studies on artificial systems with added organic matter show similar chemical changes and biological successions (Bick, 1967).

Severe organic pollution that results in extensive deoxygenation does great damage to the life of the stream and should not be permitted. Such damage is avoided by the treatment of polluted waters to reduce the content of organic compounds and suspended solids before the water enters the stream; the various forms of sewage treatment achieve a reduction in the organic content by microbial action in the same manner as is normal in a stream, but the process is accelerated and is isolated from the stream. Toxic pollutants of some types may also be broken down by bacteria, but others may require chemical treatment.

Three processes of sewage treatment are in common use and protists occur abundantly and are of importance in all three, especially the aerobic ones (Curds and Cockburn, 1970a,b; Hanel, 1979). Small-scale treatment may be carried out in a septic tank (Imhoff tank). This is a large chamber into which the polluted water flows, and within which abundant bacteria oxidise the organic matter, so that free oxygen is present only in a shallow layer at the surface. The bulk of the organic breakdown takes place under anaerobic conditions, with the use of hydrogen acceptors other than oxygen, so that much hydrogen sulphide and ammonia is usually produced; the effluent from such a tank is therefore normally only partially purified. The deeper levels of the tank contain abundant anaerobic bacteria and a number of facultatively or obligately anaerobic protists, notably those tolerant of H₂S. The ciliates *Metopus es, Trimyema compressa* and *Saprodinium putrinum* and the flagellate *Trepomonas agilis* appear to be obligate anaerobes, and other common forms in such tanks include the amoebae *Euglypha alveolata* and *Vahlkampfia guttula* and the flagellates *Cercobodo caudata, Pleuromonas jaculans* and *Hexamita inflata*.

Where sewage treatment is carried out on a larger scale, the liquor that remains after removal of solids that can be settled or skimmed off is subjected to processes which promote aerobic oxidation of the organic compounds by microorganisms (Fig. 9.15). This may be achieved either by intensive aeration of the liquor in large chambers (activated sludge tanks) or by the development of a microbial film on aerated surfaces, provided by stones in percolation filters or intermittently submerged plates in the rotating disc process. General features of these three processes will be described first and then the protozoan faunas will be considered.

In the activated sludge process settled liquor is mixed in the aeration tank with a substantial inoculum of settled sludge from the final settlement tank, and the mixture is vigorously aerated with rotating paddles or by bubbling air or oxygen through it. Solitary and filamentous bacteria from the returned sludge oxidise the organic matter in the liquor and multiply. Flocculation of the bacteria, aided probably by mucoid secretions of ciliates, removes many bacteria from suspension and the flocs provide substrata for the attachment of stalked and thigmotactic ciliates, which, like the bacteria, are aerobes. After the liquor has spent some hours in the tank it passes to a final settlement tank where the flocculated material sediments out of the effluent; some of the settled material is returned to the sludge tank and the remainder discarded.

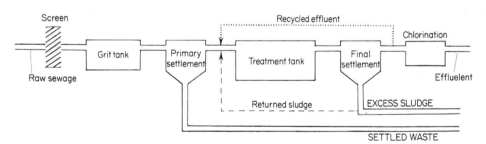

Fig. 9.15 Stages commonly involved in sewage treatment. After screening for sticks, rags, paper etc., grit is settled by slowing the flow, and organic solids settled by further reducing flow in a primary settlement tank. Different forms of aerobic treatment follow according to the type of plant, but organic solids formed during treatment are collected in a final settlement tank before the cleaned effluent is chlorinated. In some cases incompletely purified effluents may be recycled, and in the activated sludge process some settled sludge is returned to provide a substantial inoculum of microorganisms to promote aerobic oxidation in the treatment tank. Settled organic wastes may be digested anaerobically to produce methane that can provide power to operate the plant.

Percolation filters contain stones (larger ones below and smaller ones down to about 4 cm in diameter at the surface) arranged in a bed 2 m deep in such a way that air can circulate readily through spaces between the stones. A slow trickle of the sewage liquor runs down over the stones, and a bacterial film quickly appears on the stones and develops into a complex association of aerobic bacteria, fungi, blue-green algae and protozoa. Heterotrophic forms perform aerobic oxidation of organic compounds and bacteria, algae and detritus provide food for the protozoans that live in the water film which covers the stones. Larger animals like nematode and oligochaete worms and insect larvae are abundant and are important in that they burrow through the microbial film and break it up, so that fragments are washed away and the air passages kept open. Fluids drained from the filter bed are passed into a settlement tank to remove solids from the cleaned effluent.

Rotating biological discs have recently been developed as an alternative method of treating the settled liquor. A series of 4 or 5 banks of large discs, typically with a radius of 1 or 2 m, and spaced at intervals of 5 cm or so, with 20 or more discs in each bank, are mounted on a common axle over a long trough of liquor so that about 40% of each disc is immersed in the liquor and the rest is in the air. The discs are rotated slowly (1–5 rpm) and an aerobic microbial film develops on their surface. The liquor flows slowly, over a period of three hours or so, along the trough in which the discs are suspended perpendicular to the flow, and the organic compounds in the liquor are progressively oxidised. The purification of the flowing liquor is accompanied by a linear sequence of changes in the population of protozoans from stalked peritrichs in the first bank to testate amoebae, swimming ciliates and creeping flagellates in the last bank (Sudo and Aiba, 1984). The microbial film may become progressively thicker, but then its inner parts become anaerobic and the film tends to slough off and a new film forms; worms may also loosen the film in later banks. The effluent is again passed though a settlement tank to remove solids. This method of sewage purification requires little energy and is particularly suitable for small communities (Curds, personal communication).

C.R. Curds and his colleagues in Britain and R. Sudo and his colleagues in Japan have made comparative studies of protozoa in sewage treatment plants. In a survey of 52 percolating filter systems in Britain (Curds and Cockburn, 1970a), ciliates formed the most abundant group, although the testate amoeba *Arcella vulgaris* was commonly present and the flagellate *Peranema trichophorum* was frequent; the more important ciliates found in this study are listed in Table 9.6 and illustrated in Fig. 9.16. Of 56 activated sludge plants examined in the same study, three contained no ciliates, but in the others ciliates were the dominant protozoa; *Arcella* and *Peranema* were found in these also, and a somewhat different range of ciliates (Table 9.3). A similar range of species occurred in activated sludge and rotating disc plants in Japan (Sudo and Aiba, 1984).

The role of these protozoa in sewage treatment is more easily studied in activated sludge plants than percolating filters because of accessibility. In studies of small-scale experimental activated sludge systems Curds (1966) found that following the setting-up of a plant the protozoan fauna developed in a clear sequence. At first only flagellates (e.g. *Oikomonas, Bodo, Peranema*) were present, but after several days a range of free-swimming ciliates like *Paramecium*

Table 9.3 The ciliate protozoa recorded most commonly in a survey of sewage treatment plants in Britain. In each case the species are listed in order of their frequency of occurrence in plants using that type of treatment, and the percentage of cases in which the ciliates were found in very large numbers is shown alongside. Information from Curds and Cockburn (1970).

Percolating filters Species	% with high numbers	Activated-sludge plants Species	% with high numbers
Chilodonella uncinata	4	*Vorticella microstoma**	10
*Vorticella convallaria**	10	*Aspidisca costata†*	35
*Opercularia microdiscum**	44	*Trachelophyllum pusillum*	15
*Carchesium polypinum**	15	*Vorticella convallaria**	19
*Opercularia coarctata**	2	*Opercularia coarctata**	12
Aspidisca costata†	–	*Vorticella alba**	11
Cinetochilum margaritaceum	–	*Euplotes moebiusi†*	5
*Vorticella striata**	2	*Vorticella striata**	2
Trachelophyllum pusillum	–	*Vorticella fromenteli**	4
*Opercularia phryganeae**	4	*Carchesium polypinum**	8

* Peritrich ciliates.   † Hypotrich ciliates.

appeared and became dominant after about two weeks. These ciliates were replaced after about four weeks by crawling ciliates, especially species of the hypotrich **Aspidisca**, and stalked peritrich ciliates, mainly **Vorticella** spp. Finally in the mature system **Aspidisca** and **Vorticella** appeared to be competing for dominance. While the majority of the ciliates were feeding on bacteria, some, e.g. **Lionotus**, were carnivores and some, e.g. **Trachelophyllum**, were thought to ingest sludge flocs. Experimental activated sludge plants maintained free of ciliates produced turbid effluents with a high BOD (biological demand for oxygen), high quantity of suspended solids and large numbers of viable bacteria; after addition of ciliates to these plants, the clarity of the effluent improved and the BOD, suspended solids and viable count of bacteria all decreased dramatically, high numbers of ciliates being correlated with low numbers of bacteria in the effluent. The ciliates are assumed to effect this improvement primarily by feeding on the bacteria, but the enhancement of flocculation and the promotion of micro-circulation of the liquor by ciliary activity may play a part.

In the survey of activated sludge plants mentioned above it proved possible to predict the quality of effluent produced by a plant from a knowledge of the range of ciliate species present (Curds and Cockburn, 1970b). Activated sludge plants in Tokyo which produced the clearest effluent (COD, chemical oxygen demand, around $10 \, \text{mg} \, l^{-1}$) contained about 10^4 protozoa ml^{-1}, those producing poor effluent (COD around $40 \, \text{mg} \, l^{-1}$) had populations of only about 10^2 protozoa ml^{-1} (Sudo and Aiba, 1984). In the presence of larger ciliate populations the biological oxidation is enhanced and the amount of residual sludge solids is reduced; bacterial growth may be enhanced by the grazing activity of ciliates, and the latter may also reduce levels of pathogenic bacteria in the effluent.

Soil Protists

Recent reviews on this topic have been published by Stout and Heal (1967), Alabouvette *et al.* (1981) and Foissner (1987). At one time it was thought that there were no active protozoa in the soil, and that such protozoa as were found

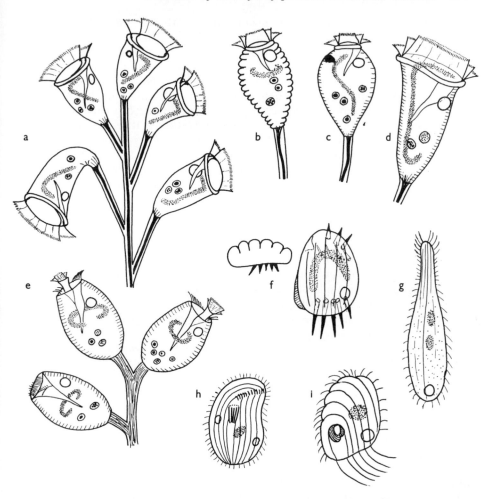

Fig. 9.16 Some ciliate protozoa commonly found in percolation filter sewage treatment systems; **a–e** are peritrichs, **a**, *Carchesium polypinum* (cell body 125 μm long): **b**, *Vorticella striata* (40 μm): **c**, *V. microstoma* (50 μm): **d**, *V. convallaria* (75 μm): **e**, *Opercularia micro-discum* (cell body 80 μm long) with non-contractile stalk: **f**, the sporadotrich hypotrich *Aspidisca costata* (30 μm): **g**, the haptorid *Trachelophyllum pusillum* (40 μm): **h**, the cyrtophorid *Chilodonella uncinata* (60 μm): **i**, the scuticociliate *Cinetochilum margaritaceum* (30 μm).

there were encysted and came from freshwater or marine sources. Early in this century, however, it was established that there was a considerable fauna of active feeding protozoa living in the water films around soil particles, where they were presumed to be feeding on soil bacteria; estimates indicate that there are between a few thousand and half a million protozoa in each gram of moist soil, along with perhaps 10^9 bacteria. From a study of soils from many parts of the world H. Sandon (1927) concluded that all soils contained protozoa, princi-pally in the surface layers, and that similar species occurred in soils at all latitudes. About 20 of the 250 protozoan species he found were only present in that habitat, while the majority of species were also present in organically pol-luted waters. In general the abundance of flagellates, ciliates and amoebae could be correlated with features associated with good bacterial growth, while the abundance of testate amoebae was correlated with the amount of organic matter, irrespective of bacterial numbers; many of the protozoa could live where the oxygen concentration was very low. It has been estimated, for example, that arable field soils may contain a protozoan biomass of about 5 g dry weight m^{-2}, a figure similar to that of earthworms, but, since protozoa reproduce much more quickly, their production is higher than that of earthworms; these soil protozoa may consume several hundred g (dry weight) of bacteria m^{-2} yr^{-1} (Stout and Heal, 1967). Other studies have confirmed these general conclusions and extended the lists of species of protozoa found in and specific to soils; though Foissner (1987) argues that more careful study of the ciliate fauna of soils reveals that only a small proportion of soil ciliates occur in fresh water. The same author estimates that 10% of the total carbon input in soils may be respired by protozoa.

There is wide agreement that the vegetative productivity of the soil is correlated with the numbers of protozoa in the soil. The principal reason for this is probably that protozoa, largely amoebae, grazing on the bacteria stimulate bacterial activity, possibly including nitrogen fixation. In addition, as in aquatic habitats (p. 259), the protozoa accelerate nutrient cycling by release of nitrogen compounds derived from bacteria, which can be used by the plants (Clarholm, 1981, 1984). Thus angiosperms growing in soils with both bacteria and ciliates grow more quickly than when only bacteria are present (Elliott *et al.*, 1984). Populations of both bacteria and protozoa are higher in the rhizosphere around plant roots than in the surrounding soil, though the same species are usually present (Darbyshire, 1966, Darbyshire and Greaves, 1973); organic compounds released from the roots presumably stimulate bacterial growth. There has also been a report that protozoa secrete plant auxins (Nikoljuk, 1969), which could influence plant growth. Protozoan grazing can reduce the numbers of harmful fungi and bacteria in the soil (Coleman *et al.*, 1977) and enhance the digestive efficiency of earthworms.

The common soil protozoa include the flagellates **Oikomonas, Cercomonas** and **Heteromita**, several species of the ciliate genus **Colpoda**, the amoebae **Hartmannella, Acanthamoeba** and **Naegleria** and species of acrasian myxomycetes. Many testate amoebae reach their greatest abundance in soils, and may be the dominant protozoa in acid soils where there are few bacteria; such genera as **Centropyxis, Nebela, Trinema** and **Euglypha** are commonly represented. It is likely that under most conditions the flagellates and small amoebae are more important than the larger amoebae or ciliates, e.g. a 1 g sample of soil from an English wheatfield contained 1500 amoebae, 32 000 flagellates and 20 ciliates, but in some soils records of thousands of ciliates and up to 70 000 testate amoebae in each g of soil indicate that these larger protozoa may sometimes be very important in the biological economy of the soil. Ciliates may be more numerous in fresh litter than in the humus beneath, while testate amoebae are more numerous in the lower layer (Foissner, pers. comm). While these protozoa are more common beneath the soil surface, the surface layers of damp soil contain photosynthetic protists of several groups, principally diatoms, xanthophytes, filamentous and coccoid chlorophytes, represented by species which, by contrast to most soil protozoa, tend to be confined to this habitat (Round, 1981).

The majority of soil protists encyst and therefore have a means of surviving dry conditions; a burst of trophic activity and reproduction follows the rewetting of dry soils. Cysts also provide a means of distribution of soil protists; it has been found that there are on average two protistan cysts m^{-3} of air, and it is thought that these minute dry cysts may be carried for some distance by air currents. A more common means of dissemination of soil species as well as freshwater protists is probably by the movements of animals, which may carry wet or dry mud, slime, etc. from place to place; a number of protists have been recovered from mud attached to the plumage and legs of birds and even from slime taken from the body of migratory water beetles.

Symbiotic relationships involving protists

Epibiotic species

Many protozoa live epibiotically on the surface of other animals and plants, frequently using these organisms only as a support; stalked and mobiline peritrich ciliates provide many examples of this mode of life in both marine and freshwater habitats and there are even epizoic peritrichs living in water films on the gills of terrestrial woodlice. A number of such epizoic forms have made the transition to parasitism, changing their mode of life to a greater or lesser extent to derive food and often other needs such as shelter and dispersal from their hosts. A range of forms may be seen among rhynchodids (p. 204) which live among the gills in the mantle cavity of bivalve molluscs; the less specialised forms feed on detritus and use the gills only as a supporting substratum, while the more advanced forms show a progressive development of a sucker or tentacle system and are thought to obtain their nourishment more directly from the host. Many species of the mobiline peritrich *Trichodina* appear to live a harmless epizoic existence, but numerous other species are parasites on the gills of their hosts, e.g. fish, and some have become endoparasites in the urinary bladder of fish. The majority of ectoparasitic protozoa are ciliates, although there are a few dinoflagellates and amoebae following this mode of life.

Endoparasitic protists

Protozoa are seldom found as parasites in terrestrial plants. Perhaps the best-known examples are the flagellates (*Phytomonas*) found in the latex channels of *Euphorbia* spp. and some other plants, and in recent years diseases of coconut and oil palms have been found to be caused by species of *Phytomonas* that live in the phloem (McGhee and McGhee, 1979); these flagellates are assumed to be transmitted by plant-sucking bugs. There have also been reports of amoebae inhabiting similar situations in euphorbias, figs and lettuce plants.

Members of several phyla of protists parasitise aquatic plants. Included amongst these are chytrids (p. 144) which attack various other protists, including flagellate and coccoid members of most photosynthetic groups, as well as higher plants. Organisms assumed to be chytrids (mostly placed in the genera *Sphaerita* and *Nucleophaga*) have been found within the cells of flagellates from termites and other hosts, in parasitic amoebae, in rumen ciliates from mammals and in gregarines from arthropods, some within the nucleus and others in the cytoplasm. Labyrinthulids (p. 138) and some oomycetes (p. 138) also parasitise aquatic plants, and plasmodiophorans and oomycetes parasitise terrestrial plants.

Multicellular animals of all phyla act as hosts of parasitic protozoa (Baker, 1969; Kreier, 1977) and provide a number of suitable habitats for them, the different locations demanding rather different means of transmission from host to host. The least specialisation is required of parasities inhabiting the alimentary canal of animals, for means of entry with the food and exit with the faeces are already provided, and the parasite does not face the immune defences of the host but only has to solve problems of avoiding desiccation before being eaten by the host and avoiding being killed and digested by the digestive juices of the host. Many gut-dwelling protozoa eat bacteria just as their free-living relatives do, so that many of them should be regarded merely as commensal endobionts rather than parasites. They survive exposure to the air by secreting protective cyst walls; these also protect them against enzymes of the host which ingests them. As with many parasites, the chances of successful transmission to a new host are enhanced by the production of enormous numbers of infective individuals – in this case cysts. It is likely that more than half of the parasitic protozoa live in such situations, inhabiting a specific part of the alimentary tract. Many of these forms attach themselves to the gut wall by means of suckers, hooks or tentacular appendages; they often grow to considerable size and many have very complex structure. While any undigested food in the gut may be ingested by many protozoa, many more feed by absorbing the food digested by the host, and a third category derive nourishment from the gut epithelia of the host, either by engulfing the cells or by extracting nutrients through appendages that penetrate into or between the cells; those of the third category are clearly the most harmful to the host.

A number of parasites that gain entry to the body through the mouth and alimentary canal proceed to invade the tissues and body cavities of the host. Thus many coccidians invade the epithelia of the alimentary tract and associated organs and release infective cysts into the gut following growth and reproduction. Some gregarines and myxosporans migrate to the body cavity and surrounding organs of the host, where they grow and produce cysts or spores. These pass out through coelomoducts, or are liberated when the host moults or lays eggs, or in some cases are released only at the death of the host. Occasionally the parasites enter the eggs of the host and are therefore passed directly to a host of the next generation (e.g. *Babesia* in the tick). Some tissue parasites gain entry to a new host through the skin, especially through gills and similar less well protected areas.

Protozoa inhabiting the blood systems of vertebrates differ from these other forms in adaptive ways related to their mode of life. These parasites generally have a simple structure, and are either very small, if they live within corpuscles, or have a small diameter, if they live free in the plasma, because of the need to pass through fine capillaries. Since the blood systems are enclosed, the parasites need a special means of entry and exit; this is frequently provided by bloodsucking vectors which transmit the protozoan from one host to another, arthropods being the usual vectors for terrestrial vertebrates and leeches for aquatic vertebrates. The parasite normally grows and reproduces in the vector also, and is a true parasite in both of its hosts. Because of the direct transmission between vertebrate and vector, these protozoa do not form resistant cyst walls. It seems likely that the relationship between the protozoa and their two hosts may have arisen differently in the sporozoans and in the flagellates. The flagellates (trypanosomids) were probably originally intestinal parasites of invertebrates, as are **Leptomonas** and **Crithidia** today, and some of these became secondarily adapted to live part of their life in vertebrate blood systems. On the contrary, it is thought that the sporozoan blood parasites were probably derived from intestinal parasites of the vertebrate hosts which became parasites in the vertebrate blood system, perhaps following an intermediate stage similar to that of the eimeriid **Schellackia**. In this parasite the sporozoites escape into the blood from the site of zygote formation in intestinal epithelial cells of a lizard and are subsequently taken up by a blood-sucking mite, the cycle being completed by a lizard eating the mite. It is important that in these less elaborate life cycles, including that of **Lankesterella**, which shows a prolonged phase in the vertebrate bloodstream, the protozoans show no changes in the invertebrate host, and it is only in more complex cycles of such genera as **Haemoproteus** and **Plasmodium** that multiplicative phases occur in the arthropod.

The immunology of parasitic infections by protozoa has been the subject of intensive study (Taliaferro and Stauber, 1969), particularly with respect to diseases caused by malaria parasites (Deans and Cohen, 1983), trypanosomes (Hajduk, 1984) and leishmanias (Mauel and Behin, 1981). The hosts possess a diversity of mechanisms that protect the body against entry of foreign organisms or which act to remove foreign organisms that gain entry into the body. Non-specific immunity is provided by such cellular mechanisms as those which isolate the parasite by encapsulation, or digest it following phagocytosis by a macrophage. Specific immunity to a particular species of parasite is somewhat more delayed and is conferred in vertebrates by the production of antibodies in response to the presence of parasite antigens. However, protozoan parasites adopt various strategies to avoid attack by host antibodies. **Trypanosoma brucei**, for example, possesses a glycoprotein cell coat in its bloodstream phase, and this glycoprotein is changed at a low frequency (Donelson and Turner, 1985), with the possibility of producing 1000 or more variant antigens, and permitting some of the parasite population to survive antibody production in response to a particular infection; trypanosomes ingested by a tsetse fly quickly lose the surface coat, but infective forms in the salivary gland possess a heterogeneous range of antigens (Cross, 1975; Hadjuk, 1984). The malarial parasite spends much of its life in the vertebrate within cells, where it is safe from direct immune attack. Both sporozoites and merozoites are, however, free in the blood for brief periods, and surface proteins of both act as antigens for antibodies that can block entry of sporozoites into hepatocytes and merozoites into erythrocytes. Mature parasites also cause the expression of malarial antigens on the membrane of the erythrocyte they inhabit; this may induce phagocytic attack by macrophages. This is a complex subject, and the range of effectiveness of the immune response of the host can be seen by the fact that some infections result in the death of the host, some in the death of the parasite and some in a chronic infection in which host and parasite live in some sort of equilibrium. Suppression of the immune response by protozoan parasites occurs widely (Krettli and Pereira, 1981). Chemotherapeutic treatments are employed to tip the balance in favour of the host.

In total there are probably about 10 000 species of parasitic protozoa, including all of the sporozoans, microsporans, myxosporans, ascetosporans, parabasalians, opalinids and some smaller groups, a substantial proportion of chytrids, oomycetes, kinetoplastids and metamonads, about one sixth of all ciliates, one eightieth of all amoebae and small numbers of several other groups.

Mutualistic symbiotic associations of protists in the guts of animals

In a number of cases the protists that live within other animals are thought to be beneficial to the host rather than harmful (Smith and Douglas, 1987). Such mutualistic associations include the ciliate populations in the alimentary canals of herbivorous mammals and the flagellate fauna of the intestine of termites (Hungate, 1966; Coleman, 1979; Breznak, 1984). In the case of cattle and sheep each ml of fluid in the rumen (anterior stomach) contains up to a million ciliate protozoa of many (often 30–50) species, mostly entodiniomorphs and trichostomes (p. 199), and perhaps 10^{10} bacteria, as well as about 8% by weight of chytrid fungi (p. 144). The ciliates are anaerobes tolerating low levels of oxygen that are transmitted from host to host in saliva; the grass from the rumen is regurgitated for chewing, and ciliates may survive for a short time in saliva on dropped grass or in drinking water. Although the ciliates thrive in the alkaline or mildly acid conditions of the rumen, they are quickly killed in the acid fluids of the posterior stomach (omasum). The entodiniomorphs ingest and break down plant fragments and both trichostomes and bacteria take up sugars from the gut contents; carbohydrates are fermented by the symbionts and the volatile fatty acids released provide two thirds of the dietary energy of the host, the cellulolytic and amylolytic activities of ciliates prolong fermentation processes and produce perhaps one third of the volatile fatty acids, while bacteria provide the rest. The ciliates contain hydrogenosomes producing hydrogen which either reacts with oxygen or is used by methanogenic bacteria (p. 18). The scavenging of oxygen by these ciliates reduces the oxygen content of the rumen from several μM before feeding to undetectable levels after feeding, and thus plays a role in stabilising fermentation in the rumen (Lloyd *et al.*, 1988). The microorganisms in the rumen also play an important role in the nitrogen metabolism of the ruminant (Rook and Thomas, 1983), but it is not yet clear how far the grazing activities of the ciliates enhance the nitrogen turnover by bacteria; some nitrogen compounds are absorbed directly in the rumen and others are released by digestion of protozoa and bacteria killed as they pass towards the intestine. However, Colombier et al. (1984) place the ciliate contribution to the microbial protein leaving the rumen at only about 5%. The large intestine of horses contains similar numbers of ciliate protozoa, but it is not known what contribution these may make to the efficiency of digestion in the host, in relation to the bacterial symbionts that accompany them; perhaps 25% of the energy needs of the horse are met by hindgut fermentation (Stevens, 1977).

Flagellate protozoa are of even greater importance to most wood-eating termites and roaches (Breznak, 1982), and it has been estimated that from one third to one seventh of the weight of a nymph of *Zootermopsis* is provided by protozoa in the hind gut. Flagellate populations of up to 10^7 ml^{-1} are accompanied by populations of various bacteria of 10^9–10^{10} ml^{-1}. The flagellates are obligate anaerobes which digest the cellulose of the wood eaten by the host insect and benefit the host because volatile fatty acids produced by anaerobic metabolism of protozoan and bacterial symbionts are absorbed by the insect as its principal source of food. In the absence of flagellates the insect starves to death, although it may continue to eat wood; the infection of termites with flagellates is normally ensured by the coprophagous habits of termites, which may also eat their own freshly cast skins after a moult in which the hind gut lining is lost. Many parabasalian and metamonad flagellates (p. 116) are found only in these insect hosts, and are as dependent upon the insects as the insects are upon them. the ecological importance of these flagellates is considerable, for in some regions of the world termites replace earthworms as major agents of humus breakdown in the soil.

Relationships in which other organisms live on or in protists

Protists are themselves subject to parasitic infection by other protists (Ball, 1969, Lee *et al*, 1985). These include small amoebae parasitic within opalinids, in foraminiferans and in other amoebae, and parasitic flagellates, such as leptomonads in ciliate nuclei (Gortz, 1983) and dinoflagellates in radiolarians, in tintinnid ciliates and even in other dinoflagellates. Chytrids occur as parasites in amoeboid, coccoid and flagellate protists. Most of the ciliates which parasitise protozoa are suctorians living in or on other ciliates, although there have also been reports of thigmotrichs living ectoparasitically on peritrichs and suctorians. A number of protozoan parasites are themselves hosts of other parasitic protozoans; microsporans of the family Metchnikovellidae are frequently found in gregarine sporozoans (p. 238), and a species of *Nosema* has been found as a parasite of myxosporans from fish.

Protozoa may act as hosts for organisms of other types. Bacteria have often been found attached to the outer surfaces of protozoa, in particular on flagellates from termites and on sand-dwelling ciliates. H. Kirby described the presence of spirochaetes and of rod-shaped and fusiformis bacteria arranged in various specific patterns on the surface of termite flagellates. Some flagellates form surface structures to which these epibionts attach. In *Mixotricha paradoxa* spirochaetes and rod-shaped bacteria are arranged in a precise pattern in relation to projecting 'brackets' on the surface of the flagellate; it has been shown that the swimming of this flagellate depends on the undulations of these spirochaetes, which move in a coordinated fashion with clear metachronal waves (Cleveland and Grimstone, 1964). Many strictly anaerobic ciliates of sulphur-rich sediments (e.g. *Plagiopyla, Metopus, Caenomorpha* and *Kentrophoros*, Fig. 9.12b) possess ecto- and/or endo-symbiotic sulphate reducing or methanogenic bacteria which presumably provide suitable anoxic electron acceptors for oxidation of products of ciliate metabolism; these ciliates lack mitochondria and cytochrome oxidase, though at least some of them contain hydrogenosomes (Fenchel *et al*., 1977), or additional symbiotic bacteria that may produce hydrogen (*see* p. 18).

Many aerobic protozoa, amoebae, ciliates and flagellates, contain bacteria, frequently gathered around or concentrated within the nucleus (Gortz, 1983). The course of development of a symbiotic association between a gram-negative rod bacterium and *Amoeba proteus*, from the stage when most amoebae were killed by the infection to a stage when the symbionts were essential to survival of the amoebae has been followed over some 200 generations by K.W. Jeon (1983). Little is known of the significance of the presence of these bacteria in most cases, but two types of example have received some attention. Rod-shaped bacteria present in the cytoplasm of some species of the flagellate *Crithidia* confer upon these hosts the ability to synthesise heme and to survive on exogenous ornithine, while flagellates deprived of symbionts require heme and arginine (McGhee and Cosgrove, 1980). Symbiotic bacteria (*Caedibacter*) have been found living in the cytoplasm of *Paramecium aurelia* (Soldo, 1974; Preer, 1981) and have been given such names as kappa, mu, pi, lambda etc.; lambda possess bacterial flagella, and several types of symbiont have bacterial cell walls. It was found many years ago that paramecia from certain stocks consistently killed paramecia from other stocks when the two were mixed, and on account of this interaction the former were called killers and the latter sensitives. Killer paramecia were found to contain kappa, but the persistence of kappa in the cytoplasm was dependent upon the presence of a dominant macronuclear gene. Only kappa bodies that contained refractile inclusions (tightly-coiled ribbons associated with virus-like granules) were able to kill, but other kappa could develop these inclusions. Other symbionts act in a comparable manner; paramecia that possess mu kill sensitives at conjugation (they are mate killers), while pi may be an inactive (mutant?) kappa, and paramecia containing lambda produce large (~ 5–10 μm) lethal granules which are released into the medium. It is assumed that these symbiotic organisms derive nutrients from the paramecia, but a killer stock has an advantage over a sensitive stock in ecological competition (Landis, 1981). Lambda apparently synthesises folic acid, and paramecia with this symbiont have no external requirement for this vitamin. Comparable examples are the range of symbionts in *Euplotes* species (Heckmann, 1983), including omikron, which is essential for survival of *E. aediculatus*, and the xenosomes of the ciliate *Parauronema* (Soldo, 1983).

Relationships between protozoa and fungi appear to be one-sided in favour of the fungi, for in all cases it appears that the fungi either parasitise protozoa or are predaceous on them – though these categories of 'attack' are not sharply separated. Soil amoebae, both naked and testate, are attacked by fungi of the Zoopagales and Hyphomycetes; these fungi either enter the cell and grow within the cytoplasm, perhaps following ingestion into food vacuoles, or entrap the amoeba with adhesive hyphae before penetrating it with hyphal roots. An infection of *Paramecium bursaria* with a yeast has been reported.

Relationships involving photosynthetic organisms

Symbiotic associations between eukaryote cells and photosynthetic cells, at first photosynthetic prokaryotes and later photosynthetic eukaryotes, have been very important in the origins and evolution of many protistan phyla (see p. 3). Such associations are doubtless of varying antiquity; some characterise whole phyla, and may be very old, whilst others are more restricted and are presumably more recent. There seems to be a continuous spectrum of associations from completely obligatory symbioses of purely plant groups, through relationships where the members of the association thrive autotrophically or heterotrophically but reproduce together so that the association remains permanent, to associations that regularly occur but require to be renewed after reproduction of the 'host' organism. The relationship of many radiolarians with their dino-flagellate symbionts is probably as obligatory for both parties as is the relation-ship of a free-swimming dinoflagellate with its plastids; such radiolarians may be effectively plants, as are adult *Convoluta*, which cease eating.

There is a considerable profusion of examples of associations of this type in which one or both members are protists (Cook, 1983, Smith and Douglas, 1987), even if we exclude those long-established associations dealt with in Chap-ter 5 (p. 100). Members of many groups are involved, and a few examples will serve to illustrate this diversity. Cyanobacterial symbionts are referred to as cyanellae, and the associations of protists with cyanellae have been termed syncyanosen. Cyanellae found externally on the chrysophycean *Oikomonas syncyanotica* appear to contribute to the nutrition of the flagellate, for motility of the colourless flagellate bearing cyanellae would occur in the absence of free oxygen, but was light-dependent. Other syncyanosen involve cyanellae that live within amoebae and flagellates, e.g. the association of the cyanobacterium *Cyanocyta korschikoffiana* within the cryptomonad *Cyanophora paradoxa*.

Symbiotic dinoflagellates, termed zooxanthellae, and usually identified as *Symbiodinium microadriaticum*, are found in a variety of marine organisms, notably radiolarians, but also foraminiferans and ciliates among protozoa, and hydroids, sea anemones, corals and clams among invertebrates. The symbionts are usually found as coccoid cells within intracellular vacuoles, but symbionts of *Tridacna* are extracellular; these coccoid cells produce gymnodinioid flagellate zoospores. Green symbionts of the genus *Chlorella*, termed zoochlorellae, occur in many freshwater protists (Christopher and Patterson, 1983), including ciliates, notably *Paramecium bursaria* and *Stentor polymorphus*, some amoebae, *Hydra viridis*, other coelenterates, flatworms and bivalves; another coccoid green alga, *Scenedesmus*, has also been found living in *P. bursaria*. The green symbionts of the marine flatworm *Convoluta roscoffensis* are prasinophytes of the genus *Tetraselmis* (= *Platymonas*), and prasinophytes also occur in some radiolaria, whilst species of *Chlamydomonas* are found in some foraminifera. Other foraminifera contain symbiotic diatoms, or sometimes unicellular red algae; symbiotic diatoms also occur in the flatworm *Convoluta convoluta*, and haptophytes in acantharians. Marine oligotrich ciliates and some foraminifera may contain isolated chloroplasts, obtained from a variety of algal groups, which appear to remain functional within the host for some weeks (Lopez, 1979; Laval-Peuto, pers. comm.).

These symbiotic protists are usually found living within vacuoles, and therefore surrounded by two membranes, within cells where contents of such vacuoles would normally be digested. Although large numbers of symbiont vacuoles may occur in a host cell, there is no clear evidence of digestion of symbionts by the host in illuminated individuals containing established symbionts. Questions therefore arise about the nature of the relationship and the mechanism of recognition of symbionts (Smith, 1981). The partnership involves transfer to the host of energy, organic carbon, oxygen and probably organic nitrogen and even phosphorus and sulphur compounds in products released by the symbiont, and the symbiont may gain carbon dioxide, inorganic nitrogen, phosphate and sulphur compounds, shelter and sometimes vitamins from the host. Zooxanthellae excrete glycerol, alanine, glucose and organic acids, among other compounds, and up to 80% of the carbon fixed by zoo chlorellae may be released as maltose, along with smaller amounts of alanine and glycollic acid. It has long been known that animals with symbionts, e.g. **Paramecium bursaria**, grow better in illuminated conditions than aposymbiotic paramecia when fed the same low concentration of bacteria; however, well-fed **P. bursaria** can show an increase in the number of zoochlorellae when kept in the dark, so there is evidently a two-way flow of nutrients. In corals another benefit to the animal of the presence of symbionts is the tenfold enhancement of deposition of the calcareous skeleton in the light, compared with dark deposition – presumably because the removal of CO_2 for photosynthesis aids the precipitation of $CaCO_3$; a similar enhancement of calcification occurs in foraminiferans. The autotrophic symbiont also benefits from the association, for records of very high rates of carbon fixation have been obtained for the red ciliate **Mesodinium** (*see* Lindholm, 1985) and for radiolaria (*see* Anderson, 1983).

The problem of recognition of symbionts by their hosts has been studied in examples where aposymbiotic hosts have been allowed to ingest algal cells of the type normally symbiotic within them. Many of the ingested algae are digested, but some of them escape digestion, and may be carried to the base of the digestive cells of **Hydra** or to peripheral cytoplasm of **P.bursaria**, where they may remain in stable symbiosis. It has been shown that while normal food vacuoles in **P. bursaria** fuse with lysosomes, vacuoles with stable symbionts do not – the presence of symbionts somehow inhibits the fusion of lysosomes with the vacuole membrane (Karakashian and Rudzinska, 1981). It is also known that **Chlorella** cells ingested by **Hydra** secrete maltose, perhaps stimulated by acid pH, and the release of maltose seems essential for recognition of the symbiont by the **Hydra** cells and inhibition of lysosome fusion (McAulay and Smith, 1982).

Appendix

Methods of obtaining, culturing and preserving protists

Protozoa may be collected from the natural sources indicated in many parts of this book, or cultures may be bought from suppliers and culture collections. Many biological supply houses list a range of the more commonly studied species of free-living protists in their catalogues, and some suppliers have quite extensive lists, e.g. Sciento Educational Services, 61 Bury Old Road, White-field, Manchester, M25 5TB, UK. A wider range of species is available in some countries from specialised culture collections, (see p. 288), and a list of culture sources and methods has been published in the Journal of Protozoology (**5**, p. 1, 1958).

A very wide range of culture techniques and recipes for media have been used and will be found in the literature; many of these can be obtained from the culture list in the Journal of Protozoology, in advice booklets from Culture Centres, e.g. Page (1981), Cote (1984), as well as in their culture lists (*see* p. 288), in appropriate books, e.g. Galtsoff *et al.*, (1959), Mackinnon and Hawes (1961) or Lee *et al.*, (1985), and in accounts of studies on the particular species involved. A beginners' guide to the culture of freshwater protozoa (and to their collection, isolation and identification) has been published by Finlay *et al*, (1988). Culture methods for parasitic protozoa are necessarily rather specific (e.g. Taylor and Baker, 1978); they will not be mentioned further here, but may be traced from sources mentioned above. The use of two or three of the culture techniques listed below will usually provide large numbers of a variety of species of free-living protists for study from samples obtained from soil or marine or freshwater habitats. The same culture methods may be used to maintain cultures of selected species isolated from these collections. The maintenance of pure cultures depends on nutritional requirements, especially for protozoa, where one has often to be content with cultures in which the desired organism is a dominant member of the community of organisms in the culture – most protozoa feed on other living organisms and seldom is a single food organism as satisfactory as a mixed diet. Axenic or monoxenic culture methods have been developed for some species.

Samples obtained from freshwater or marine sources contain more species if they include tufts of filamentous plants or sediment containing decaying organic matter; a sample of water lacking any form of solid substratum is seldom very productive unless the water has a high organic content. It is often worthwhile to make several collections from one body of water, keeping separate containers for samples of mud, weeds of different sorts and the surface film (especially if there are floating weeds or scum). Similarly with terrestrial samples – dry soil contains spores, but a wider variety of species may usually be obtained from damp soil, particularly if humus is present, and especially if there is a good growth of vegetation. Methods for soil protozoa have been described by Heal (1971).

In order to encourage the multiplication of a variety of species from the samples, it is advisable to separate a sample into several containers and treat each differently. One should be put in a well-illuminated place out of direct sunlight to encourage growth of photosynthetic species and those that feed upon them; the growth of autotrophs may be promoted by addition of a suitable mixture of inorganic salts – obtained most simply by autoclaving 1 kg of garden soil with 1 l of water for 30 min, adding a little $CaCO_3$ and filtering, or by dissolving the following in 1 l of glass distilled water: $NaNO_3$ 250 mg, $CaCO_3$ 25 mg, $MgSO_4.7H_2O$ 85 mg, K_2HPO_4 85 mg, KH_2PO_4 185 mg, $NaCl_2$ 25 mg and a trace of $FeCl_3$. For other cultures less light is advisable but darkness is not normally necessary; in these a variety of bacteria may be encouraged to multiply by providing different food in the different dishes – for example, the following provide rather different mixtures of protists from the same samples: 0.5% proteose peptone, enough milk to impart a faint turbidity to the water, a pinch of dried yeast, a few wheat (or other cereal or pulse) grains that have been boiled to kill the seeds, 20% of hay infusion. The last is made by boiling 10 g of chopped hay in 1 l of water and allowing the mixture to stand for 24 hours in a flask stoppered with cotton cool before decanting the infusion fluid. The water used for culturing may be autoclaved (or twice boiled) tapwater if this is known not to be toxic, but filtered and autoclaved rain or pond water may be substituted, or a solution like Chalkley's medium (below) may be used. For marine species seawater that has been briefly autoclaved may be used, or artificial seawater may be reconstituted from commercial sea-salt or made by dissolving the following in 1 l of distilled water: $NaCl$ 27.7 g, KCl 0.7 g, $CaCl_2$ 1.03 g, $MgCl_2$ 2.2 g, $MgSO_4$ 2.6 g, $NaHCO_3$ 0.18 g, adjusted to pH 8.0 with NaOH. Ideally water should be filtered through a 0.2 μm pore size membrane filter before autoclaving. Some protists grow best in tall jars, others in shallow dishes, and some thrive better if a thin layer of nonnutrient agar is provided at the base of the container.

The resulting cultures will contain a diversity of amoebae and ciliates, as well as flagellates and non-flagellated algae, which may be visible under a dissecting microscope, but usually require a good compound microscope, preferably with phase contrast or best of all differential interference contrast (Nomarski) optics, for adequate study and subsequent identification. The time during which protists may be studied on a slide may be prolonged if a thin ring of vaseline is applied around the margin of the cover glass before placing it over the culture. Most free-living protists are best studied alive, and where there are ciliates that move too quickly to be seen, it is advisable to place a drop of culture within a ring of 2% methyl cellulose and gently lower a cover glass, so that the ciliates will swim slowly in the viscous material. 0.01% Nickel sulphate is an effective anaesthetic for many ciliate protozoa. Many flagellates may be successfully studied after being killed with Lugol's iodine solution (dissolve 4 g of iodine in 100 ml of a 6% solution of potassium iodide); the same method may be used for other protists, though protozoa fixed in this way deteriorate unless transferred to something like Da Fano's fluid (15 ml 40% formalin, 1 g $NaCl$, 1 g $CoNO_3$ in

100 ml distilled water) (Legner, 1977). There are special staining methods for parasitic forms, e.g. Giemsa's stain for blood parasites (*see* Baker, 1969), and silver staining techniques for revealing the infraciliature of ciliates, included in lists of fixation and staining techniques given by Mackinnon and Hawes (1961) and Lee *et al.*, (1985). Fixation methods for electron microscopy commonly use a combination of glutaraldehyde and osmic acid in a suitable buffer, but special techniques of freeze substitution (Barlow and Sleigh, 1979) or rapid chemical fixation (Parducz, 1967) may be used in preparing specimens for scanning microscopy.

The correct identification of protists is seldom an easy matter, but some information is to be found in many books. General books on the identification of organisms from particular habitats may have sections on algae or protozoa, and several chapters in the second edition of Ward and Whipple's Freshwater Biology (Edmondson, 1959) may be found especially helpful. Books specifically concerned with protists, such as Kudo (1966), Prescott (1970), Belcher and Swale (1976), Jahn *et al.*, (1979) and Lee *et al.*, (1985), and the beautifully illustrated colour atlas of Hausmann and Patterson (1983), are also valuable, but it must be recognised that such books only describe some representative members of the groups concerned. Kahl (1930–5), Curds (1982) and Curds *et al.*, (1983) describe many genera and species of ciliates, and Borror (1973) gives a key to the families of marine ciliates, but the bibliographies given by Borror, Curds, Corliss (1979) and Small and Lynn (1985) should be consulted for recent references to species descriptions for correct identification of ciliates. New works are appearing for many groups; for example, Page (1976, 1987) provides the information to partially replace the classic books of Cash and Wailes (1905–21) on rhizopods and heliozoa, and references to many books and reviews on flagellate protists can be found in Chapter 5. References to works on parasitic protozoa can be found in Kreier (1977) and Kreier and Baker (1987). References to systematic studies on most groups will be found in appropriate chapters of Lee *et al.*, (1985) and very useful lists of sources are given in Kerrich *et al.*, (1978) and Sims (1980).

The culture techniques mentioned above are usually satisfactory for short-term maintenance of mixed cultures of protists, but it is advisable to remove predators and competitors in order to prolong the life of cultures. The following examples of culture techniques that may be used on five common species indicate methods that may be modified for culture of many other protists, or to provide food organisms for such predatory species as **Didinium**, **Actinosphaerium** or suctorians.

Amoeba proteus is grown in 10–15 cm diameter glass bowls with covers, containing Chalkley's medium (below) to a depth of about 3 cm with 6 boiled wheat grains and 3 boiled 5 cm lengths of Timothy hay stalks. The dish is seeded with amoebae, together with flagellates and ciliates for food, from an existing culture, and then put in a dimly illuminated place. The amoebae are found mainly on the glass surface, and will persist for many months if more wheat and grass are added monthly. (Chalkley's medium contains the following: NaCl 80 mg, $NaHCO_3$ 4 mg, KCl 4 mg, $CaCl_2$ 4 mg, $CaH_4(PO_4)_2$. H_2O 1.6 mg dissolved in 1 l of glass distilled water).

Chlamydomonas spp. may be grown in tubes or petri dishes containing the following medium: Proteose peptone (e.g. Difco) 1 g, KNO_3 0.2 g, K_2HPO_4 20 mg, $MgSO_4$ $7H_2O$ 20 mg, Agar 10 g dissolved in 1 l of glass distilled water and autoclaved; the agar may be omitted for liquid cultures. A simple medium made by boiling a little garden soil is often effective with many photosynthetic forms, after experimenting with appropriate concentrations for your own soil. All photosynthetic forms should be placed in a well illuminated position out of direct sunlight.

Euglena spp. are grown in jam jars with glass covers, three quarters filled with autoclaved pond water or Chalkley's medium, and with 20 boiled wheat grains; cultures require good illumination, but not direct sunlight. Wheat grains may be added monthly to maintain the cultures.

Paramecium caudatum are grown in jam jars with glass covers, three quarters filled with Chalkley's medium or autoclaved pond water, and with 7–12 drops of skimmed milk (reconstituted dried milk works well) added each week. The jars are kept in a dark corner away from direct lighting. Wheat grains or hay extract also provide suitable food for bacteria that *Paramecium* eat.

Tetrahymena spp. are kept in axenic cultures in rimless 30 ml test tubes with metal caps (or autoclavable culture bottles) which are autoclaved after addition of 15 ml of peptone medium (10 g proteose peptone (e.g. Difco), 5 g NaCl in 1 l of distilled water); subcultures are made every 3–4 weeks using sterile pipettes and normal sterile precautions, and the cultures are kept in the dark. Axenic cultures of the flagellates *Astasia* and *Chilomonas* may be maintained in more dilute peptone media in which NaCl is replaced with sodium acetate.

Lists of cultured protists may be obtained from the following:

American Type Culture Collection, 12301 Parklawn Drive, Rockville, MD 20852, USA.

The Culture Collection of Algae and Protozoa (Natural Environment Research Council), Freshwater Biological Association, The Ferry House, Ambleside, Cumbria, LA22 OLP, UK.

The Microbial Culture Collection, The National Institute for Environmental Studies, Japan Environment Agency, Yatabe-machi, Tsukuba, Ibaraki 305, Japan.

Algensammlung, Pflanzenphysiologisches Institut, Gottingen, West Germany.

The Curator of the Culture Collection of Algae, Botany Department, Indiana University, Bloomington, Indiana 47401, USA.

References

Aiello, E. and Sleigh, M.A. (1972). The metachronal wave of lateral cilia of *Mytilus edulis. J. Cell Biology*, **54**, 493–506.

Aikawa, M. (1971). *Plasmodium*: The fine structure of malarial parasites. *Experimental Parasitology*, **30**, 284–320.

Aikawa, M., Udeinya, I.J., Rabbege, J., Dayan, M., Leech, J.H., Howard, R.J. and Miller, L.H. (1985). Structural alteration of the membrane of erythrocytes infected with *Plasmodium falciparum. Journal of Protozoology,* **32**, 424–429.

Ainsworth, G.C., Sparrow, F.K. and Sussman, A.S. (1973). *The Fungi.* **IV**, *Taxonomic review with keys*. Academic Press, New York.

Alabouvette, C., Couteaux, M.M., Old, K.M., Pussard, M., Reisinger, O. and Toutain, F. (1981). Les protozoaires du sol: Aspects écologiques et méthodologiques. *Année Biologique*, **20**, 255–303.

Albergoni, V. and Piccinni, E. (1983). Biological response to trace metals and their biochemical effects. In *Trace Element Speciation in Surface Waters and its Ecological Implications*, G.G. Leppard (ed.) 159–175, Plenum, New York.

Alberts, B., Bray, D., Lewis, J., Raff, M., Roberts, K. and Watson, J.D. (1983). *Molecular Biology of the Cell*. Garland, New York.

Al-Khazzar, A.R., Earnshaw, M.J., Butler, R.D. Emes, M.J. and Sigee, D.C. (1984). Tentacle contraction in *Discophrya collini*: The effects of the ionophore A 23187 and ruthenium red on Ca^{2+}-induced contraction and uptake of extracellular calcium. *Protoplasma*, **122**, 125–131.

Allen, H.L. (1971). Primary productivity, chemo-organotrophy and nutritional interactions of epiphytic algae and bacteria on macrophytes in the littoral of a lake. *Ecological Monographs*, **41**, 97–127.

Allen, Richard D. (1967). Fine structure, reconstruction and possible functions of components of the cortex of *Tetrahymena pyriformis. Journal of Protozoology*, **14**, 553–565.

Allen, Richard D. (1969). The morphogenesis of basal bodies and accessory structures of the cortex of the ciliated protozoon *Tetrahymena pyriformis. Journal of Cell Biology*, **40**, 716–733.

Allen, Richard D. (1971). Fine structure of membranous and microfibrillar systems in the cortex of *Paramecium caudatum. Journal of Cell Biology*, **49**, 1–20.

Allen, Richard D. (1978). Membranes of ciliates: ultrastructure, biochemistry and fusion. In *Membrane Fusion*, G. Poste and G.L. Nicholson, (eds.) 657–763, Elsevier/North Holland.

Allen, Richard D. (1984). *Paramecium* phagosome membrane: From oral region to cytoproct and back again. *Journal of Protozoology,* **31**, 1–6

Allen, Robert D. (1961). A new theory of ameboid movement and protoplasmic streaming. *Experimental Cell Research, Supplement* **8**, 17–31.

Allen, Robert D. (1973). Biophysical aspects of pseudopodium formation and retraction. In *The Biology of Amoeba*, K.W. Jeon, (ed.) 201–247, Academic Press, New York.

Allen S.L. and Gibson, I. (1973). Genetics of *Tetrahymena*. In *Biology of Tetrahymena*, A.M. Elliott (ed.) 307–373, Dowden, Hutchinson and Ross, Stroudsburg, Pa.

Amos, W.B. (1975). Contraction and calcium binding in the vorticellid ciliates. In *Molecules and Cell Movement*, S. Inoue and R.E. Stephens (eds.) 411–436, Raven Press, New York.

Amos, W.B., Grimstone, A.V., Rothschild, L.J. and Allen, R.D. (1979). Structure, protein composition and birefringence of the costa: A motile flagellar root fibre in the flagellate *Trichomonas*. *Journal of Cell Science*, **35**, 139–164.

Andersen, R.A. (1987). Synurophyceae Classis Nov., a new class of algae. *American Journal of Botany*, **74**, 337–353.

Anderson, J.M. (1981). *Ecology for Environmental Sciences: Biosphere, Ecosystems and Man*. Edward Arnold, London.

Anderson, O.R. (1981). Radiolarian fine structure and silica deposition. In *Silicon and Siliceous Structures in Biological Systems*, T.L. Simpson and B.E. Volcani, (eds.) 347–380, Springer-Verlag, New York.

Anderson, O.R. (1983). *Radiolaria*. Springer-Verlag, New York.

Anderson, O.R. and Botfield, M. (1983). Biochemical and fine structure evidence for cellular specialization in a large spumellarian radiolarian *Thalassicolla nucleata*. *Marine Biology*, **72**, 235–241.

Anderson, O.R., Spindler, M., Be, A., and Hemleben, C.H. (1979). Trophic activity of planktonic foraminifera. *Journal of the Marine Biological Association*, UK, **59**, 791–799.

Baker, J.R. (1969). *Parasitic Protozoa*. Hutchinson, London.

Balamuth, W., Bradbury, P.C. and Schuster, F.L. (1983). Ultrastructure of the amoeboflagellate *Tetramitus rostratus*. *Journal of Protozoology*., **30**, 445–455.

Ball G.H. (1969). Organisms living on and in Protozoa. In *Research in Protozoology 3*, T.-T. Chen, (ed.) 565–718, Pergamon Press, Oxford.

Bannister, L.H. (1972). The structure of trichocysts in *Paramecium caudatum*. *Journal of Cell Science*, **11**, 899–929.

Bannister, L.H. (1977). The invasion of red cells by *Plasmodium*. *Symposium of the British Society of Parasitology*, **15**, 27–55.

Bannister, L.H. and Tatchell, E.C. (1968). Contractility and the fibre systems of *Stentor coeruleus*. *Journal of Cell Science*, **3**, 295–308.

Bardele, C.F. (1971). Microtubule model systems: cytoplasmic transport in the suctorian tentacle and the centrohelidian axopod. *Proceedings of the 29th Annual meeting of the Electron Microscopy Society of America*, 334–335.

Bardele, C.F. (1972a). A microtubule model for ingestion and transport in the suctorian tentacle. *Zeitschrift für Zellforschung*, **126**, 116–134.

Bardele, C.F. (1972b). Cell cycle, morphogenesis and ultrastructure in the pseudoheliozoan *Clathrulina elegans*. *Zeitschrift für Zellforschung*, **130**, 219–242.

Bardele, C.F. (1976). Particle movement in heliozoan axopods associated with lateral displacement of highly ordered membrane domains. *Zeitschrift für Naturforschung*, **31C**, 190–194.

Bardele, C.F. (1981). Functional and phylogenetic aspects of the ciliary membrane: A comparative freeze-fracture study. *BioSystems*, **14**, 403–421.

Bardele, C.F. (1987). Auf neuen Wegen zu alten Zielen – moderne Ansatze zur Verwandtschaftsanalyse der Ciliaten. *Verhandlungen de Deuktschen Zoologischen Gesellschaft*, **80**, 59–75.

Bardele, C.F. and Grell, K.G. (1967). Elektronenmikroskopische Beobachtungen zur Nahrungsaufnahme bei dem Suktor *Acineta tuberosa* Ehrenberg. *Zeitschrift für Zellforschung*, **80**, 108–123.

Bark, A.W. (1981). The temporal and spatial distribution of planktonic and benthic protozoan communities in a small productive lake. *Hydrobiologia*, **85**, 239–255.

Barlow, D.I. and Sleigh, M.A. (1979). Freeze-substitution for preservation of ciliated surfaces for scanning electron microscopy. *Journal of Microscopy*, **115**, 81–95.

Barr, D.J.S. (1980). An outline for the reclassification of the Chytridiales, and for a new order, the Spizellomycetales. *Canadian Journal of Botany*, **58**, 2380-2394.

Barr, D.J.S. (1981). The phylogenetic and taxonomic implications of flagellar rootlet morphology among zoosporic fungi. *BioSystems*, **14**, 359-370.

Barr, D.J.S. and Allan, P.M.E. (1985). A comparison of the flagellar apparatus in *Phytophthora, Saprolegnia, Thraustochytrium* and *Rhizidiomyces*. *Canadian Journal of Botany*, **63**, 138-154.

Beale, G.H. (1954). *The Genetics of Paramecium aurelia*. Cambridge University Press.

Beam, C.A. and Himes, M. (1980). Sexuality and meiosis in dinoflagellates. In *Biochemistry and Physiology of Protozoa*, Second edition, M. Levandovsky and S.H. Hutner, (eds) **2**, 171-206, Academic Press, New York.

Beam, C.A. and Himes, M. (1984). Dinoflagellate genetics. In *Dinoflagellates*, D.L. Spector, (ed.) 263-298, Academic Press, New York.

Bean, B. (1984). Microbial geotaxis. In *Membranes and Sensory Transduction*, G. Colombetti and F. Lenci, (eds) 163-198, Plenum Press, New York.

Belar, K. (1926). Der Formwechsel der Protistenkerne. *Ergebnisse der Zoologie*, **6**, 235-654.

Belcher, J.H. and Miller, J.D.A. (1960). Studies on the growth of Xanthophyceae in pure culture. IV Nutritional types amongst the Xanthophyceae. *Archiv für Mikrobiologie* **36**, 219-228.

Belcher, H. and Swale, E. (1976). *A Beginners Guide to Freshwater Algae*. Institute of Terrestrial Ecology, Cambridge, UK.

Berninger, U.-G., Finlay, B.J. and Canter, H.M. (1986). The spatial distribution and ecology of zoochlorella-bearing ciliates in a productive pond. *Journal of Protozoology*, **33**, 557-563.

Bick, H. (1973). Population dynamics of protozoa associated with the decay of organic materials in fresh water. *American Zoologist*, **13**, 149-160.

Bisalputra, T. (1974). Plastids. In *Algal Physiology and Biochemistry*, W.D.P. Stewart, (ed.) 124-160, Blackwell, Oxford.

Blake, J.R. and Sleigh, M.A. (1974). Mechanics of ciliary locomotion. *Biological Reviews*, **49**, 85-125.

Bloodgood, R.A. (1981). Flagella-dependent gliding motility in *Chlamydomonas*. *Protoplasma*, **106**, 183-192.

Bomford, R. (1966). The syngens of *Paramecium bursaria*: new mating types and intersyngenic mating reactions. *Journal of Protozoology*, **13**, 497-501.

Borror, A.C. (1972). Revision of the order Hypotrichida (Ciliophora, Protozoa). *Journal of Protozoology*, **19**, 1-23.

Borror, A.C. (1973). *Marine Flora and Fauna of the Northeastern United States. Protozoa: Ciliophora*. NOAA Technical Report, National Marine Fisheries Service, Circ-378, Seattle, Wa.

Bouck, G.B. (1971). The structure, origin, isolation and composition of the tubular mastigonemes of the *Ochromonas* flagellum. *Journal of Cell Biology*. **50**, 362-384.

Bourrelly, P. (1970). *Les Algues D'eau Douce*, III. *Les Algues Bleues et Rouges*, Boubee, Paris.

Bowers, B. and Olszewski, T.I. (1972). Pinocytosis in *Acanthamoeba castellanii*. Journal of Cell Biology, **53**, 681-694.

Breznak, J.A. (1982). Intestinal microbiota of termites and other xylophagous insects. *Annual Review of Microbiology*, **36**, 323-343.

Breznak, J.A. (1984). Biochemical aspects of symbiosis between termites and their intestinal microbiota. In *Invertebrate-Microbial Interactions*, J.M. Anderson, A.D.M. Rayner and D.W.H. Walton, (eds) 173-204, Cambridge University Press, UK.

Brokaw, C.J. (1965). Non-sinusoidal bending waves of sperm flagella. *Journal of Experimental Biology*, **43**, 155-169.

Brokaw, C.J., Luck, D.J.L. and Huang, B. (1982). Analysis of the movement of *Chlamydomonas* flagella: the function of the radial-spoke system is revealed by comparison of wild-type and mutant flagella. *Journal of Cell Biology*, **92**, 722–732.

Brook, A.J. (1981). *The Biology of Desmids*. Blackwell, Oxford.

Brown, I.D., Connolly, J.G. and Kerkut , G.A. (1981). Galvanotaxic response of *Tetrahymena vorax*. *Comparative Biochemistry and Physiology*, **69C**, 281–291.

Brugerolle, G. (1973). Étude ultrastructurale du trophozoite et du kyste chez le genre *Chilomastix*, Alexeieff, 1910 (Zoomastigophorea, Retortamonadida, Grassᵉ, 1952). *Journal of Protozoology*, **20**, 574–585.

Brugerolle, G. (1975). Étude de la cryptopleuromitose et de la morphogénèse de division chez *Trichomonas vaginalis* et chez plusieurs genres de trichomonadines primitives. *Protistologica*, **11**, 457–468.

Brugerolle, G. (1982). Caractères ultrastructuraux d'une mastigamibe: *Mastigina hylae* (Frenzel). *Protistologica*, **18**, 227–235.

Brugerolle, G. (1985a). Des trichocystes chez les bodonides, un caractère phylogénétique supplémentaire entre Kinetoplastida et Euglenida. *Protistologica*, **21**, 339–348.

Brugerolle, G. (1985b). Ultrastructure d'*Hedriocystis pellucida* (Heliozoa Desmothoracida) et de sa forme migratrice flagellée. *Protistologica*, **21**, 259–265.

Brugerolle, G. (1986). Séparation des genres *Trimitus* (Diplomonadida) et *Tricercomitus* (Trichomonadida) d'après leur ultrastructure. *Protistologica*, **22**, 31–37.

Brugerolle, G. and Joyon, L. (1975). Étude cytologique ultrastructurale des genres *Proteromonas* et *Karotomorpha* (Zoomastigophorea, Proteromonadida, Grassé, 1952). *Protistologica*, **11**, 531–546.

Brugerolle, G., Lom, J., Nohynkova, E. and Joyon, L. (1979). Comparaison et évolution des structures cellulaires chez plusiers espèces de bodonidés et cryptobiidés appartenant aux genres *Bodo, Cryptobia* et *Trypanoplasma*. *Protistologica*, **15**, 197–221.

Brugerolle, G. and Mignot, J.-P. (1979a). Observations sur le cycle, l'ultrastructure et la position systématique de *Spiromonas perforans* (*Bodo perforans* Hollande 1938), flagellé parasite de *Chilomonas paramecium*: ses relations avec les dinoflagellés et sporozoaires. *Protistologica*, **15**, 183–196.

Brugerolle, G. and Mignot, J.-P. (1979b). Distribution et organisation de l'ADN dans le complexe kinétoplaste-mitochondrie chez un Bodonidé, protozoaire kinétoplastidé: variation au cours du cycle cellulaire. *Biologie Cellulaire*, **35**, 111–114.

Brugerolle, G. and Mignot, J.-P. (1983). Caractéristiques ultrastructurales de l'hélioflagellé *Tetradimorpha* (Hsiung) et leur intérêt pour l'etude phyletique des héliozoaires. *Journal of Protozoology*, **30**, 473–480.

Brugerolle, G. and Mignot, J.-P. (1984). Les caractéristiques ultrastructurales de l'hélioflagellé *Dimorpha mutans* Gruber (Sarcodina-Actinopoda) et leur intérêt phyletique. *Protistologica*, **20**, 97–112.

Brugerolle, G. and Taylor, F.J.R. (1979) . Taxonomy, cytology and evolution of the Mastigophora. *Proceedings of the 5th International Congress on Protozoology*, New York, 14–28.

Buetow, D.E. (1967–1982). *The Biology of Euglena* (3 vols). Academic Press, New York.

Burzell, L.A. (1973). Observations on the proboscis-cytopharynx and flagella of *Rhynchomonas metabolita* Pshenin 1964 (Zoomastigophorea, Bodonidae). *Journal of Protozoology*, **20**, 385–393.

Butzel, H.M. (1974). Mating type determination and development in *Paramecium aurelia*. In *Paramecium, A Current Survey*, W.J. van Wagtendonk, (ed.) 91–130, Elsevier, Amsterdam.

Cachon, J. and Cachon–Enjumet, M. (1967a). *Cymbiodinium elegans* nov. gen., nov. sp., péridinien Noctilucidae Saville-Kent. *Protistologica*, **3**, 313–318.

Cachon, J. and Cachon–Enjumet, M. (1967b). Contribution a l'étude des Noctilucidae (Saville-Kent), I. Les Kofoidininae (Cachon, J. et M.), évolution morphologique et systématique. *Protistologica*, **3**, 427–444.

Cachon, J. and Cachon, M. (1969). Contribution a l'étude des Noctilucidae Saville-Kent. Evolution morphologique, cytologie, systématique. II. Les Leptodiscinae.

Protistologica, **5**, 11-33.

Cachon, J. and Cachon, M. (1973). Systèms microtubulaires de l'astropyle et des parapyles de phaeodariés. *Archiv für Protistenkunde*, **115**, 324-335.

Cachon, J. and Cachon, M. (1981). Movement by non-actin filament mechanisms. *Bio-Systems*, **14**, 313-326.

Cachon, J. and Cachon, M. (1985). Class Acantharea Haeckel 1881, 274-282; Class Polycystinea Ehrenberg 1838, Emend Riedel 1967, 283-295; Class Phaeodarea Haeckel, Emend Haeckel, Reschetnjak, 295-302. In *An Illustrated Guide to the Protozoa*, J.J. Lee, S.H. Hutner and E.C. Bovee, (eds) Society of Protozoologists, Lawrence, Kansas.

Cachon, J., Cachon, M. and Boillot, A. (1983). Flagellar rootlets as myonemal elements for pusule contractility in dinoflagellates. *Cell Motility*, **3**, 61-77.

Canella, M.F. and Rocchi–Canella, I, (1976). Biologie des Ophryoglenina. *Annali Dell'Università di Ferrara (Nuova Serie) Sezione III-Biologia Animale* **3**, Supplemento **2**, 1-510.

Cann, J.P. (1986). Ultrastructural observations of taxonomic importance on the euglenoid genera **Gyropaigne** Skuja, **Parmidium** Christen and **Rhabdospira** Pringsheim. *Archiv für Protistenkunde*, **132**, 395-401.

Canning, E.U. and Lom, J. (1986). *The Microsporidia of Vertebrates*. Academic Press, London.

Caron, D.A., Davis, P.G., Madin, L.P. and Sieburth, J. McN. (1982). Heterotrophic bacteria and bacterivorous protozoa in oceanic microaggregates. *Science*, **218**, 795-797.

Cash, J. and Wailes, G.H. (1905-1921). *The British Freshwater Rhizopoda and Heliozoa*. (5 vols.) Ray Society, London.

Cavalier–Smith, T. (1981). Eukaryote kingdoms: seven or nine? *BioSystems*, **14**, 461-481.

Cavalier–Smith, T. (1987). The simultaneous symbiotic origin of mitochondria, chloroplasts and microbodies. *Annals of the New York Academy of Sciences*, **503**, 55-71.

Chang, K.P. (1983). Cellular and molecular mechanisms of intracellular symbiosis in leishmaniasis. *International Review of Cytology*, Supplement **14**, 267-305.

Chapman–Andresen, C. (1973). Endocytic processes. In *The Biology of Amoeba*, K.W. Jeon, (ed.) 319-348, Academic Press, New York.

Chapman, D.V., Dodge, J.D. and Heaney, S.I. (1982). Cyst formation in the freshwater dinoflagellate **Ceratium hirundinella** (Dinophyceae). *Journal of Phycology*, **18**, 121-129.

Chatton, E. and Lwoff, A. (1935). La constitution primitive de la strie ciliaire des Infusoires. La desmodexie. *Comptes rendus des Séances de la Société de Biologie*, Paris, **118**, 1068-1071.

Clarholm, M. (1981). Protozoan grazing of bacteria in soil – impact and importance. *Microbial Ecology*, **7**, 343-350.

Clarholm, M. (1984). Microbes as predators or prey. Heterotrophic, free-living protozoa: neglected microorganisms with an important task in regulating bacterial populations. In *Current Perspectives in Microbial Ecology*, M.J. Klug and C.A. Reddy, (eds) 321-326, *American Microbiology Society*, Washington.

Cleveland, L.R. (1950). Hormone-induced sexual cycles of flagellates. II Gametogenesis, fertilization and one-division meiosis in **Oxymonas**. *Journal of Morphology*, **86**, 185-214.

Cleveland, L.R. (1956). Brief accounts of the sexual cycles of the flagellates of **Cryptocercus**. *Journal of Protozoology*, **3**, 161-180.

Cleveland, L.R. and Grimstone, A.V. (1964). The fine structure of the flagellate **Mixotricha paradoxa** and its associated micro-organisms. *Proceedings of the Royal Society of London*. series B, **159**, 668-685.

Coleman, D.C. Cole, C.V., Anderson, R.V., Blaha, M., Campion, M.K., Clarholm, M., Elliott, E.T., Hunt, H.W., Shaefer, B. and Sinclair, J. (1977). An analysis of rhizosphere-saprophage interactions in terrestrial ecosystems. *Ecological Bulletin* (Stockholm), **25**, 299–309.

Coleman, G.S. (1979). Rumen ciliate protozoa. In *Biochemistry and Physiology of Protozoa*, Second edition, M. Levandowsky and S.H. Hutner, (eds) 381–408, Academic Press, New York.

Collombier, J., Grolière, C.A. Senaud, J., Jouany, J.P., Grain, J. and Thivend, P. (1984). Etude du rôle des protozoaires ciliés du rumen dans l'apport d'azote microbien entrant dans le duodénum du ruminant, par l'estimation directe de la quantité de ciliés sortant du rumen. *Protistologica*, **20**, 431–436.

Connolly, J.G. and Kerkut, G.A. (1983). Ion regulation and membrane potential in *Tetrahymena* and *Paramecium*. *Comparative Biochemistry and Physiology*, **76A**, 1–16.

Cook, C.B. (1983). Metabolic interchange in algae-invertebrate symbiosis. *International Review of Cytology, Supplement* **14**, 177–210.

Corliss, J.O. (1961). *The Ciliated Protozoa: Characterization, Classification, and Guide to the Literature*. Pergamon Press, Oxford.

Corliss, J.O. (1973). Guide to the literature on Tetrahymena: A companion piece to Elliott's 'General Biography'. *Transactions of the American Microscopical Society,* **92**, 468–491.

Corliss, J.O. (1979). *The Ciliated Protozoa: Characterization, Classification, and Guide to the Literature*, Second edition. Pergamon Press, Oxford.

Corliss, J.O. (1981). What are the taxonomic and evolutionary relationships of the Protozoa to the Protista? *BioSystems*, **14**, 445–459.

Corliss, J.O. (1984). The Kingdom Protista and its 45 phyla. *BioSystems*, **17**, 87–126.

Corliss, J.O. (1985). Consideration of taxonomic-nomenclatural problems posed by report of myxosporidians with a two-host life cycle. *Journal of Protozoology*, **32**, 589–591.

Corliss, J.O. and Esser, S.C. (1974). Comments on the role of the cyst in the life cycle and survival of free-living protozoa. *Transactions of the American Microscopical Society*, **93**, 578–593.

Cote, R. (1984). *ATTC Media Handbook*. American Type Culture Collection, Rockville, Maryland.

Crawford, R.M. (1973). The protoplasmic ultrastructure of the vegetative cell of *Melosira varians* C.A. Agardh. *Journal of Phycology*, **9**, 50–61.

Crawford, R.M. (1981). Valve formation in diatoms and the fate of the silicalemma and plasmalemma. *Protoplasma*, **106**, 157–166.

Cross, G.A.M. (1975). Identification, purification and properties of clone-specific glycoprotein antigens constituting the surface coat of *Trypanosoma brucei. Parasitology*, **71**, 393–417.

Creutz, C. and Diehn, B. (1976). Motor response to polarised light and gravity sensing in *Euglena. Journal of Protozoology*, **23**, 552–556.

Curds, C.R. (1966). An ecological study of the ciliated protozoa in activated sludge. *Oikos*, **15**, 282–289.

Curds, C.R. (1975). A guide to the species of the genus *Euplotes* (Hypotrichida Ciliatea). *Bulletin of the British Museum of Natural History (Zoology)*, **28**, 1–61.

Curds, C.R. (1982). *British and Other Freshwater Ciliated Protozoa. Part I. Ciliophora: Kinetofragminophora*. University Press, Cambridge.

Curds, C.R. and Cockburn, A. (1968). Studies on the growth and feeding of *Tetrahymena pyriformis* in axenic and monoxenic culture. *Journal of general Microbiology*, **54**, 343–358.

Curds, C.R. and Cockburn, A. (1970a). Protozoa in biological sewage-treatment processes – I. A survey of the protozoan fauna of British percolating filters and activated-sludge plants. *Water Research*, **4**, 225–236.

Curds, C.R. and Cockburn, A. (1970b). Protozoa in biological sewage-treatment processes – II. Protozoa as indicators in the activated-sludge process. *Water Research*, **4**, 237–249.

Curds, C.R. and Cockburn, A. (1971). Continuous monoxenic culture of *Tetrahymena pyriformis. Journal of general Microbiology*, **66**, 95–108.

Curds, C.R., Gates, M.A. and Roberts, D. McL. (1983). *British and Other Freshwater Ciliated Protozoa. Part II. Ciliophora: Oligohymenophora and Polyhymenophora.* University Press, Cambridge.

Curds C.R. and Vandyke, J.M. (1966). The feeding habits and growth rates of some fresh-water ciliates found in activated-sludge plants. *Journal of applied Ecology*, **3**, 127–137.

Daniels, E.W. (1973). Ultrastructure. In *The Biology of Amoeba*, K.W. Jeon, (ed.) 125–169, Academic Press, New York.

Darbyshire, J.F. (1966). Protozoa in the rhizosphere of **Lolium perenne L.** *Canadian Journal of Microbiology*, **12**, 1287–1289.

Darbyshire, J.F. and Greaves, M.P. (1973). Bacteria and protozoa in the rhizosphere. *Pesticide Science*, **4**, 349–360.

Danforth, W.F. (1967). Respiratory metabolism. In *Research in Protozoology* 1, T.-T. Chen, (ed.) 201–306, Pergamon Press, Oxford.

Davies, S.F.M., Joyner, L.P. and Kendall, B.S. (1963). *Coccidiosis.* Oliver and Boyd, Edinburgh.

Deans, J.A. and Cohen, S. (1983). Immunology of Malaria. *Annual Review of Microbiology*, **37**, 25–49.

Deflandre, G. (1952). Classe des Ebriédiens. In *Traité de Zoologie*, **I**(i), P.-P. Grassé, (ed.) 407–424, Masson, Paris.

Deitmer, J.W., Machemer, H. and Martinac, B. (1984). Motor control in three types of ciliary organelles in the ciliate **Stylonychia.** *Journal of Comparative Physiology A*, **154**, 113–120.

Dentler, W.L. (1980). Structures linking the tips of ciliary and flagellar microtubules to the membrane. *Journal of Cell Science*, **42**, 207–220.

Desportes, I. (1981). Étude ultrastructurale de la sporulation de **Paramyxa paradoxa** Chatton (Paramyxida) parasite de l'annelide polychéte **Poecilochaetus serpens.** *Protistologica*, **17**, 365–386.

Desportes, I. (1984). The Paramyxea Levine 1979: an original example of evolution towards multicellularity. *Origins of Life*, **13**, 343–352.

Desportes, I. and Nashed, N.N. (1983). Ultrastructure of sporulation in **Minchinia dentali** (Arvy), an haplosporean parasite of **Dentalium entale** (Scaphopoda, Mollusca): taxonomic implications. *Protistologica*, **19**, 435–460.

De Terra, N. (1970). Cytoplasmic control of macronuclear events in the cell cycle of **Stentor.** *Symposium of the Society for Experimental Biology*, **24**, 345–368.

De Terra, N. (1977). Control of cell and nuclear division by cortical pattern change during the cell cycle of **Stentor.** *Fifth International Congress on Protozoology*, New York, Abstract **52a**.

Diehn, B. and Kint, B. (1970). The flavin nature of the photoreceptor molecule for phototaxis in **Euglena.** *Physiological Chemistry and Physics*, **2**, 483–488.

Diller, W.F. (1975). Nuclear behaviour and morphogenetic changes in fission and conjugation in **Aspidisca costata** Dujardin. *Journal of Protozoology*, **22**, 221–229.

Dingle, A.D. and Fulton, C. (1966). Development of the flagellar apparatus of **Naegleria.** *Journal of Cell Biology*, **31**, 43–54.

Dixon, P.S. (1973). *Biology of the Rhodophyta.* Oliver and Boyd, Edinburgh.

Dodge, J.D. (1965). Chromosome structure in the dinoflagellates and the problem of the mesokaryotic cell. In *Progress in Protozoology*, 264–265, Excerpta Medica, London.

Dodge, J.D. (1969). The ultrastructure of **Chroomonas mesostigmatica** Butcher (Cryptophyceae). *Archiv für Mikrobiologie*, **69**, 266–280.

Dodge, J.D. (1971). Fine structure of the Pyrrophyta. *Botanical Review*, **37**, 484–508.

Dodge, J.D. (1972). The ultrastructure of the dinoflagellate pusule: a unique osmoregulatory organelle. *Protoplasma*, **75**, 285–302.

Dodge, J.D. (1973). *The Fine Structure of Algal Cells.* Academic Press, London.

Dodge, J.D. (1975). A survey of chloroplast ultrastructure in the Dinophyceae. *Phycologia*, **14**, 253–263.

Dodge. J.D. (1979). The phytoflagellates: fine structure and phylogeny. In *Biochemistry and Physiology of Protozoa*, second edition, S.H. Hutner and M. Levandowsky, (eds) **1**, 7–57, Academic Press, New York.

Dodge, J.D. (1982). *Marine Dinoflagellates of the British Isles.* H.M. Stationery Office, London.

Dodge, J.D. (1985). *Atlas of Dinoflagellates.* Farrand Press, London.

Dodge, J.D. and Crawford, R.M. (1968). Fine structure of the dinoflagellate *Amphidinium carteri* Hulbert. *Protistologica.* **4**, 231–242.

Dodge, J.D. and Crawford, R.M. (1969). Observations on the fine structure of the eyespot and associated organelles in the dinoflagellate *Glenodinium foliaceum. Journal of Cell Science*, **5**, 479–493.

Dodge, J.D. and Crawford, R.M. (1970a). The morphology and fine structure of *Ceratium hirundinella* (Dinophyceae). *Journal of Phycology*, **6**, 137–149.

Dodge, J.D. and Crawford, R.M. (1970b). A survey of thecal fine structure in the Dinophyceae. *Botanical Journal of the Linnean Society*, **63**, 53–67.

Dodge, J.D. and Crawford, R.M. (1971). Fine structure of the dinoflagellate *Oxyrrhis marina. Protistologica*, **7**, 295–304.

Dodge, J.D. and Lee, J.J. (1985). Order Dinoflagellida Butschli, 1885. In *An Illustrated Guide to the Protozoa*, J.J. Lee, S.H. Hutner and E.C. Bovee, (eds) 22–41, Society of Protozoologists, Lawrence, Kansas.

Dogiel, V.A. (1965). *General Protozoology.* Revised by J.I. Poljanskij and E.M. Chejsin, Second edition, University Press, Oxford.

Donelson, J.E. and Turner, M.J. (1985). How the trypanosome changes its coat. *Scientific American*, **252** (2), 32–39

Dragesco, J. (1963a). Compléments à la connaissance des ciliés mésopsammiques de Roscoff. I. Holotriches, *Cahiers de Biologie marine*, **4**, 91–119.

Dragesco, J. (1963b). Compléments à la connaissance des ciliés mésopsammiques de Roscoff. II. Hétérotriches, III. Hypotriches. *Cahiers de Biologie marine*, **4**, 251–275.

Droop, M. (1974). Heterotrophy of carbon. In *Algal Physiology and Biochemistry*, W.D.P. Stewart, (ed.) 530–559, Blackwell, Oxford.

Dryl, S. (1963). Oblique galvanotaxis in *Stylonychia mytilus. Journal of Protozoology*, **10**, Supplement 26.

Dunlap, J.C., Taylor, W. and Hastings, J.W. (1981). The control and expression of bioluminescence in dinoflagellates. In *Bioluminescence: Current Perspectives*, K.H. Nealson, (ed.) 108–124, Burgess, Minneapolis, Minnesota.

Eckert, B.A. and McGee-Russell, S.M. (1973). The patterned organisation of thick and thin microfilaments in the contracting pseudopod of *Difflugia. Journal of Cell Science*, **13**, 727–739.

Edds, K.T. (1975). Motility in *Echinosphaerium nucleofilum* II. Cytoplasmic contractility and its molecular basis. *Journal of Cell Biology*, **66**, 156–164.

Edgar, L.A. and Pickett–Heaps, J.D. (1983). The mechanism of diatom locomotion. I. An ultrastructural study of the motility apparatus. *Proceedings of the Royal Society of London*, Series B, **218**, 331–343.

Edmondson, W.T. (1959). *Freshwater Biology*, Second edition, Wiley, New York.

Elliott, A.M. (1973). *Biology of Tetrahymena.* Dowden, Hutchinson and Ross, Stroudsburg, Pennsylvania.

Elliott, E.T., Coleman, D.C. Ingham, R.E. and Trofymow, J.A. (1984). Carbon and energy flow through microflora and microfauna in the soil subsystem of terrestrial ecosystems. In *Current Perspectives in Microbial Ecology*, M.J. Klug and C.A. Reddy, (eds) 424–433, American Society of Microbiology, Washington.

Eppley, R.W. and Peterson, B.J. (1979). Particulate organic matter flux and plank-tonic new production in the deep ocean. *Nature*, **282**, 677–680.

Ettienne, E.M. (1970). Control of contractility in *Spirostomum* by dissociated calcium ions. *Journal of general Physiology*, **56**, 168–179.

Faure-Fremiet, E. (1950). Morphologie comparée et systématique des ciliés. *Bulletin de la Societé Zoologique de France*, **75**, 109–122.

Faure-Fremiet, E. (1951a). Associations infusoriennes à *Beggiatoa*. *Hydrobiologia*, **3**, 65–71.

Faure-Fremiet, E. (1951b). The marine sand-dwelling ciliates of Cape Cod. *Biological Bulletin of the Marine Biological Laboratory*, Woods Hole, **100**, 59–70.

Febvre, J. (1981). The myoneme of the Acantharia (Protozoa): A new model of cellular motility. *BioSystems*, **14**, 327–336.

Febvre-Chevalier, C. (1980). Behaviour and cytology of *Actinocoryne contractilis* nov. gen., nov. sp., a new stalked heliozoan (Centrohelida): comparison with the other related genera. *Journal of the marine biological Association* UK, **60**, 909–928.

Febvre-Chevalier, C. (1985). Heliozoea Haeckel 1866. In *An Illustrated Guide to the Protozoa*, J.J. Lee, S.H. Hutner and E.C. Bovee, (eds) 302–317, Society of Proto-zoologists, Lawrence, Kansas.

Fenchel, T. (1969). The ecology of marine microbenthos. IV. Structure and function of the benthic ecosystem. *Ophelia*, **6**, 1–182.

Fenchel, T. (1980a). Relation between particle size selection and clearance in suspen-sion-feeding ciliates. *Limnology and Oceanography*, **25**, 733–738.

Fenchel, T. (1980b). Suspension feeding in ciliated protozoa: structure and function of feeding organelles. *Archiv für Protistenkunde*, **123**, 239–260.

Fenchel, T. (1980c). Suspension feeding in ciliated protozoa: functional response and particle size selection. *Microbial Ecology*, **6**, 1–11.

Fenchel, T. (1980d). Suspension feeding in ciliated protozoa: feeding rates and their ecological significance. *Microbial Ecology*, **6**, 13–25.

Fenchel, T. (1982a). Ecology of heterotrophic microflagellates. I. Some important forms and their functional morphology. *Marine Ecology Progress Series*, **8**, 211–223.

Fenchel, T. (1982b). Ecology of heterotrophic microflagellates. II. Bioenergetics and growth. *Marine Ecology Progress Series*, **8**, 225–231.

Fenchel, T. (1982c). Ecology of heterotrophic microflagellates. III. Adaptations to heterogeneous environments. *Marine Ecology Progress Series*, **9**, 25–33.

Fenchel, T. (1982d). Ecology of heterotrophic microflagellates. IV. Quantitative occur-rence and importance as bacterial consumers. *Marine Ecology Progress Series*, **9**, 35–42.

Fenchel, T. (1986). Protozoan filter feeding. *Progress in Protistology*, **1**, 65–113.

Fenchel, T. (1987). *Ecology of Protozoa*. Science Technical, Madison, Wisconsin.

Fenchel, T. and Finlay, B. (1983). Respiration rates in heterotrophic, free-living pro-tozoa. *Microbial Ecology*, **9**, 99–122.

Fenchel, T. and Finlay, B. (1984). Geotaxis in the ciliated protozoon *Loxodes*. *Journal of experimental Biology*, **110**, 17–33.

Fenchel, T. and Finlay, B. (1986a). The structure and function of Muller vesicles in loxodid ciliates. *Journal of Protozoology*, **33**, 69–79.

Fenchel, T. and Finlay, B. (1986b). The responses to light and to oxygen in the ciliated protozoon *Loxodes striatus*. *Journal of Protozoology*, **33**, 139–145.

Fenchel, T. and Harrison, P. (1976). The significance of bacterial grazing and mineral cycling for the decomposition of particulate detritus. In *The Role of Terrestrial and Aquatic Organisms in Decomposition Processes*, J.M. Anderson and A. Mac-Fadyen, (eds) 285–299, Blackwell, Oxford.

Fenchel, T. and Patterson, D.J. (1986). *Percolomonas cosmopolitus* (Ruinen) n. gen., a new type of filter feeding flagellate from marine plankton. *Journal of marine biological Association*, UK, **66**, 465–482.

Fenchel, T., Perry, T. and Thane, A. (1977). Anaerobiosis and symbiosis with bacteria in free-living ciliates. *Journal of Protozoology*, **24**, 154–163.

Fenchel, T. and Reidl, R.J. (1970). The sulfide system: a new biotic community underneath the oxidised layer of marine sand bottoms. *Marine Biology*, **7**, 255–268.

Finlay, B.J. (1982). Effects of seasonal anoxia on the community of benthic ciliated protozoa in a productive lake. *Archiv für Protistenkunde*, **125**, 215–222.

Finlay. B.J. (1985). Nitrate respiration by protozoa (*Loxodes* spp.) in the hypolimnetic nitrite maximum of a productive freshwater pond. *Freshwater Biology*, **15**, 333–346.

Finlay, B.J. and Fenchel, T. (1986). Photosensitivity in the ciliate protozoon *Loxodes*: pigment granules, absorption and action spectra, blue light perception and ecological significance. *Journal of Protozoology*, **33**, 534–542.

Finlay, B.J., Rogerson, A. and Cowling, A.J. (1988). A beginners' guide to the collection, isolation, cultivation and indentification of freshwater Protozoa. *Culture Collection of Algae and Protozoa*, Freshwater Biological Association, Ambleside, UK.

Fitch, C.D. (1983). Mode of action of antimalarial drugs. *Ciba Foundation Symposium*, **94**, 222–232.

Fjerdingstad, E. (1971). Microbial criteria of environmental qualities. *Annual Review of Microbiology*, **25**, 563–582.

Flickinger, C.J. (1973). Cellular membranes in amoebae. In *The Biology of Amoeba*, K.W. Jeon, (ed.) 171–199, Academic Press, New York.

Foissner, W. (1987). Soil protozoa: fundamental problems, ecological significance, adaptations in ciliates and testaceans, bioindicators and guide to the literature. *Progress in Protistology*, **2**, 69–212.

Foster, K.W. and Smyth, R.D. (1980). Light antennas in phototactic algae. *Microbiological Reviews*, **44**, 572–630.

Frankel, J. (1967). Studies on the maintenance of oral development in *Tetrahymena pyriformis* GL-C. II. The relationship of protein synthesis to cell division and oral organelle development. *Journal of Cell Biology*, **34**, 841–858.

Furness, D.N. and Butler, R.D. (1985). The cytology of sheep rumen ciliates. II. Ultrastructure of *Eudiplodinium maggii*. *Journal of Protozoology*, **32**, 205–214.

Gaines, G. and Taylor, F.J.R. (1984). Extracellular digestion in marine dinoflagellates. *Journal of Plankton Research*, **6**, 1057–1061.

Gallo, J.-M. and Schrevel, J. (1982). Euglenoid movement in *Distigma proteus*. I. Cortical rotational motion. *Biology of the Cell*, **44**, 139–148.

Galtsoff, P.S., Lutz, F.E., Welch, P.S. and Needham, J.G. (1959). *Culture Methods for Invertebrate Animals*. A.A.A.S., Dover, New York.

Gantt, E. (1980). Photosynthetic cryptophytes. In *Phytoflagellates*, E.R. Cox, (ed.) 381–405, Elsevier North Holland, New York.

Gao, Q.-R. (1980). Preliminary study of chromatin structure of *Oxyrrhis marina*. II. Electron microscopic observations on nucleosome-like structures. *Acta Biologicae Experimentalis Sinica*, **13**, 469–470.

Garnham, P.C.C. (1966). *Malaria Parasites and other Haemosporidia*. Blackwell, Oxford.

Gause, G.F. (1934). *The Struggle for Existence*. Williams and Wilkins, Baltimore, Maryland.

Gause, G.F. (1935). Experimental demonstrations of Volterra's periodic oscillations in the number of animals. *Journal of Experimental Biology*, **12**, 44–48.

Gayral, P. and Fresnel, J. (1983). Description, sexualité et cycle de développment d'une nouvelle Coccolithophoracée (Prymnesiophyceae): *Pleurochrysis pseudoroscoffensis* sp. nov. *Protistologica*, **19**, 245–261.

Geller, A. and Muller, D.G. (1981). Analysis of the flagellar beat pattern of male *Ecto-*

carpus siliculosus gametes (Phaeophyta) in relation to chemotactic stimulation by female cells. *Journal of Experimental Biology*, **92**, 53–66.

Gerassimova, Z.P. and Seravin, L.N. (1976). Ectoplasmic fibrillar system of Infusoria and its role for the understanding of their phylogeny. *Zoologicheskii Zhurnal*, **55**, 645–656, (In Russian with English summary).

Gibbons, I.R. (1981). Cilia and flagella of eukaryotes. *Journal of Cell Biology*, **91**, 107s–124s.

Gibbons, I.R. (1983). Conclusion to dynein workshop. *Journal of Submicroscopic Cytology*, **15**, 243–245.

Gibbs, S.P. (1981). The chloroplasts of some algal groups may have evolved from eukaryotic algae. *Annals of the New York Academy of Science*, **361**, 193–208.

Giese, A.C. (1973). *Blepharisma: The Biology of a Light-Sensitive Protozoan*. Stanford University Press, Stanford, California.

Gitelman, S.E. and Witman, G.B. (1980). Purification of calmodulin from *Chlamydomonas*: Calmodulin occurs in cell bodies and flagella. *Journal of Cell Biology*, **98**, 764–770.

Gliddon, R. (1966). Ciliary organelles and associated fibre systems in *Euplotes eurystomus* (Ciliata, Hypotrichida). I. Fine structure. *Journal of Cell Science*, **1**, 439–448.

Gojdics, M. (1953). *The Genus Euglena*. University of Wisconsin Press, Madison.

Goldacre, R.J. and Lorch, I.J. (1950). Folding and unfolding of protein molecules in relation to cytoplasmic streaming, amoeboid movement and osmotic work. *Nature*, **166**, 497–500.

Goldstein, S.F. (1974). Isolated, reactivated and laser-irradiated cilia and flagella. In *Cilia and Flagella* M.A. Sleigh, (ed.) 111–130, Academic Press, London.

Gooday, A.J. and Nott, J.A. (1982). Intracellular barite crystals in two xenophyophores, **Aschemonella ramuliformis** and **Galatheammina** sp. (Protozoa, Rhizopoda) with comments on the taxonomy of **A. ramuliformis**. *Journal of the marine biological Association*, UK, **62**, 595–605.

Goodfellow, L.P. Belcher, J.H. and Page, F.C. (1974). A light- and electron-microscopical study of **Sappinia diploidea**, a sexual amoeba. *Protistologica*, **10**, 207–216.

Gortz, H.-D. (1983). Endonuclear symbionts in ciliates. *International Review of Cytology*, Supplement **14**, 145–176.

Grain, J. (1969). Le cinétosome et ses dérivés chez les ciliés. *Annals of Biology,* **8**, 54–97.

Grassé, P.-P. (1952). *Traité de Zoologie* **1** (i) and (ii). Masson, Paris.

Grassé, P.-P. (1952). Les Ellobiopsidae. In *Traité de Zoologie* P.-P. Grassé, (ed.) **1** (i), 1023–1030, Masson, Paris.

Grebecki, A. (1982). Supramolecular aspects of amoeboid movement. Proceedings of the Sixth International Congress of Protozoology, special Congress Volume, *Acta Protozoologica*, **I**, 117–130.

Grebecki, A. (1986a). Adhesion-dependent movements of the cytoskeletal cylinder of amoebae. *Acta Protozoologica*, **25**, 255–268.

Grebecki, A. (1986b). Two-directional pattern of movements on the cell surface of **Amoeba proteus.** *Journal of Cell Science,* **83**, 23–25.

Grebecki, A. and Klopocka, W. (1981). Functional interdependence of pseudopodia in **Amoeba proteus** stimulated by light-shade difference. *Journal of Cell Science*, **50**, 245–258.

Grell, K.G. (1967). Sexual reproduction in protozoa. In *Research in Protozoology*, T.-T. Chen, (ed.) **2**, 147–213, Pergamon Press, Oxford.

Grell, K.G. (1973). *Protozoology*. Springer-Verlag. Berlin.

Grell, K.G. and Meister, A. (1983). Die ultrastruktur von **Hypocoma acinetarum** Collin. *Protistologica*, **19**, 51–72.

Greuet, C. (1972). La nature trichoplastaire du cnidoplaste dans le complexe cnidoplaste-nematocyste du **Polykrikos schwartzi** Butschli. *Compte rendu Hebdomadaire des Séances de L'Académie des Sciences*, Paris, Series D, **275**, 1239–1242.

Greuet, C. (1977). Évolution structurale et ultrastructurale de l'ocelloide d'*Erythropsidinium pavillardi* Kofoid et Swezy au cours des divisons binaire et palintomiques. *Protistologica*, **13**, 127–143.

Greuet, C. (1982). Photorécepteurs et phototaxie des flagellés et des stades unicellulaires d'organismes inferieurs. *Année Biologique*, **21**, 97–141.

Greuet, C., Gayol, P., Salvano, P. and Laval–Peuto, M. (1986). Preliminary report on the ultrastructural organization of the contractile appendix of **Tontonia appendiculariformis** (Ciliophora, Oligotrichina). *Cell Motility and the Cytoskeleton*, **6**, 217–224.

Hajduk, S.L. (1984). Antigenic variation during the developmental cycle of **Trypanosoma brucei**. *Journal of Protozoology*, **31**, 41–47.

Hall, R.P. (1967). Nutrition and growth of protozoa. In *Research in Protozoology*, T.-T. Chen, (ed.) **1**, 337–404, Pergamon Press, Oxford.

Hall, S.L. and Fisher, F.M. (1985). Annual productivity and extracellular release of dissolved organic compounds by the epibenthic algal community of a brackish marsh. *Journal of Phycology*, **21**, 277–281.

Halldal, P. (1964). Phototaxis in protozoa. In *Biochemistry and Physiology of Protozoa*, S.H. Hutner, (ed.) **3**, 277–296, Academic Press, New York.

Hanel, K. von (1979). Systematics and ecology of colourless flagellates in sewage. *Archiv für Protistenkunde*, **121**, 73–137.

Hanson, E.D. (1977). *The Origin and Early Evolution of Animals*. Wesleyan University, Middleton, Connecticut.

Hardham, A.R. (1987). Microtubules and the flagellar apparatus in zoospores and cysts of the fungus **Phytophthora cinnamomi**. *Protoplasma*, **137**, 109–124.

Hartmann, M. and Nagler, K. (1908). Copulation bei **Amoeba diploidea** mit Selbststandigbleiben der Gametenkerne wahrend des ganzen Lebenscyclus. *Sitzungsberichte der Gesellschaft naturforschender Freunde zu Berlin*, **5**, 112–125.

Hauser, M. and Hausmann, K. (1982). Electron microscopic localization of ATPase activity in the cytopharyngeal basket of the ciliate **Pseudomicrothorax dubius**. *Differentiation*, **22**, 67–72.

Hauser, M., Hausmann, K. and Jockusch, B. (1980). Demonstration of tubulin, actin and α-actinin by immunofluorescence in the microtubule-microfilament complex of the cytopharyngeal basket of the ciliate **Pseudomicrothorax dubius**. *Experimental Cell Research*, **125**, 265–274.

Hausmann, K. (1978). Extrusive organelles in protists. *International Review of Cytology*, **52**, 197–276.

Hausmann, K, and Patterson, D.J. (1983). *Taschenatlas der Einzeller*. Kosmos Franckh, Stuttgart.

Hausmann, K. and Peck, R.K. (1978). Microtubules and microfilaments as major components of a phagocytic apparatus: the cytopharyngeal basket of the ciliate **Pseudomicrothorax dubius**. *Differentiation*, **11**, 157–167.

Hausmann, K. and Peck, R.K. (1979). The mode of function of the cytopharyngeal basket of the ciliate **Pseudomicrothorax dubius**. *Differentiation*, **14**, 147–158.

Hausmann, K., Stockem, W. and Wohlfarth–Bottermann, K.E. (1972a). Cytological studies on trichocysts. I. The fine structure of the discharged spindle trichocyst of **Paramecium caudatum**. *Cytobiologie*, **5**, 208–227.

Hausmann, K., Stockem, W. and Wohlfarth–Bottermann, K.E. (1972b). Cytological studies on trichocysts. II. The fine structure of resting and inhibited spindle trichocysts in **Paramecium caudatum**. *Cytobiologie*, **5**, 228–246.

Hawes, R.S.J. (1963). The emergence of asexuality in protozoa. *Quarterly Review of Biology*, **38**, 234–242.

Heal, O.W. (1967). Quantitative feeding studies on the soil amoebae. In *Progress in Soil*

Biology, O. Graff and J.E. Satchell (eds) 120–126, Elsevier North Holland, Amsterdam.

Heal, O.W. (1971). Protozoa. In *Quantitative Soil Ecology*, J. Phillipson, (ed.) 51–71, Blackwell, Oxford.

Heath, I.B. (1980). Variant mitoses in lower eukaryotes: indicators of the evolution of mitosis? *International Review of Cytology*, **64**, 1–80.

Heckmann, K. (1983). Endosymbionts of *Euplotes*. *International Review of Cytology Supplement*, **14**, 111–144.

Hedley, R.H. and Ogden, C.G. (1973). Biology and fine structure of *Euglypha rotunda* (Testacea, Protozoa). *Bulletin of the British Museum of Natural History (Zoology)*, **25**, 121–137.

Hedley, R.H., Parry, D.M. and Wakefield, J.St.J. (1968). Reproduction in *Boderia turneri* (Foraminifera). *Journal of Natural History*, **2**, 147–151.

Hedley, R.H. and Wakefield, J.St.J. (1969). Fine structure of *Gromia oviformis* (Rhizopodea, Protozoa). *Bulletin of the British Museum of Natural History (Zoology)*, **18**, 69–89.

Hellebust, J.A. and Lewin, J. (1977). Heterotrophic nutrition. In *The Biology of Diatoms*, D. Werner (ed.) 169–197, Blackwell, Oxford.

Henebry, M.J. and Cairns, J. (1980). Monitoring of stream pollution using protozoan communities on artificial substrates. *Transactions of the American Microscopical Society*, **99**, 151–160.

Heywood, P. (1980). Chloromonads, In *Phytoflagellates*, E.R. Cox, (ed.) 351–379, Elsevier North Holland, New York.

Heywood, P. (1988). Ultrastructure of *Chilomonas paramecium* and the phylogeny of the cryptoprotists. *BioSystems*, **21**, 293–298.

Heywood, P. and Leedale, G.F. (1985). Raphidomonadida Heywood and Leedale, 1983. In *An Illustrated Guide to the Protozoa*, J.J. Lee, S.H. Hutner and E.C. Bovee, (eds) 70–74, Society of Protozoologists, Lawrence, Kansas.

Heywood, P. and Rothschild, L.J. (1987). Reconciliation of evolution and nomenclature among the higher taxa of protists. *Biological Journal of the Linnean Society,* **30**, 91–98.

Hibberd, D.J. (1975). Observations on the ultrastructure of the choanoflagellate *Codosiga botrytis* (Ehr.) Saville Kent with special reference to the flagellar apparatus. *Journal of Cell Science*, **17**, 191–219.

Hibberd, D.J. (1976). Observations on the ultrastructure of three new species of *Cyathobodo* Petersen et Hansen (*C. salpinx, C. intricatus* and *C. simplex*) and on the external morphology of *Pseudodendromonas vlkii* Bourrelly. *Protistologica*, **12**, 249–261.

Hibberd, D.J. (1979). The structure and phylogenetic significance of the flagellar transition region in the chlorophyll c-containing algae. *BioSystems*, **11**, 243–261.

Hibberd, D.J. (1980a). Xanthophytes. In *Phytoflagellates*, E.R. Cox (ed.) 243–271, Elsevier North Holland, New York.

Hibberd, D.J. (1980b). Prymnesiophytes (= Haptophytes). In *Phytoflagellates*, E.R. Cox, (ed.) 273–317. Elsevier North Holland, New York.

Hibberd, D.J. (1980c). Eustigmatophytes. In *Phytoflagellates*, E.R. Cox, (ed.) 319–334, Elsevier North Holland, New York.

Hibberd, D.J. (1983). Ultrastructure of the colonial colourless zooflagellates *Phalansterium digitatum* Stein (Phalansteriida ord. nov.) and *Spongomonas uvella* Stein (Spongomonadida ord. nov.). *Protistologica*, **19**, 523–535.

Hibberd, D.J. and Leedale, G.F. (1985). Prymnesiida. In *An Illustrated Guide to the Protozoa* J.J. Lee, S.H. Hutner and E.C. Bovee, (eds.) 74–88, Society of Protozoologists, Lawrence, Kansas.

Hiramoto, Y. (1974). Mechanics of ciliary movement. In *Cilia and Flagella*, M.A. Sleigh (ed.) 177–196, Academic Press, London.

Hitchen, E.J. and Butler, R.D. (1973). Ultrastructural studies of the commensal suctorian *Choanophrya infundibulifera* Hartog. I. Tentacle structure, movement and feeding. *Zeitschrift für Zellforschung*, **144**, 37–57.

Hitchen, E.T. and Butler, R.D. (1974). The ultrastructure and function of the tentacle in *Rhyncheta cyclopum* Zenker (Ciliatea, Suctorida). *Journal of Ultrastructure Research*, **46**, 279–295.

Hoffman, E.K. and Rasmussen, L. (1972). Phenylalanine and methionine transport in *Tetrahymena pyriformis*: characteristics of a concentrating, inducible transport system. *Biochimica Biophysica Acta*, **266**, 206–216.

Hoham, R.W. (1980). Unicellular chlorophytes – snow algae. In *Phytoflagellates*, E.R. Cox, (ed.) 61–84, Elsevier North Holland, New York.

Holberton, D.V. (1973). Fine structure of the ventral disk apparatus and the mechanism of attachment in the flagellate *Giardia muris*. *Journal of Cell Science*, **13**, 11–41.

Hollande, A., Cachon, J., Cachon–Enjumet, M. and Valentin, J. (1967). Infrastructure des axopodes et organisation générale de *Sticholonche zanclea* Hertwig (Radiolaire Sticholonchidea). *Protistologica*, **3**, 155–166.

Hollande, A. and Carruette–Valentin, J. (1971). Les atractophores, l'induction du fuseau et la division cellulaire chez le hypermastigines: étude infrastructurale et revision systématique des trichonymphines et des spirotrichonymphines. *Protistologica*, **7**, 5–100.

Hollande, A. and Valentin, J. (1968). Morphologie infrastructurale de *Trichomonas (Trichomitopsis* Kofoid et Swezy 1919) *termopsidis*, parasite intestinal de *Termopsis angusticollis* Walk. Critique de la notion de centrosome chez les polymastigines. *Protistologica*, **4**, 127–140.

Hollande, A. and Valentin, J. (1969a). Appareil de Golgi, pinocytose, lysosomes, mitochondries, bactéries symbiontiques, atractophores et pleuromitose chez les hypermastigines du genre *Joenia*. Affinités entre Joenides et Trichomonadines. *Protistologica*, **5**, 39–86.

Hollande, A. and Valentin, J. (1969b). La cinétide et ses dépendances dans le genre *Macrotrichomonas* Grassi. Considerations générales sur la sous-famille des Macrotrichomonadinae. *Protistologica*, **5**, 335–343.

Holwill. M.E.J. (1964). The motion of *Strigomonas oncopelti*. *Journal of Experimental Biology*, **42**, 125–137.

Holwill, M.E.J. (1966). Physical aspects of flagellar movement. *Physiological Reviews*, **46**, 696–785.

Holwill, M.E.J. (1974). Hydrodynamic aspects of ciliary and flagellar movement. In *Cilia and Flagella*, M.A. Sleigh, (ed.) 143–175, Academic Press, London.

Holwill, M.E.J. and McGregor, J.L. (1976). Effects of calcium on flageller movement in the trypanosome *Crithidia oncopelti*. *Journal of Experimental Biology*, **65**, 229–242.

Holwill, M.E.J. and Sleigh, M.A. (1967). Propulsion by hispid flagella. *Journal of Experimental Biology*, **47**, 267–276.

Holwill, M.E.J. and Sleigh, M.A. (1969). Energetics of ciliary movement in *Sabellaria* and *Mytilus*. *Journal of Experimental Biology*, **50**, 733–743.

Holz, G.G. (1960). Structural and functional changes in a generation in *Tetrahymena*. *Biological Bulletin. Marine Biological Laboratory, Woods Hole*, **118**, 84–95.

Honigberg, B.M., Balamuth, W., Bovee, E.C., Corliss, J.O. Gojdics, M., Hall R.P., Kudo, R.R., Levine, N.D., Loeblich, A.R., Weiser, J. and Wenrich, D.H. (1964). A revised classification of the Phylum Protozoa. *Journal of Protozoology*, **11**, 7–20.

Hoops H.J. and Witman, G.B. (1985). Basal bodies and associated structures are not required for normal flagellar motion or phototaxis in the green alga *Chlorogonium elongatum*. *Journal of Cell Biology,* **100**, 297–309.

Hori, H. and Osawa, S. (1986). Evolutionary change in 5S rRNA secondary structure and a phylogenetic tree of 352 5S rRNA species. *BioSystems*, **19**, 163–172.

Hori, H. and Osawa, S. (1987). Origin and evolution of organisms as deduced from 5S ribosomal RNA sequences. *Molecular Biology and Evolution*, **4**, 445–472.

Huang, B. and Mazia, D. (1975). Microtubules and filaments in ciliate contractility. In *Molecules and Cell Movement*, S. Inoue and R.E. Stephens (eds) 389–409, Raven Press, New York.

Huang, B. and Pitelka, D.R. (1973). The contractile process in the ciliate *Stentor coeruleus*. I. The role of microtubules and filaments. *Journal of Cell Biology*, **57**, 704–728.

Hungate, R.E. (1966). *The Rumen and its Microbes*. Academic Press, New York.

Hutchison, W.M., Dunachie, J.F., Siim, J.C. and Work, K. (1970). Coccidian-like nature of *Toxoplasma gondii*. *British Medical Journal*, **1**, 142–144.

Hyams, J. (1982). The *Euglena* paraflagellar rod: structure, relationship to other flagellar components and preliminary biochemical characterisation. *Journal of Cell Science*, **55**, 199–210.

Hyams, J.S. and Borisy, G.G. (1978). Isolated flagellar apparatus of *Chlamydomonas*: characterization of forward swimming and alteration of waveform and reversal of motion by calcium ions *in vitro*. *Journal of Cell Science*, **33**, 235–253.

Hyman, L.H. (1940). *The Invertebrates Vol. 1 Protozoa through Ctenophora*. McGraw Hill, New York.

Hynes, H.B.N. (1960). *The Biology of Polluted Water*. Liverpool University Press.

Ibanez. F. and Rassoulzadegan, F. (1977). A study of the relationships between pelagic ciliates (Oligotrichina) and planktonic flagellates of the neritic ecosystem of the Bay of Villefranche-Sur-Mer. Analysis of chronological series. *Annales de l'Institut Oceanographique*, Paris, **53**, 17–30.

Inouye, I. and Pienaar, R.N. (1984). New observations on the coccolithophorid *Umbilicosphaera sibogae*, var. *foliosa* (Prymnesiophyceae) with reference to cell covering, cell structure and flagellar apparatus. *British Phycological Journal*, **19**, 357–369.

Irvine, D.E.G. and Price, J.H. (1978). *Modern approaches to the taxonomy of red and brown algae*. Academic Press, London.

Iwasaki, H. (1979). Physiological ecology of red tide flagellates. In *Physiology and Biochemistry of Protozoa*, Second edition, M. Levandowsky and S.H. Hutner, 357–393, Academic Press, London.

Jahn, T.L., Bovee, E.C. and Jahn, F.F. (1979). *How to Know the Protozoa*. Second edition. Brown, Dubuque, Iowa.

Jamieson, G.A., Vanaman, T.C. and Blum, J.J. (1979). Presence of calmodulin in *Tetrahymena*. *Proceedings of the National Academy of Sciences of the United States of America*, **76**, 6471–6475.

Jankowski, A.W. (1967). A new system of ciliate protozoa (Ciliophora) [in Russian]. *Academia Nauk SSSR, Trudy Zoological Institute*, **43**, 3–54.

Jankowski, A.W. (1973). Taxonomic revision of subphylum Ciliophora Doflein, 1901 [in Russian]. *Zoologicheskii Zhurnal*, **52**, 165–175.

Jeon, K.W. (1983). Integration of bacterial endosymbionts in amoebae. *International Review of Cytology* Supplement, **14**, 29–47.

Johannes, R.E. (1965). Influence of marine protozoa on nutrient regeneration. *Limnology and Oceanography*, **10**, 434–442.

Jorgen Hansen, H. (1972). Pore pseudopodia and sieve plates of *Amphistegina*. *Micropalaeontology*, **18**, 223–230.

Jurand, A. and Selman, G.G. (1969). *The anatomy of Paramecium aurelia*. Macmillan, London.

Kahl, A. (1930–1935). Urtiere oder Protozoa. 1: Wimpertiere oder Ciliata (Infusoria), eine Bearbeitung der freilebenden und ectocommensalen Infusorien der Erde, unter Ausschluss der marinen Tintinnidae. In *Die Tierwelt Deutschlands*, F. Dahl, (ed.) parts 18, 21, 25, 30, p. 1–886. G. Fischer, Jena.

Kaneshiro, E.S., Holz, G.G. and Dunham, P.B. (1969). Osmoregulation in a marine ciliate, *Miamiensis avidus*. II. Regulation of intracellular free amino acids. *Biological Bulletin of the Marine Biological Laboratory*, Woods Hole, **137**, 161–169.

Karakashian, S.J. and Rudzinska, M.A. (1981). Inhibition of lysosomal fusion with symbiont-containing vacuoles in *Paramecium bursaria*. *Experimental Cell Research*, **131**, 387–393.

Kazama, F.Y. (1972). Ultrastructure and phototaxis of the zoospores of *Phlyctochytrium* sp., an estuarine chytrid. *Journal of General Microbiology*, **71**, 555–566.

Kennedy, J.R. (1965). The morphology of *Blepharisma undulans* Stein. *Journal of Protozoology*, **12**, 542–561.

Kerrich, G.J., Hawksworth, D.L. and Sims, R.W. (1978). *Key Works to the Fauna and Flora of the British Isles and Northwestern Europe*. Academic Press, London.

Kimball, R. (1942). The nature and inheritance of mating types in *Euplotes patella*. *Genetics*, **27**, 269–285.

King, C.A. (1981). Cell surface interaction of the protozoan *Gregarina* with concanavalin A beads – implications for models of gregarine gliding. *Cell Biology International Reports*, **5**, 297–305.

Kitamura, A. and Hiwatashi, K. (1978). Are sugar residues involved in the specific cell recognition of mating in *Paramecium*? *Journal of Experimental Zoology*, **203**, 99–108.

Kitching, J.A. (1951). The physiology of contractile vacuoles. VII. Osmotic relations in a suctorian, with special reference to the mechanism of control of vacuolar output. *Journal of Experimental Biology*, **28**, 203–214.

Kitching, J.A. (1956). Food vacuoles of protozoa. In *Protoplasmatologia*, L.V. Heilbrunn and F. Weber, (eds) **III** D, 3b, 1–54. Springer-Verlag, Vienna.

Kitching, J.A. (1967). Contractile vacuoles, ionic regulation and excretion. In *Research in Protozoology*, T.-T. Chen, (ed.) **I**, 307–336, Pergamon, Oxford.

Kjelleberg, S. and Hermansson, M. (1987). Short-term responses to energy fluctuations by marine heterotrophic bacteria. In *Microbes in the Sea*, M.A. Sleigh (ed.) 203–219, Ellis Horwood, Chichester, UK.

Klopocka, W. (1983). The question of geotaxis in *Amoeba proteus*. *Acta Protozoologica*, **22**, 211–217.

Knight–Jones, W. (1954). Relation between metachronism and direction of the ciliary beat in metazoa. *Quarterly Journal of Microscopical Science*, **95**, 503–521.

Kolkwitz, R. and Marsson, M. (1908). Okologie der pflanzliche Saprobien. *Bericht der Deutschen Botanischen Gesellschaft*, **26**, 505–519.

Kolkwitz, R. and Marsson, M. (1909). Okologie der tierischen Saprobien. *Internationale Revue der gesamten Hydrobiologie*, **2**, 1–126.

Kreier, J.P. (1977). *Parasitic Protozoa* (4 vols). Academic Press, London.

Kreier, J.P. and Baker, J.R. (1987). *Parasitic Protozoa*. Allen and Unwin, London.

Krettli, A.U. and Pereira, F.E.L. (1981). Immunosuppression in protozoal infections. In *Biochemistry and Physiology of Protozoa*, M. Levandovsky and S.H. Hutner, (eds) **4**, 431–462, Academic Press, New York.

Kudo, R.R. (1966). *Protozoology*, Fifth edition Thomas, Springfield, Illinois.

Kugrens, P. and Lee, R.E. (1987). An ultrastructural survey of cryptomonad periplasts using quick-freezing freeze fracture techniques. *Journal of Phycology*, **23**, 365–376.

Kung, C., Chang, S.-Y., Satow, Y., Houten, J. van and Hansma, H. (1975). Genetic dissection of behavior in *Paramecium*. *Science*, **188**, 898–904.

Lake, J.A. Henderson, E., Oakes, M. and Clark, M.W. (1984). Eocytes, a new ribosome structure indicates a kingdom with a close relationship to eukaryotes. *Proceedings of the National Academy of Science of the United States of America*, **81**, 3786–3790.

Landis, W.G. (1981). The ecology, role of killer trait, and interactions of five species of the *Paramecium aurelia* complex inhabiting the littoral zone. *Canadian Journal of Zoology*, **59**, 1734–1743.

Langreth, S.G. (1976). Feeding mechanisms in extracellular *Babesia microti* and *Plasmodium lophurae*. *Journal of Protozoology*, **23**, 215–223.

Lapage, G. (1925). Notes on the choanoflagellate *Codosiga botrytis* Ehrbg. *Quarterly Journal of Microscopical Science*, **69**, 471–508.

Lapidus, R. and Levandowsky, M. (1981). Mathematical models of behavioral responses to sensory stimuli by protozoa. In *Biochemistry and Physiology of Protozoa*, M. Levandowsky and S.H. Hutner, (eds) **4**, 235–260. Academic Press, New York.

Larsson, R. (1986). Ultrastructure, function and classification of Microsporidia. *Progress in Protistology*, **1**, 325–390.

Laybourn, J. (1976). Energy budgets for *Stentor coeruleus* Ehrenberg (Ciliophora). *Oecologia*, **22**, 431–437.

Laybourn-Parry, J. (1984). *A Functional Biology of Free-Living Protozoa*. Croom Helm, London.

Leadbeater, B.S.C. (1971). Observations by means of cine photography on the behaviour of the haptonema in planktonic flagellates of the Class Haptophyceae. *Journal of the Marine Biological Association*, UK, **51**, 207–217.

Leadbeater, B.S.C. (1972a). Fine-structural observations on some marine choanoflagellates from the coast of Norway. *Journal of the Marine Biological Association*, UK, **52**, 67–79.

Leadbeater, B.S.C. (1972b). *Paraphysomonas cylicophora* sp. nov., a marine species from the coast of Norway. *Norwegian Journal of Botany*, **19**, 179–185.

Leadbeater, B.S.C. (1983). Life history and ultrastructure of a new marine species of *Proterospongia* (Choanoflagellida). *Journal of the Marine Biological Association*. UK, **63**, 135–160.

Leadbeater, B.S.C. (1985). Choanoflagellida Kent, 1880. In *An Illustrated Guide to the Protozoa*, J.J. Lee, S.H. Hutner and E.C. Bovee, (eds) 106–116, Society of Protozoologists, Lawrence, Kansas.

Lee, J.J. (1980). Nutrition and physiology of the Foraminifera. In *Biochemistry and Physiology of Protozoa*, M. Levandowsky and S.H. Hutner, (eds) **3**, 43–66, Academic Press, New York.

Lee, J.J. (1985). Ebriida Poche. In *Illustrated Guide to the Protozoa*, J.J. Lee, S.H. Hutner and E.C. Bovee, (eds) 140–141, Society of Protozoologists, Lawrence, Kansas.

Lee, J.J. and Corliss, J.O. (1985). A Symposium on Symbiosis in Protozoa. *Journal of Protozoology*, **32**, 371–403.

Lee, J.J., Hutner, S.H. and Bovee, E.C. (1985). *An Illustrated Guide to the Protozoa*, Society of Protozoologists, Lawrence, Kansas.

Lee, J.J., Soldo, A.T., Reisser, W., Lee, M.J., Jeon, K.W. and Gortz, H.-D. (1985). The extent of algal and bacterial endosymbioses in Protozoa, *Journal of Protozoology*, **32**, 391–403.

Leedale, G.F. (1967a). *Euglenoid Flagellates*. Prentice-Hall, Englewood Cliffs, New Jersey.

Leedale, G.F. (1971). *The Euglenoids*. Oxford University Press, Oxford.

Leedale, G.F. (1978). Phylogenetic criteria in euglenoid flagellates. *BioSystems*, **10**, 183–187.

Leedale, G.F. (1985). Euglenida Butschli, 1884. In *An Illustrated Guide to the Protozoa*, J.J. Lee, S.H. Hutner and E.C. Bovee, (eds) 41–54, Society of Protozoologists, Lawrence, Kansas.

Legner, M. (1977). A simple method for the preservation of quantitative samples of free-living ciliates. *Journal of Protozoology*, **24**, 52A.

Levine, N.D. (1978). *Perkinsus* gen. n. and other new taxa in the protozoan phylum Apicomplexa. *Journal of Parasitology*, **64**, 549.

Levine, N.D. (1985). Phylum Apicomplexa Levine, 1970. In *An Illustrated Guide to the Protozoa*, J.J. Lee, S.H. Hutner and E.C. Bovee, (eds) 322–374, Society of Protozoologists, Lawrence, Kansas.

Levine, N.D., Corliss, J.O., Cox, F.E.G., Deroux, G., Grain, J., Honigberg, B.M., Leedale, G.F., Loeblich, A.R., Lom, J., Lynn, D., Merinfeld, E.G., Page, F.C. Poljansky, G. Sprague, V., Vavra, J. and Wallace, F.G. (1980). A newly revised classification of the Protozoa. *Journal of Protozoology*, **27**, 37–59.

Li, G.-W. and Volcani, B.E. (1987). Four new apochlorotic diatoms, *British Phycological Journal*, **22**, 375–382.

Lighthill, J. (1976). Flagellar hydrodynamics. *SIAM Review*, **18**, 161–230.

Lipscomb, D.L. and Corliss, J.O. (1982). *Stephanopogon*, a phylogenetically important 'ciliate' shown by ultrastructural studies to be a flagellate. *Science*, **215**, 303–304.

Lindholm, T. (1985). *Mesodinium rubrum* – a unique photosynthetic ciliate. *Advances in Aquatic Microbiology*, **3**, 1–48.

Lloyd, D., Hillman, K., Yarlett, N. and Williams. A.G. (1988). Hydrogen production by rumen holotrich protozoa: effects of oxygen and implications for metabolic control by *in situ* conditions. *Journal of Protozoology*, in Press.

Lloyd, F.E. (1926). Some behaviors of *Vampyrella lateritia* and the response of *Spirogyra* to its attack. *Papers of the Michigan Academy of Science, Arts and Letters*, **7**, 395–416.

Loeblich, A.R. (1984). Dinoflagellate evolution. In *Dinoflagellates*, D.L. Spector, (ed.) 482–522, Academic Press, New York.

Lom, J. and Corliss, J.O. (1971). Morphogenesis and cortical ultrastructure of *Brooklynella hostilis*, a Dysteriid ciliate ectoparasitic on marine fishes. *Journal of Protozoology*, **18**, 261–281.

Lom, J. and Dykova, I. (1985). *Hoferellus cyprini* Doflein, 1898 from carp kidney: a well established myxosporean species or a sequence in the developmental cycle of *Sphaerospora renicola* Dykova and Lom, 1982? *Protistologica*, **21**, 195–206.

Lom, J., Dykova, I. and Lhotakova, S. (1982). Fine structure of *Sphaerospora renicola* Dykova and Lom, 1982, a myxosporean from carp kidney and comments on the origin of pansporoblasts. *Protistologica*, **18**, 489–502.

Lopez, E. (1979). Algal chloroplasts in the protoplasm of three species of benthic Foraminifera: taxonomic affinity, viability and persistence. *Marine Biology*, **53**, 201–211.

Loubès, C. (1979). Recherches sur la méiose chez les microsporidies: consequences sur la cycle biologique. *Journal of Protozoology*, **26**, 200–208.

Luck, D.J.L. (1984). Genetic and biochemical dissection of the eucaryotic flagellum. *Journal of Cell Biology*, **98**, 789–794.

Luporini, P. and Miceli, C. (1986). Mating pheromones. In *The Molecular Bilogy of Ciliated Protozoa*, J.G. Gall, (ed.) 263–288, Academic Press, New York.

Lwoff, A. (1950). *Problems of Morphogenesis in Ciliates*. Wiley, New York.

Lynn, D.H. (1981). The organisation and evolution of microtubular organelles in ciliated protozoa. *Biological Reviews*, **56**, 243–292.

Lynn, D.H. (1982). Dimensionality and contractile vacuole function in ciliated protozoa. *Journal of Experimental Zoology*, **233**, 219–229.

Lynn, D.H. and Small, E.G. (1981). Protist kinetids: structural conservatism, kinetid structure and ancestral states. *BioSystems*, **14**, 377–385.

Machemer, H. (1972a). Properties of polarised ciliary beat in *Paramecium. Acta Protozoologica*, **11**, 295–300.

Machemer, H. (1972b). Ciliary activity and the origin of metachrony in *Paramecium*: effects of increased viscosity. *Journal of Experimental Biology*, **57**, 239–259.

Machemer, H. (1974). Ciliary activity and metachronism in Protozoa. In *Cilia and Flagella*, M.A. Sleigh, (ed.) 199–286, Academic Press, London.

Machemer, H. (1976). Interactions of membrane potential and cations in regulation of ciliary activity in *Paramecium. Journal of Experimental Biology*, **65**, 427–448.

Machemer, H. (1977). Motor activity and bioelectric control of cilia. *Fortschritte der Zoologie*, **24**, 195–210.

Machemer, H. (1986). Electromotor coupling in cilia. *Fortschritte der Zoologie*, **33**, 205–250.

Machemer, H. (1987). From structure to behaviour. *Stylonychia* as a model system for cellular physiology. *Progress in Protistology*, **2**, 213–330.

Machemer, H. and De Peyer, J. (1977). Swimming sensory cells: electrical membrane parameters, receptor properties and motor control in ciliated Protozoa. *Verhandlungen der Deutschen Zoologischen Gesellschaft*, 1977, 86–100.

Mackinnon, D.L. and Hawes, R.S.J. (1961). *An Introduction to the Study of Protozoa*. University Press, Oxford.

Maly, E.J. (1978). Stability of the interaction between *Didinium* and *Paramecium*: effects of dispersal and predator time lag. *Ecology*, **59**, 733–741.

Manton, I. and Leedale, G.F. (1961). Observations on the fine structure of *Paraphysomonas vestita*, with special reference to the Golgi apparatus and the origin of scales. *Phycologia*, l, 37–57.

Manton, I. and Leedale, G.F. (1969). Observations on the microanatomy of *Coccolithus pelagicus* and *Cricosphaera carterae*, with special reference to the origin and nature
of the coccoliths and scales. *Journal of the Marine Biological Association*, UK, **49**, 1–16.

Margulis, L. (1981). *Symbiosis in Cell Evolution*. Freeman, San Francisco.

Margulis, L. and Sagan, D. (1985). Order amidst animalcules: the protoctista kingdom and its undulipodiated cells. *BioSystems*, **18**, 141–147.

Markus, M.B., Killick–Kendrick, R. and Garnham, P.C.C. (1974). The coccidial nature and life-cycle of *Sarcocystis*. *Journal of Tropical Medicine and Hygiene*, **77**, 248–259.

Maruyama, T. (1982). Fine structure of the longitudinal flagellum in *Ceratium tripos*, a marine dinoflagellate. *Journal of Cell Science*, **58**, 109–123.

Mast, S.O. (1926). Structure, movement and locomotion and stimulation in *Amoeba*. *Journal of Morphology*, **41**, 347–425.

Mato, J.M. and Konijn, T.M. (1979). Chemosensory transduction in *Dictyostelium discoideum*. In *Biochemistry and Physiology of Protozoa*, Second edition, M. Levandowsky and S.H. Hutner, (eds) **2**, 181–219, Academic Press, New York.

Mauel, J. and Behin, R. (1981). Immunology of Leishmaniasis. In *Biochemistry and Physiology of Protozoa*, Second edition, M. Levandowsky and S.H. Hutner, (eds) **4**, 385–429. Academic Press, New York.

McAuley, P.J. and Smith, D.C. (1982). The green hydra symbiosis. V. Stages in the intracellular recognition of algal symbionts by digestive cells. *Proceedings of the Royal Society of London*, series B, **216**, 7–23.

McGhee, B. and Cosgrove, W.B. (1980). Biology and physiology of the lower Trypanosomatidae. *Microbiological Reviews*, **44**, 140–173.

McGhee, R.B. and McGhee, A. (1979). Biology and structure of *Phytomonas staheli* sp. n., a trypanosomatid located in sieve tubes of coconut and oil palms. *Journal of Protozoology*, **26**, 348–351.

Medlin, L.K., Crawford, R.M. and Andersen, R.A. (1986). Histochemical and ultrastructural evidence for the function of the labiate process in the movement of centric diatoms. *British Phycological Journal*, **21**, 297–301.

Melkonian, M. (1977). The flagellar root system of zoospores of the green alga *Chlorosarcinopsis* (Chlorosarcinales) as compared with *Chlamydomonas* (Volvocales). *Plant Systematics and Evolution*, **128**, 79–88.

Melkonian, M. (1984). Flagellar apparatus ultrastructure in relation to green algal classification. In *Systematics of the Green Algae*, D.E.G. Irvine and D.M. John, (eds) 73–120, Academic Press, London.

Melkonian, M. and Robenek, H. (1984). The eyespot apparatus of flagellated green algae: a critical review. *Progress in Phycological Research*, **3**, 193–268.

Mignot, J.-P. (1963). Quelques particularités de l'ultrastructure d'*Entosiphon sulcatum*. *Compte rendu hebdomadaire des Séances de L'Académie des Sciences*, **257**, 2530–2533.

Mignot, J.-P. (1976). Compléments a l'étude des Chloromonadines: ultrastructure de

Chattonella subsala, Biecheler, flagellé d'eau saumatre. *Protistologica*, **12**, 279–293.

Mignot, J.-P. (1985). Les coenobes chez les Volvocales: un example du passage des unicellulaires aux pluricellulaires. *Année Biologique*, **24**, 1–26.

Mignot, J.-P. and Brugerolle, G. (1975a). Étude ultrastructurale de *Cercomonas* Dujardin (*Cercobodo* Krassilstchick), protiste flagellé. *Protistologica*, **11**, 547–554.

Mignot, J.-P. and Brugerolle, G. (1975b). Étude ultrastructurale du flagellé phago-trophe *Colponema loxodes* Stein. *Protistologica*, **11**, 429–444.

Miyake, A. (1981). Cell interaction by gamones in *Blepharisma*. In *Sexual Interactions in Eukaryote Microbes*, D.H. O'Day and P.A. Horgen, (eds) 95–129, Academic Press, New York.

Moestrup, O. (1982). Flagellar structure in algae: a review, with new observations parti-cularly on the Chrysophyceae, Phaeophyceae (Fucophyceae), Euglenophyceae and *Reckertia*. *Phycologia*, **21**, 427–528.

Moestrup, O. and Ettl, H. (1979). A light and electron microscopical study of the flagel-late *Nephroselmis olivacea* (Prasinophyceae). *Opera Botanica a Societate Botanica Lundensi*, **49**, 1–39.

Moestrup, O. and Thomsen, H.A. (1976). Fine structural studies on the flagellate genus *Bicoeca*. I. *Bicoeca maris* with particular emphasis on the flagellar apparatus. *Pro-tistologica*, **12**, 101–120.

Molyneux, D. and Ashford, R.W. (1983). *The Biology of Trypanosoma and Leish-mania, Parasites of Man and Domestic Animals*. Taylor and Francis, London.

Morris, H.R., Taylor, G.W., Masento, M.S., Jermyn, K.A. and Kay, R.R. (1987). Chemical structure of the morphogen differentiation inducing factor from *Dictyo-stelium discoideum*. *Nature*, **328**, 811–814.

Moss, S.T. (1985). An ultrastructural study of taxonomically significant characters of the Thraustochytriales and the Labyrinthulales. *Botanical Journal of the Linnean Society*, **91**, 329–357.

Muller, M. (1975). Biochemistry of protozoan microbodies (peroxysomes, α-glycero-phosphate bodies and hydrogenosomes). *Annual Review of Microbiology*, **29**, 467–483.

Muller, M. (1980). The hydrogenosome. *Symposia of the Society of General Microbio-logy*, **30**, 127–142.

Muller, M. (1985). Search for cell organelles in Protozoa. *Journal of Protozoology*, **32**, 559–563.

Munch, R. (1970). Endocytosenachweis bei *Opalinen*. *Cytobiologie*, **2**, 108–122.

Murray, J.W. (1971). *An Atlas of British Recent Foraminiferids*. Heinemann, London.

Naitoh, Y. and Eckert, R. (1969). Ciliary orientation: controlled by cell membrane or by intracellular fibrils? *Science*, **166**, 1633–1635.

Naitoh, Y. and Eckert, R. (1974). The control of ciliary activity in Protozoa. In *Cilia and Flagella*, M.A. Sleigh, (ed.) 305–352, Academic Press, London.

Naitoh, Y. and Kaneko, H. (1973). Control of ciliary activities by adenosinetriphosphate and divalent cations in triton-extracted models of *Paramecium caudatum*. *Journal of Experimental Biology*, **58**, 657–676.

Neff, R.J. and Neff, R.H. (1969). The biochemistry of amoebic encystment. *Symposia of the Society for Experimental Biology*, **23**, 51–81.

Netzel, H. (1971). Die schalenbildung bei den thekamoben gattung *Arcella* (Rhizopoda, Testacea). *Cytobiologie*, **3**, 89–92.

Netzel, H. and Durr, G. (1984). Dinoflagellate cell cortex. In *Dinoflagellates*, D.L. Spector, (ed.) 43–105, Academic Press, New York.

Ng, S.F. (1986). The somatic function of the micronucleus of ciliated protozoa. *Progress in Protistology*, **1**, 215–286.

Nikoljuk, V.F. (1969). Some aspects of the study of soil protozoa. *Acta Protozoologica*, **7**, 99–109.

Nilsson, J.R. (1979). Phagotrophy in *Tetrahymena*. In *Biochemistry and Physiology of Protozoa*, Second edition, M. Levandowsky and S.H. Hutner, (eds) **2**, 339–379, Academic Press, New York.

Nilsson, J.R. and Van Deurs, B. (1983). Coated pits and pinocytosis in *Tetrahymena*. *Journal of Cell Science*, **63**, 209–222.

Nisbet, B. (1974). An ultrastructural study of the feeding apparatus of *Peranema trichophorum*. *Journal of Protozoology*, **21**, 39–48.

Noirot-Timothee, C. (1959). Recherches sur l'ultrastructure d'*Opalina ranarum*. *Annales des Sciences Naturelles, Zoologie*, 12 serie, **1**, 265–281.

Norris, R.E. (1980). Prasinophytes. In *Phytoflagellates*, E.R. Cox, (ed.) 85–145, Elsevier North Holland, New York.

Nozawa, Y. and Thompson, G.A. (1979). Lipids and membrane organisation in *Tetrahymena*. In *Biochemistry and Physiology of Protozoa*, Second edition, M. Levandowsky and S.H. Hutner, (eds) **2**, 275–338, Academic Press, New York.

O'Day, D.H. and Lewis, K.E. (1981). Pheromonal interactions during mating in *Dictyostelium*. In *Sexual Interactions in Eukaryotic Microbes*, D.H. O'Day and P.A. Horgen, (eds) 199–221, Academic Press, New York.

Ogden, C.G. (1979). An ultrastructural study of division in *Euglypha* (Protozoa: Rhizopoda). *Protistologica*, **15**, 541–556.

Ogura, A. and Machemer, H. (1980). Distribution of mechanoreceptor channels in the *Paramecium* surface membrane. *Journal of Comparative Physiology, A,* **135**, 233–242.

Okagaki, T. and Kamiya, R. (1986). Microtubule sliding in mutant *Chlamydomonas* axonemes devoid of outer or inner dynein arms. *Journal of Cell Biology*, **103**, 1895–1902.

O'Kelly, C.J. and Floyd, G.L. (1984). The absolute configuration of the flagellar apparatus in zoospores from two species of Laminariales (Phaeophyceae). *Protoplasma*, **123**, 18–25.

Omoto, C.K. and Kung, C. (1980). Rotation and twist of the central pair microtubules in the cilia of *Paramecium*. *Journal of Cell Biology*, **87**, 33–46.

Omoto, C.K. and Witman, G.B. (1981). Functionally significant central-pair rotation in a primitive eukaryotic flagellum. *Nature*, **290**, 708–710.

Opperdoes, F.R. and Borst, P. (1977). Localisation of nine glycolytic enzymes in a microbody-like organelle in *Trypanosoma brucei*: the glycosome. *F.E.B.S. Letters*, **80**, 360–364.

Orias, E. and Rasmussen, L. (1977). Dual capacity for nutrient uptake in *Tetrahymena*. II. Role of the two systems in vitamin uptake. *Journal of Protozoology*, **24**, 507–511.

Page, F.C. (1976). *An Illustrated Key to Freshwater and Soil Amoebae*. F.B.A., Ambleside, UK.

Page, F.C. (1981). *The Culture and Use of Free-living Protozoa in Teaching*. Institute of Terrestrial Ecology, Cambridge, UK.

Page, F.C. (1987). The classification of 'naked' amoebae (Phylum Rhizopoda). *Archiv für Protistenkunde*, **133**, 199–217.

Page, F.C. and Blanton, R.L. (1985). The Heterolobosea (Sarcodina: Rhizopoda), a new class uniting the Schizopyrenida and the Acrasidae (Acrasida). *Protistologica*, **21**, 121–132.

Parducz, B. (1967). Ciliary movement and coordination in ciliates. *International Review of Cytology*, **21**, 91–128.

Parke, M., Manton, I. and Clarke, B . (1955). Studies on marine flagellates. II. Three new species of *Chrysochromulina*. *Journal of the Marine Biological Association*, UK, **34**, 579–609.

Patterson, D.J. (1977). On the behaviour of contractile vacuoles and associated structures of *Paramecium caudatum* (Ehrbg). *Protistologica*, **13**, 205–212.

Patterson, D.J. (1978). Membranous sacs associated with cilia of *Paramecium*. *Cytobiologie*, **17**, 107–113.

Patterson, D.J. (1979). On the organization and classification of the protozoon, *Actinophrys sol* Ehrenberg 1830. *Microbios*, **26**, 165–208.

Patterson, D.J. (1980). Contractile vacuoles and associated structures: their organization and function. *Biological Reviews*, **55**, 1–46.

Patterson D.J. (1983). On the organization of the naked filose amoeba, *Nuclearia moebiusi* Frenzel, 1897 (Sarcodina, Filosea) and its implications. *Journal of Protozoology*, **30**, 301–307.

Patterson, D.J. (1985a). The fine structure of *Opalina ranarum* (Family Opalinidae): Opalinid phylogeny and classification. *Protistologica*, **21**, 413–428.

Patterson, D.J. (1985b). On the organization and affinities of the amoeba, *Pompholyxophrys punicea* Archer, based on ultrastructural examination of individual cells from wild material. *Journal of Protozoology*, **32**, 241–246.

Patterson, D.J. and Fenchel, T. (1985). Insights into the evolution of heliozoa (Protozoa, Sarcodina) as provided by ultrastructural studies on a new species of flagellate from the genus *Pteridomonas*. *Biological Journal of the Linnean Society*, **34**, 381–403.

Patterson, D.J. and Hausmann, K. (1981). Feeding by *Actinophrys sol* (Protista, Heliozoa): I. Light microscopy. *Microbios*, **31**, 39–55.

Patterson, D.J., Larsen, J. and Corliss, J.O. (1988). The ecology of heterotrophic flagellates and ciliates living in marine sediments. *Progress in Protistology*, **3**, In Press.

Patterson, D.J. and Sleigh, M.A. (1976). Behaviour of the contractile vacuole of *Tetrahymena pyriformis* W: A redescription with comments on the terminology. *Journal of Protozoology*, **23**, 410–417.

Patterson, M., Woodworth, F., Marciano–Cabral, F. and Bradley, S.G. (1981). Ultrastructure of *Naegleria fowleri* enflagellation. *Journal of Bacteriology*, **147**, 217–226.

Peck, R.K. (1977). Cortical ultrastructure of the scuticociliates *Dexiotricha media* and *Dexiotricha colpidiopsis* (Hymenostomatida). *Journal of Protozoology*, **24**, 122–134.

Perkins, F.O. (1973). Observations of thraustochytriaceous (Phycomycetes) and labyrinthulid (Rhizopodea) ectoplasmic nets on natural and artificial substrates – an electron microscope study. *Canadian Journal of Botany*, **51**, 485–491.

Perkins, F.O. (1976). Zoospores of the oyster pathogen *Dermocystidium marinum* . I. Fine structure of the conoid and other sporozoan-like organelles. *Journal of Parasitology*, **62**, 959–974.

Perkins, F.O. and Amon, J.P. (1969). Zoosporulation in *Labyrinthula* sp.; an electron microscopic study. *Journal of Protozoology*, **16**, 235–257.

Peters, W. (1970). *Chemotherapy and Drug Resistance in Malaria*. Academic Press, London.

Pfiester, L.A. (1984). Sexual reproduction. In *Dinoflagellates,* D.L. Spector, (ed.) 181–199, Academic Press, New York.

Piccinni, E. and Omodeo, P. (1975). Photoreceptors and phototactic programs in Protista. *Bollettino di Zoologia, Pubblicato dall'Unione Zoologica Italiana*, **42**, 57–79.

Picken, L.E.R. (1937). The structure of some protozoan communities. *Journal of Ecology*, **25**, 368–384.

Pickett-Heaps, J. (1982). *New light on the green algae*. Carolina Biology Supply Company, Burlington N. Carolina.

Pienaar, R.N. (1980). Chrysophytes. In *Phytoflagellates*, E.R. Cox, (ed.) 213–242, Elsevier North Holland, New York.

Pitelka, D.R. (1963). *The electron microscopic structure of Protozoa*. Pergamon Press, Oxford.

Pitelka, D.R. (1965). New observations on cortical ultrastructure in *Paramecium*. *Journal de Microscopie*, **4**, 373–394.

Pitelka, D.R. (1969). Fibrillar systems in Protozoa. In *Research in Protozoology*, T.-T. Chen, (ed.) **3**, 280–388, Pergamon Press, Oxford.

Poisson, R. (1953). Sous-embranchement des Cnidosporidies. In *Traité de Zoologie*, P.-P. Grassé, (ed.) **1**, ii, 1006–1088, Masson, Paris.

Pokorny, K.S. (1985). Phylum Labyrinthomorpha. In *An Illustrated Guide to the Protozoa*, J.J. Lee, S.H. Hutner and E.C. Bovee, (eds) 318–321, Society of Protozoologists, Lawrence, Kansas.

Pollard, T.D., Aebi, V., Cooper, J.A., Fowler, W.E., Kiehart, D.P., Smith, P.R. and Tseng, P.C. (1982). Actin and myosin function in *Acanthamoeba*. *Philosophical Transactions of the Royal Society of London*, Series B, **229**, 237–245.

Pomeroy, L.R. (1980). Detritus and its role as a food source. In *Fundamentals of Aquatic Ecosystems*, R.S.K. Barnes and K.H. Mann, (eds) 84–102, Blackwell, Oxford.

Pommerville, J. (1978). Analysis of gamete and zygote motility in *Allomyces*. *Experimental Cell Research*, **113**, 161–172.

Preer, L.B. (1981). Prokaryotic symbionts of Paramecium. In *The Prokaryotes – A Handbook of Habitats, Isolation and Identification of Bacteria*, M.P. Starr, H. Stolp. H.G. Truger, A. Balows and H.G. Schlegel, (eds) **2**, 2127–2136, Springer-Verlag. Berlin.

Prescott, D.M. (1970). *How to Know the Freshwater Algae*. Brown, Dubuque, Iowa.

Prescott, D.M. and Stone, G.E. (1967). Replication and function of the protozoan nucleus. In *Research in Protozoology*, T.-T. Chen, (ed.) **2**, 117–146, Pergamon Press, Oxford.

Preston, T.M. (1985). A prominent microtubule cytoskeleton in *Acanthamoeba*. *Cell Biology International Reports*, **9**, 307–314.

Puytorac, P. De (1954). Contribution à l'étude cytologique et taxonomique des infusoires astomes. *Annals des Sciences Naturelles (Zoologie), 11 serie,* **16**, 85–270.

Puytorac, P. De Batisse, A., Bohatier, J., Corliss, J.O. Deroux, G., Didier, P., Dragesco, J., Fryd–Versavel, G., Grain, J., Grolière, C.-A., Hovasse, R., Iftode, F., Laval, M., Roque, M., Savoie, A. and Tuffrau, M. (1974). Proposition d'une classification du phylum Ciliophora Doflein, 1901 *Compte Rendu Hebdomadaire des Séances de L'Académie des Sciences*, Paris, **278**, 2799–2802.

Qu, L.-H., Perasso, B., Baroin, A., Brugerolle, G., Bachellerie, J.-P. and Adoutte, A. (1988). Molecular evolution of the 5–terminal domain of large-subunit rRNA from lower eukaryotes. A broad phylogeny covering photosynthetic and non-photosynthetic protists. *BioSystems*, **21**, 203–208.

Ragan, M.A. (1988). Ribosomal RNA and the major lines of evolution: a perspective. *BioSystems*, **21**, 177–187.

Ragan, M.A. and Chapman, D.J. (1978). *A Biochemical Phylogeny of the Protists*. Academic Press, New York.

Raikov, I.B. (1969). The macronucleus of ciliates. In *Research in Protozoology*, T.-T. Chen, (ed.) **3**, 1–128, Pergamon Press, Oxford.

Raikov, I.B. (1978). Ultrastructure du cytoplasme et des nématocystes du cilié **Remanella multinucleata** Kahl (Gymnostomata, Loxodidae). Existence de nématocystes chez les ciliés. *Protistologica*, **14**, 413–432.

Raikov, I.B. (1982). *The Protozoan Nucleus, Morphology and Evolution*. Springer-Verlag, Vienna.

Redfield, A.C. (1958). The biological control of chemical factors in the environment. *American Scientist*, **46**, 205–221.

Rees, A.J.J. and Leedale, G.F. (1980). The dinoflagellate transverse flagellum: three-dimensional reconstructions from serial sections. *Journal of Phycology*, **16**, 73–80.

Riddick, D.H. (1968). Contractile vacuole in the amoeba *Pelomyxa carolinensis*. *American Journal of Physiology*, **215**, 736–740.

Rikmenspoel, R. and Sleigh, M.A. (1970). Bending moments and elastic constants in cilia. *Journal of Theoretical Biology*, **28**, 81–100.

Robenek, H. and Melkonian, M. (1979). Rhizoplast-membrane associations in the flagellate *Tetraselmis cordiformis* Stein (Chlorophyceae) revealed by freeze etching and thin sections. *Archiv für Protistenkunde*, **122**, 340–351.

Roberts, A.M. (1981). Hydrodynamics of protozoan swimming. In *Biochemistry and Physiology of Protozoa*, Second edition, M. Levandowsky and S.H. Hutner, (eds) **4**, 6–66, Academic Press, New York.

Roberts, K. and Hyams, J.S. (1979). *Microtubules*. Academic Press, London.

Roberts, K.R. (1985). The flagellar apparatus of *Oxyrrhis marina* (Pyrrophyta). *Journal of Phycology*, **21**, 641–655.

Roberts, K.R., Stewart, K.D. and Mattox, K.R. (1981). The flagellar apparatus of *Chilomonas paramecium* (Cryptophyceae) and its comparison with certain zooflagellates. *Journal of Phycology*, **17**, 159–167.

Rogerson, A. (1981). The ecological energetics of *Amoeba proteus* (Protozoa). *Hydrobiologia*, **85**, 117–128.

Rook, J.A.F. and Thomas, P.C. (1983). *Nutritional Physiology of Farm Animals*. Longman, London.

Round, F.E. (1981). *The Ecology of Algae,* University Press, Cambridge.

Rudzinska, M.A. (1965). The fine structure and function of the tentacle in *Tokophrya infusionum*. *Journal of Cell Biology,* **25**, 459–477.

Rudzinska, M., Spielman, A., Lewengrub, S., Piesman, J. and Karakashian, S. (1984). The sequence of developmental events of *Babesia microti* in the gut of *Ixodes dammini*. Protistologica, **20**, 649–663.

Sale, W.S. and Satir, P. (1977). Direction of active sliding of microtubules in *Tetrahymena* cilia. *Proceedings of the National Academy of Sciences of the United States of America*, **74**, 2045–2049.

Salisbury, J.L. (1983). Contractile flagellar roots: the role of calcium. *Journal of Submicroscopic Cytology*, **15**, 105–110.

Sandon, H. (1972). *The Composition and Distribution of the Protozoan Fauna of the Soil*. Oliver and Boyd, Edinburgh.

Santore, U.J. and Leedale, G.F. (1985). Cryptomonadida Senn 1900. In *An Illustrated Guide to the Protozoa*, J.J. Lee, S.H. Hutner and E.C. Bovee, (eds) 19–22, Society of Protozoologists, Lawrence, Kansas.

Satir, B., Schooley, C. and Satir, P. (1973). Membrane fusion in a model system. Mucocyst secretion in *Tetrahymena*. *Journal of Cell Biology*, **56**, 153–176.

Schindler, D.W. (1977). Evolution of phosphorus limitation in lakes. *Science*, **195**, 260–262.

Schmidt-Nielsen, B. and Schrauger, C.R. (1963). *Amoeba proteus*: studying the contractile vacuole by micropuncture. *Science*, **139**, 606–607.

Schnepf, E., Deichgraber, G., Roderer, G. and Herth, W. (1977). The flagellar root apparatus, the microtubular system and associated organelles in the chrysophycean flagellate *Poteriochromonas malhamensis* Peterfi (syn. *Poteriochromonas stipitata* Scherffel and *Ochromonas malhamensis* Pringsheim). *Protoplasma*, **92**, 87–107.

Seravin, L.N. and Gerassimova, Z.P. (1978). A new macrosystem of ciliates. *Acta Protozoologica*, **17**, 399–418.

Seravin, L.N. and Goodkov, A. (1984). The main types and forms of agamic cell fusion in protozoa. *Cytologia*, **26**, 123–132.

Sherr, B.F., Sherr, E.B. and Berman, T. (1983). Grazing, growth and ammonium excretion rates of a heterotrophic microflagellate fed with four species of bacteria. *Applied Environmental Microbiology*, **45**, 1196–1201.

Shigenaka, Y. (1976). Microtubules in protozoan cells. II. Heavy metal ion effects on degradation and stabilization of the heliozoan microtubules. *Annotationes Zoological Japonenses*, **49**, 164–176.

Shilo, M. (1967). Formation and mode of action of algal toxins. *Bacteriological Reviews*, **31**, 180–193.

Sieburth, J.McN. (1987). Contrary habitats for redox-specific processes: methano genesis in oxic waters and oxidation in anoxic waters. In *Microbes in the Sea*, M.A. Sleigh, (ed.) 11–38, Ellis Horwood, Chichester, UK.

Siegel, R.W. (1967). Genetics of ageing and the life cycle in ciliates. *Symposia of the Society for Experimental Biology,* **21**, 127–148.

Sigee, D.C. (1986). The dinoflagellate chromosome. *Advances in Botanical Research,* **12**, 205–264.

Silvester, N.R. and Sleigh, M.A. (1985). The forces on microorganisms at surfaces in flowing water. *Freshwater Biology,* **15**, 433–448.

Sims, R.W. (1980). *Animal Identification. A Reference Guide.* 2 vols. Wiley, Chichester, UK.

Singer, S.J. and Nicholson, G.L. (1972). The fluid mosaic model of the structure of cell membranes. *Science,* **175**, 720–731.

Sleigh, M.A. (1966). Striated and microtubular kinetodesmata in ciliates. *Journal of Protozoology,* **13**, Supplement 29.

Sleigh, M.A. (1968). Patterns of ciliary beating. *Symposia of the Society for Experimental Biology,* **22**, 131–150.

Sleigh, M.A. (1974). *Cilia and Flagella.* Academic Press, London.

Sleigh, M.A. (1979). Radiation of the eukaryote protista. In *The Origin of Major Invertebrate Groups*, M.R. House, (ed.) 23–54, Academic Press, London.

Sleigh, M.A. (1981). Flagellar beat patterns and their possible evolution. *BioSystems,* **14**, 423–431.

Sleigh, M.A. (1984a). The integrated activity of cilia: function and coordination. *Journal of Protozoology,* **31**, 16–21.

Sleigh, M.A. (1984b). Motile systems of ciliates, *Protistologica,* **20**, 299–305.

Sleigh, M.A. (1986a). Evolution of cells. In *Cell Biology*, B. King, (ed.) 30–68, Allen and Unwin, London.

Sleigh, M.A. (1986b). The origin of flagella – autogenous or symbiontic? *Cell Motility and the Cytoskeleton,* **6**, 96–98.

Sleigh, M.A. (1988). Flagellar root maps allow speculative comparisons of root patterns and of their ontogeny. *BioSystems,* **21**, 277–282.

Sleigh, M.A. and Barlow, D. (1976). Collection of food by **Vorticella**. *Transactions of the American Microscopical Society,* **95**, 482–486.

Sleigh, M.A. and Barlow, D.I. (1982a). The use of cilia by protozoa. Progress in Protozoology, *Proceedings of the VI International Congress of Protozoology*, Special Congress Volume *Acta Protozoologica,* **I**, 141–147.

Sleigh, M.A. and Barlow, D.I. (1982b). How are different ciliary beat patterns produced? *Symposia of the Society for Experimental Biology,* **35**, 139–157.

Sleigh, M.A., Dodge, J.D. and Patterson, D.J. (1984), Kingdom Protista. In *A Synoptic Classification of Living Organisms*, R.S.K. Barnes, (ed.) 25–88, Blackwell, Oxford.

Sleigh, M.A. and Holwill, M.E.J. (1969). Energetics of ciliary movement in *Sabellaria* and *Mytilus*. Journal of Experimental Biology,. **50**,. 733–743.

Sleigh, M.A. and Pitelka, D.R. (1974). Processes of contractility in protozoa. *Actualités Protozoologiques,* **1**, 293–306, Universite de Clermont.

Sleigh, M.A. and Silvester, N.R. (1983). Anchorage functions of the basal apparatus of cilia. *Journal of Submicroscopic Cytology,* **15**, 101–104.

Sluiman, H.J. (1983). The flagellar apparatus of the zoospore of the filamentous green alga **Coleochaete pulvinata**: absolute configuration and phylogenetic significance. *Protoplasma,* **115**, 160–175.

Sluiman, H.J., Roberts, K.R., Stewart, K.D. and Mattox, K.R. (1980). Comparative cytology and taxonomy of the Ulvaphyceae. I. The zoospore of *Ulothrix zonata* (Chlorophyta). *Journal of Phycology,* **16**, 537–545.

Small, E.G. and Lynn, D.H. (1981). A new macrosystem for the phylum Ciliophora Doflein, 1901. *BioSystems,* **14**, 387–401.

Small, E.G. and Lynn, D.H. (1985). Phylum Ciliophora Doflein, 1901. In *An Illustrated*

Guide to the Protozoa, J.J Lee, S.H. Hutner and E.C. Bovee, (eds) 393–575, Society of Protozoologists, Lawrence, Kansas.

Small, E.G. and Marszalek, D.S. (1969). Scanning electron microscopy of fixed, frozen and dried protozoa. *Science*, **163**, 1064–1065.

Smith, D.C. (1981). The role of nutrient exchange in recognition between symbionts. *Berichte der Deutschen Botanischen Gesellschaft*, **94**, Supplement 517–528.

Smith, D.C. and Douglas, A.E. (1987). *The Biology of Symbiosis*. Edward Arnold, London.

Smith, R.McK. and Patterson, D.J. (1986). Analyses of heliozoan interrelationships: an example of the potentials and limitations of ultrastructural approaches to the study of protistan phylogeny. *Proceedings of the Royal Society of London*, Series B, **227**, 325–366.

Sogin, M.L., Elwood, H.J. and Gunderson, J.H. (1986). Evolutionary diversity of eukaryotic small-subunit rRNA genes. *Proceedings of the National Academy of Sciences of the United States of America*, **83**, 1383–1387.

Soldo, A.T. (1974). Intracellular particles in *Paramecium*. In *Paramecium, A Current Survey*, W.J. Van Wagtendonk, (ed.) 377–432, Elsevier, Amsterdam.

Soldo, A.T. (1983). The biology of the xenosome, an intracellular symbiont. *International Review of Cytology*, Supplement **14**, 79–109.

Solomon, J.A. Walne, P.A. and Kivic, P. (1987). *Entosiphon sulcatum* (Euglenophyceae): flagellar roots of the basal body complex and reservoir region. *Journal of Phycology*, **23**, 85–98.

Song, P.-S. and Walker, E.B. (1981). Molecular aspects of photoreceptors in protozoa and other microorganisms. In *Biochemistry and Physiology of Protozoa*, M. Levandowsky and S.H. Hutner, (eds) Second edition, **4**, 199–233, Academic Press, New York.

Sonneborn, T.M. (1957). Breeding systems, reproductive methods and species problems in protozoa. In *The Species Problem*, E. Mayr, (ed.) 155–325, American Association for the Advancement of Science, Washington.

Sonneborn, T.M. (1975). The *Paramecium aurelia* complex of fourteen sibling species. *Transactions of the American Microscopical Society*, **94**, 155–178.

Sorokin, Y.I. (1977). The heterotrophic phase of plankton succession in the Japan Sea. *Marine Biology*, **41**, 107–117.

Sparrow, F.K. (1973). Chytridiomycetes, Hyphochytridiomycetes. In *The Fungi*, G.C. Ainsworth, F.K. Sparrow and A.S. Sussman, (eds) **IVB**, 85–110, Academic Press, New York.

Spector, D.L. (1984). *Dinoflagellates*. Academic Press, New York.

Spero, H.J. (1982). Phagotrophy in *Gymnodinium fungiforme* (Pyrrophyta): the peduncle as an organelle of ingestion. *Journal of Phycology*, **18**, 356–360.

Sprague, V. (1979). Classification of the Haplosporidia. *Marine Fisheries Review*, **41**, 40–44.

Steidinger, K.A. and Baden, D.G. (1984). Toxic marine dinoflagellates. In *Dinoflagellates*, D.L. Spector, (ed.) 201–261, Academic Press, New York.

Stevens, C.E. (1977). Comparative physiology of the digestive system. In *Dukes' Physiology of Domestic Animals*, M.J. Swanson, (ed.) 216–232, Cornell University Press, Ithaca, New York.

Stewart, K.D. and Mattox, K.R. (1978). Structural evolution in the flagellated cells of green algae and land plants. *BioSystems*, **10**, 145–152.

Stockem, W. (1977). Endocytosis. In *Mammalian Cell Membranes*, G.A. Jamieson and D.M. Robinson, (eds) **5**, 151–195, Butterworths, London.

Stockem, W., Hoffmann, H.U. and Gawlitta, W. (1982). Spatial organisation and fine structure of the cortical filament layer in normal locomoting *Amoeba proteus*. *Cell and Tissue Research*, **221**, 505–519.

Stoner, L.C. and Dunham, P.B. (1970). Regulation of cellular osmolarity and volume in *Tetrahymena*. *Journal of Experimental Biology*, **53**, 391–399.

Stout, J.D. and Heal, O.W. (1967). Protozoa. In *Soil Biology*, N.A. Burges and F. Raw, (eds) 149–195, Academic Press, London.

Stuart, V., Lucas, M.I. and Newell, R.C. (1981). Heterotrophic utilisation of particulate matter from the kelp *Laminaria pallida*. *Marine Ecology Progress Series*, **4**, 337–348.

Sudo, R. and Aiba, S. (1984). Role and function of protozoa in the biological treatment of polluted waters. *Advances in Biochemical Engineering/Biotechnology*, **29**, 117–141.

Surek, B. and Melkonian, M. (1986). A cryptic cytostome is present in *Euglena*. *Protoplasma*, **133**, 39–49.

Suzaki, T. and Williamson, R.F. (1986a). Pellicular ultrastructure and euglenoid movement in *Euglena enhrenbergii* Klebs and *Euglena oxyuris* Schmarda. *Journal of Protozoology*, **33**, 165–171.

Suzaki, T. and Williamson, R.F. (1986b). Ultrastructure and sliding of pellicular structures during euglenoid movement in *Astasia longa* Pringsheim (Sarcomastigophora, Euglenida). *Journal of Protozoology*, **33**, 179–184.

Taliaferro, W.H. and Stauber, L.A. (1969). Immunology of protozoan infections. In *Research in Protozoology*, T.-T. Chen, (ed.) **3**, 505–564, Pergamon Press, Oxford.

Tamm, S.L. and Horridge, G.A. (1970). The relation between the orientation of the central fibrils and the direction of beat in cilia of *Opalina*. *Proceedings of the Royal Society of London*, Series B, **175**, 219–233.

Tamm, S.L., Sonneborn, T.M. and Dippell, R.V. (1975). The role of cortical orientation in the control of the direction of ciliary beat in *Paramecium*. *Journal of Cell Biology*, **64**, 98–112.

Tamm, S.L. and Tamm, S. (1974). Direct evidence for fluid membranes. *Proceedings of the National Academy of Sciences of the United States of America*, **71**, 4589–4593.

Tartar, V. (1961). *The Biology of Stentor*. Pergamon Press, Oxford.

Tartar, V. (1966). Stentors in dilemmas. *Zeitschrift für Allgemeine Mikrobiologie,* **6**, 125–134.

Tartar, V. (1967). Morphogenesis in protozoa. In *Research in Protozoology*, T.-T. Chen, (ed.) **2**, 1–116, Pergamon Press, Oxford.

Tartar, V. (1968). Micrurgical experiments on cytokinesis in *Stentor coeruleus*. *Journal of Experimental Zoology*, **167**, 21–36.

Taylor, A.E.R. and Baker, J.R. (1978). *Methods of Cultivating Parasites in Vitro*. Academic Press, London.

Taylor, D.L. and Condeelis, J.S. (1979). Cytoplasmic structure and contractility in amoeboid cells. *International Review of Cytology*, **56**, 57–144.

Taylor, F.J.R. (1980). On dinoflagellate evolution. *BioSystems*, **13**, 65–108.

Taylor, F.J.R. (1987). *The Biology of Dinoflagellates*. Blackwell, Oxford.

Taylor. G.T. (1982). The role of pelagic heterotrophic protozoa in nutrient cycling: a review. *Annales de L'Institut oceanographique*, Paris, **58**, 227–241.

Taylor, W.D. and Lean, D.R.S. (1981). Radiotracer experiments on phosporus uptake and release by limnetic microzooplankton. *Canadian Journal of Fishery and Aquatic Sciences*, **38**, 1316–1321.

Tendal, O.S. (1972). A monograph of the Xenophyophorea (Rhizopodea, Protozoa). *Galathea Reports*, **12**, 7–103.

Thompson, J.E. and Pauls, K.P. (1980). Membranes of small ameobae. In *Biochemistry and Physiology of Protozoa*, Second edition, M. Levandowsky and S.H. Hutner, (eds) **3**, 207–253. Academic Press, New York.

Tokuyasu, K. and Scherbaum, O.H. (1965). Ultrastructure of mucocysts and pellicle of *Tetrahymena pyriformis*. *Journal of Cell Biology*, **27**, 67–81.

Travis, J.L. and Bowser, S.S. (1986). Microtubule-dependent reticulopodial motility: is there a role for actin? *Cell Motility and the Cytoskeleton*, **6**, 146–152.

Travis, J.L., Kenealy, J.F.X. and Allen, R.D. (1983). Studies on the motility of the Foraminifera II. The dynamic microtubular cytoskeleton of the reticulopodial network of *Allogromia laticollaris*. *Journal of Cell Biology*, **97**, 1668–1676.

Triemer, R.E. (1982). A unique mitotic variation in the marine dinoflagellate *Oxyrrhis marina* (Pyrrophyta). *Journal of Phycology*, **18**, 399–411.

Triemer, R.E. and Fritz, L. (1984). Cell cycle and mitosis. In *Dinoflagellates*, D.L. Spector, (ed.) 149–179, Academic Press, New York.

Triemer, R.E. and Fritz, L. (1987). Structure and operation of the feeding apparatus in a colourless euglenoid *Entosiphon sulcatum*. *Journal of Protozoology*, **34**, 39–47.

Tucker, J.B. (1967). Changes in nuclear structure during binary fission in the ciliate *Nassula*. *Journal of Cell Science*, **2**, 481–498.

Tucker, J.B. (1968). Fine structure and function of the cytopharyngeal basket in the ciliate *Nassula*. *Journal of Cell Science*, **3**, 493–514.

Tucker, J.B. (1970). Morphogenesis of a large microtubular organelle and its association with basal bodies in the ciliate *Nassula*. *Journal of Cell Science*, **6**, 385–429.

Tucker, J.B. (1971). Microtubules and a contractile ring of microfilaments associated with a cleavage furrow. *Journal of Cell Science*, **8**, 557–571.

Tucker, J.B. (1972). Microtubule-arms and propulsion of food particles inside a large feeding organelle in the ciliate *Phascolodon vorticella*. *Journal of Cell Science*, **10**, 883–903.

Tucker, J.B. Beisson, J. Roche, D.L.J. and Cohen, J. (1980). Microtubules and control of macronuclear amitosis in *Paramecium*. *Journal of Cell Science*, **44**, 135–151.

Tuffrau, M. (1970). Nouvelles observations sur l'origine du primordium buccal chez les hypotriches. *Compte Rendu Hebdomadaire des Séances de l'Académie des Sciences*, Paris, **270**, 104–107.

Tuttle, J.H., Wirsen, C.O. and Jannasch, H.W. (1983). Microbial activities in the emitted hydrothermal waters of the Galapagos rift vents. *Marine Biology*, **73**, 293–299.

Uspenskaya, A.V. (1976). The nuclear cycle of Myxosporidia according to cytophotometry. *Acta Protozoologica*, **15**, 301–313.

Van Houten, J., Hauser, D.C.R. and Levandowsky, M. (1981). Chemosensory behavior in protozoa. In *Biochemistry and Physiology of Protozoa*, Second edition, M. Levandowsky and S.H. Hutner, (eds) **4**, 67–124, Academic Press, New York.

Van Valkenburg, S.D. (1980). Silicoflagellates. In *Phytoflagellates*, E.R. Cox, (ed.) 335–350, Elsevier North Holland, New York.

Van Wagtendonk, W.J. (1974). *Paramecium: A Current Survey*. Elsevier, New York.

Vesk, M. and Moestrup, O. (1987). The flagellar root system in *Heterosigma akashiwo* (Raphidophyceae). *Protoplasma*, **137**, 15–28.

Vickerman, K. (1969). On the surface coat and flagellar adhesion in trypanosomes. *Journal of Cell Science*, **5**, 163–194.

Vickerman, K. (1974). The ultrastructure of pathogenic flagellates. *Ciba Foundation Symposium*, **20**, 171–198.

Vickerman, K. (1976). The diversity of the kinetoplastid flagellates. In *Biology of the Kinetoplastida*, W.H.R. Lumsden and D.A. Evans, (eds) **1**, 1–34, Academic Press, London.

Vickerman, K., Darbyshire, J.F. and Ogden, C.G. (1974). *Apusomonas proboscidea* Alexeieff 1924, an unusual phagotrophic flagellate from soil. *Archiv für Protistenkunde*, **116**, 254–269.

Vincent, W.F. (1980). The physiological ecology of a *Scenedesmus* population in the hypolimnion of a hypertrophic pond. II. Heterotrophy. *British Phycological Journal*, **15**, 35–41.

Vivier, E. (1979). Données nouvelles sur les microsporidies. Ultrastructure – cycles – systématique. *Bulletin de la Société Zoologique de France*, **104**, 281–396.

Vivier, E. (1982). Réflexions et suggestions a propos de la systématique des sporozoaires: création d'une classe des Hematozoa. *Protistologica*, **18**, 449–457.

Volcani, B.E. (1981). Cell wall formation in diatoms: morphogenesis and biochemistry. In *Silicon and Siliceous Structures in Biological Systems*, T.L. Simpson and B.E. Volcani, (eds) 157–200, Springer-Verlag, New York.

Vossbrinck, C.R. and Woese, C.R. (1986). Eukaryotic ribosomes that lack a 5.8S RNA. *Nature*, **320**, 287–288.

Vossbrinck, C.R., Maddox, J.V., Friedman, S., Debrunner-Vossbrinck, B.A. and Woese, C.R. (1987). Ribosomal RNA sequence suggests Microsporidia are extremely ancient eukaryotes. *Nature*, **326**, 411–414.

Walker, M.H., Mackenzie, C., Bainbridge, S.P. and Orme, C. (1979). A study of the structure and gliding movement of *Gregarina garnhami*. *Journal of Protozoology*, **26**, 566–574.

Walne P.L. (1980). Euglenoid flagellates. In *Phytoflagellates*, E.R. Cox, (ed.) 165–212, Elsevier North Holland, New York.

Warner, F.D. and Mitchell, D.R. (1978). Structural conformation of ciliary dynein arms and the generation of sliding forces in *Tetrahymena* cilia. *Journal of Cell Biology*, **76**, 261–277.

Watanabe, M.M., Takeda, Y., Sasa, T., Inouye, I., Suda, S., Sawaguchi, T. and Chihara, M. (1987). A green dinoflagellate with chlorophylls a and b: morphology, fine structure of the chloroplast and chlorophyll composition. *Journal of Phycology*, **23**, 382–389.

Webb, M. (1956). An ecological study of brackish water ciliates. *Journal of Animal Ecology*, **25**, 148–175.

Wehland, J., Weber, K., Gawlitta, W. and Stockem, W. (1979). Effects of the actin-binding protein DNAase I on cytoplasmic streaming and ultrastructure of *Amoeba proteus*. An attempt to explain amoeboid movement. *Cell and Tissue Research*, **199**, 353–372.

Weiser, J. (1985). Microspora Sprague, 1969, and Myxozoa Grasse, 1970. In *An Illustrated Guide to the Protozoa*, J.J. Lee, S.H. Hutner and E.C. Bovee, (eds) 375–383 and 384–392, Society of Protozoologists, Lawrence, Kansas.

Werner, D. (1977). *The Biology of Diatoms*. Blackwell, Oxford.

Wessenberg, H. (1961). Studies on the life cycle and morphogenesis of *Opalina*. *University of California Publications in Zoology*, **61**, 315–370.

Wessenberg, H. (1966). Observations on cortical ultrastructure in *Opalina*. *Journal de Microscopie*, **5**, 471–492.

Wessenberg, H. and Antipa, G. (1970). Capture and ingestion of *Paramecium* by *Didinium nasutum*. *Journal of Protozoology*, **17**, 250–270.

Whatley, J.M., John, P. and Whatley, F.R. (1979). From extracellular to intracellular: the establishment of mitochondria and chloroplasts. *Proceedings of the Royal Society of London*, Series B, **204**, 165–187.

Wichterman, R. (1953). *The Biology of Paramecium*. Blakiston, New York.

Wichterman, R. (1986). *The Biology of Paramecium*. Second edition Plenum Press, New York.

Woese, C.R. (1981). Archaebacteria. *Scientific American*, **244**(6), 98–122.

Woese, C.R. (1987). Bacterial evolution. *Microbiological Reviews*, **51**, 221–271.

Wohlfarth-Bottermann, K.E. (1974). Plasmalemma invaginations as characteristic constituents of plasmodia of *Physarum polycephalum*. *Journal of Cell Science*, **16**, 23–37.

Wolf, K. and Markiw, M.E. (1984). Biology contravenes taxonomy in the Myxozoa: new discoveries show alternation of invertebrate and vertebrate hosts. *Science*, **225**, 1449–1452.

Wright, M.A. Moisand, A. and Mir, L. (1979). The structure of the flagellar apparatus of the swarm cells of *Physarum polycephalum*. *Protoplasma*, **100**, 231–250.

Yamada, K. and Asai, H. (1982). Extraction and some properties of the protein spastin B from the spasmoneme of *Carchesium polypinum. Journal of Biochemistry*, **91**, 1187–1195.

Yang, D., Oyaizu, Y., Oyaizu, H., Olsen, G.J. and Woese, C.R. (1985). Mitochondrial origins. *Proceedings of the National Academy of Sciences of the United States of America*, **82**, 4443–4447.

Zeuthen, E. (1971). Synchronization of cell division and replication using temperature shocks (*Tetrahymena pyriformis*). *Experimental Cell Research*, **67**, 26.

Zeuthen, E. and Rasmussen, L. (1972). Synchronised cell division in Protozoa. In *Research in Protozoology*, T.-T. Chen, (ed.) **4**, 9–145, Pergamon Press, Oxford.

Zeuthen, E. and Scherbaum, O. (1954). Synchronous divisions in mass cultures of the ciliate protozoon *Tetrahymena pyriformis*, as induced by temperature changes. *Colston Papers*, **7**, 141–157.

Index

stomatogenesis, 209
   Tetrahymena, 91, 197, 209
   types, 197
   Stentor, 94
stream protists, 247, 267, 271
striated fibres,
   flagellar roots, 150
Strigomonas (*Crithidia*) flagellar
   propulsion, 107
Strobilidium, 217, 265
stroboscope, 33
Strombidium, 214–5, 219
strontium sulphate, in Acantharia, 180,
   183
studying protists, 286–7
Stylochona, 219
Stylonychia, 216–7, 219–20
   ciliary control, 41, 217
   galvanotaxis, 41
Suctoria (-ida), 205–7, 218–21
   feeding, 63–4
   tentacles, 64
sulphate-reducing bacteria, 265, 281
sulphide zone, 265
sulphur bacteria, 267, 269–70, 281
survival through adverse conditions,
   259–63
suspension feeding, 252–6
Syllis, host of *Urosporidium*, 238
Symbiodinium, 112, 283
symbionts,
   algae, 264, 282–4
   bacteria, 281–2
   chloroplast, 20, 103, 283
   ciliates, 209, 217
   diatoms, 172
   flagellates, 182–3
   mitochondrial, 3, 17
   nutrient cycling and, 259
     exchange, 283–4
   recognition and establishment, 283–4
synaptinemal complex, 72
synchronous division, 92
   Astasia, 260
   Tetrahymena, 92, 260
synchrony of ciliary beat, 38
Synchytrium, 146, 155
syncilia, 202
syncyanosen, 283
Syndiniales, 154
syndiniids, 112, 114
Syndinium, 154
syngens, 84–5
Synhymeniida, 202, 218–9

synistosome, 139
synthesis, of DNA, 44–5
   of proteins, 51–3
syntomy, 55
Synura, 132–3, 154
   and carbon dioxide, 261
   in freshwater plankton, 267
Synurales, 154
Synurophyceae, 98–9, 133, 152, 154
   scales, 133
syzygy, 77, 226

Tachyblaston, 205
Taxopodida, 179–80, 185
Technitella, 172
tectin,
   in shells, 54
telokinetal stomatogenesis, 197
telotroch, 215
Telotrochidium, 215
temperature, and cell division, 92
   and lives of protists, 260
Tenebrio, host of *Gregarina*, 226
Terebella, host of *Selenidium*, 227
termite flagellates, 75, 280–1
   and humus breakdown, 281
Testaceafilosea, 185
Testacealobosea, 185
testate amoebae, 162–3
   acid soils, 261
   binary fission, 163, 166, 168
   habitats, 162, 167
   pseudopodia, 160, 162
   soil fauna, 275–7
tests, 54
Tetradimorpha, 177, 185
Tetrahymena, 8, 205, 219
   amicronucleate, 85
   buccal cavity, 197
   cell cycle, 45, 90–4
   ciliary components, 29
   culture, 288
   division synchrony, 92, 94, 260
   DNA synthesis, 45, 91–2
   feeding on bacteria, 253–4
   galvanotaxis, 41
   growth curve, 252
     efficiency, 250
     rate and food, 253–4
   heat shocks, 92
   life cycle, 90–4
   macronucleus, 187
   mouth formation, 91
   nutrition, 24, 57, 61, 252–4